Practical Field Ecology:
A Project Guide

Practical Field Ecology: A Project Guide

C. Philip Wheater[1], James R. Bell[2] and Penny A. Cook[3]

[1]Department of Environmental and Geographical Sciences, Manchester Metropolitan University, Manchester, UK

[2]Plant & Invertebrate Ecology, Rothamsted Research, Harpenden, UK

[3]Faculty of Health and Applied Social Sciences, Liverpool John Moores University, Liverpool, UK

A John Wiley & Sons, Ltd., Publication

This edition first published [2011] © [2011] by [John Wiley & Sons Ltd]

Wiley-Blackwell is an imprint of John Wiley & Sons, formed by the merger of Wiley's global Scientific, Technical and Medical business with Blackwell Publishing.

Registered office: John Wiley & Sons, Ltd, The Atrium, Southern Gate, Chichester, West Sussex, PO19 8SQ, UK

Editorial offices: 9600 Garsington Road, Oxford, OX4 2DQ, UK
The Atrium, Southern Gate, Chichester, West Sussex, PO19 8SQ, UK
111 River Street, Hoboken, NJ 07030-5774, USA

For details of our global editorial offices, for customer services and for information about how to apply for permission to reuse the copyright material in this book please see our website at www.wiley.com/wiley-blackwell.

The right of the author to be identified as the author of this work has been asserted in accordance with the UK Copyright, Designs and Patents Act 1988.

All rights reserved. No part of this publication may be reproduced, stored in a retrieval system, or transmitted, in any form or by any means, electronic, mechanical, photocopying, recording or otherwise, except as permitted by the UK Copyright, Designs and Patents Act 1988, without the prior permission of the publisher.

Designations used by companies to distinguish their products are often claimed as trademarks. All brand names and product names used in this book are trade names, service marks, trademarks or registered trademarks of their respective owners. The publisher is not associated with any product or vendor mentioned in this book. This publication is designed to provide accurate and authoritative information in regard to the subject matter covered. It is sold on the understanding that the publisher is not engaged in rendering professional services. If professional advice or other expert assistance is required, the services of a competent professional should be sought.

Library of Congress Cataloging-in-Publication Data

Wheater, C. Philip, 1956-
 Practical field ecology : a project guide / C. Philip Wheater, James R. Bell, Penny A. Cook.
   p. cm.
 Includes bibliographical references and index.
 ISBN 978-0-470-69428-2 (cloth) – ISBN 978-0-470-69429-9 (pbk.)
 1. Ecology–Research–Methodology. 2. Ecology–Fieldwork. I. Bell, James R. (James Robert), 1969- II. Cook, Penny A., 1971- III. Title.
 QH541.2.W54 2011
 577.072'3–dc22

2010047197

A catalogue record for this book is available from the British Library.

This book is published in the following electronic formats: ePDF [978-0-470-97506-0]; ePub [978-0-470-97670-8]

Set in 10.5/12.5pt Minion by Thomson Digital, Noida, India

First Impression   2011

Contents

Tables xi
Figures xiii
Boxes xvii
Case Studies xix
Preface xxi
Acknowledgements xxiii

1 Preparation 1
Choosing a topic for study 3
Ecological research questions 4
    Monitoring individual species and groups of species ... 4
    Monitoring species richness 5
    Monitoring population sizes and density 5
    Monitoring community structure 6
    Monitoring behaviour 6
    A note of caution 6
Creating aims, objectives and hypotheses 7
Reviewing the literature 8
    Primary literature 9
    Secondary literature 9
    Other sources of information 9
    Search terms 10
    Reading papers 10
Practical considerations 11
    Legal aspects 11
    Health and safety issues 12
    Implementation 13
    Time management 16
Project design and data management 18
    Designing and setting up experiments and surveys .. 20
    Types of data 20
    Sampling designs 22
    Planning statistical analysis 28
Choosing sampling methods 33
Summary 34

2 Monitoring site characteristics ... 35
Site selection ... 35
Site characterisation ... 36
    Habitat mapping ... 36
    Examination of landscape scale ... 41
    Measuring microclimatic variables ... 42
    Monitoring substrates ... 45
    Monitoring water ... 51
    Other physical attributes ... 54
    Measuring biological attributes ... 56
    Identification ... 58

3 Sampling static organisms ... 67
Sampling techniques for static organisms ... 70
    Quadrat sampling ... 73
    Pin-frames ... 83
    Transects ... 84
    Distribution of static organisms ... 88
    Forestry techniques ... 90

4 Sampling mobile organisms ... 95
General issues ... 96
    Distribution of mobile organisms ... 97
    Direct observation ... 97
    Behaviour ... 98
    Indirect methods ... 103
    Capture techniques ... 104
    Marking individuals ... 106
    Radio-tracking ... 108
Invertebrates ... 111
    Direct observation ... 112
    Indirect methods ... 114
    Capture techniques ... 115
    Marking individuals ... 117
Capturing aquatic invertebrates ... 121
    Netting ... 123
    Suction sampling ... 127
    Benthic coring ... 128
    Drags, dredges and grabs ... 128
    Wet extraction ... 129
    Artificial substrate samplers ... 131
    Baited traps ... 132
Capturing soil-living invertebrates ... 133
    Dry sieving ... 133
    Floatation and phase-separation ... 134
    Tullgren funnels as a method of dry extraction ... 134

- Chemical extraction 137
- Electrical extraction 138
- Capturing ground-active invertebrates 139
  - Pitfall traps...139
  - Suction samplers 148
  - Emergence traps..150
- Capturing invertebrates from plants.............................. 152
  - Pootering 153
  - Sweep netting 154
  - Beating 156
  - Fogging 156
- Capturing airborne invertebrates 158
  - Sticky traps..161
  - Using attractants 162
  - Refuges 165
  - Flight interception (window and malaise) traps 165
  - Light traps 167
  - Rotary traps 172
  - Water traps 172
- Fish 174
  - Direct observation 176
  - Indirect methods 176
  - Capture techniques..177
  - Marking individuals..181
- Amphibians.. 184
  - Direct observation 186
  - Indirect methods 187
  - Capture techniques..187
  - Marking individuals..192
- Reptiles 193
  - Direct observation 193
  - Indirect methods 194
  - Capture techniques..195
  - Marking individuals..199
- Birds 200
  - Direct observation 201
  - Indirect methods 209
  - Capture techniques..211
  - Marking individuals..213
- Mammals... 216
  - Direct observation 217
  - Indirect methods 219
  - Capture techniques..225
  - Marking individuals..233

5 Analysing and interpreting information 235

Keys to tests 238
Exploring and describing data 244
    Transforming and screening data 244
    Spatial and temporal distributions 252
    Population estimation techniques: densities and population sizes 252
    Richness and diversity 256
    Similarity, dissimilarity and distance coefficients 258
    Recording descriptive statistics 261
Testing hypotheses using basic statistical tests and simple general
linear models 261
    Differences between samples 265
    Relationships between variables 269
    Associations between frequency distributions 274
More advanced general linear models for predictive analysis 276
    Multiple regression 276
    Analysis of covariance and multivariate analysis of variance 277
    Discriminant function analysis 279
Generalized linear models 280
    Extensions of the generalized linear model 285
Statistical methods to examine pattern and structure in communities:
classification, indicator species and ordination 286
    Classification 287
    Indicator species analysis 292
    Ordination 294

6 Presenting the information 305

Structure 306
    Title 307
    Abstract 307
    Acknowledgements 308
    Contents 308
    Introduction 309
    Methods 310
    Results 311
    Discussion 316
    References 317
    Appendices 321
Writing style 321
    Tense 324
    Numbers 324
    Abbreviations 325
    Punctuation 327
    Choice of font 328
    Common mistakes 329

Computer files 331
Summary. 331

References 333
Appendix 1   Glossary of statistical terms 345
Index. 351

Tables

Table 1.1	Example time-scales for a short research project	17
Table 1.2	Random numbers	27
Table 1.3	Common statistical tests	30
Table 2.1	Common factors influencing living organisms	37
Table 2.2	Major taxonomic groups	63
Table 2.3	Major divisions of the Raunkiær plant life-form system	66
Table 3.1	DAFOR, Braun–Blanquet and Domin scales for vegetation cover	77
Table 3.2	Abundance (ESACFORN) scales for littoral species	79
Table 3.3	Recommended quadrat sizes for various organisms	80
Table 4.1	Some considerations in the choice of radio-tracking equipment	109
Table 4.2	Summary of killing and preservation techniques for commonly studied invertebrates	116
Table 4.3	Factors to consider when using pitfall traps	146
Table 4.4	Examples of baits and target insect groups	162
Table 4.5	Factors to consider when choosing light traps to collect moths	171
Table 4.6	Summary of different types of net	179
Table 4.7	Example of timed species counts	205
Table 4.8	Comparison of bat detector systems	224
Table 5.1	Abundance of invertebrates in ponds	249
Table 5.2	Summary of commonly used methods of population estimation based on mark-release-recapture techniques	255
Table 5.3	Common diversity and evenness indices	257
Table 5.4	Commonly used similarity measures	259
Table 5.5	Statistics that should be reported for difference tests	267
Table 5.6	Statistics that should be reported for relationship tests	270
Table 5.7	Statistics that should be reported for tests used to examine associations between two frequency distributions	274
Table 5.8	Using dummy variables	276
Table 5.9	A spider indicator species analysis	293
Table 5.10	Types of stress measure for computing MDS solutions	299
Table 6.1	Mean number of individuals of invertebrate orders found in polluted and clean ponds	312
Table 6.2	Uses of different types of graphs	314
Table 6.3	Examples of words used unnecessarily when qualifying terms	323
Table 6.4	SI units of measurement	325
Table 6.5	Conventions for the use of abbreviations	326
Table 6.6	Examples of Latin and foreign words and their emphasis	329

Figures

Figure 1.1	Flowchart of the planning considerations for research projects	4
Figure 1.2	Example time-scales for a medium-term research project	17
Figure 1.3	Example of a data-recording sheet for an investigation into the distribution of woodland birds	22
Figure 1.4	Examples of sampling designs	26
Figure 1.5	Latin square design for five different treatments	28
Figure 1.6	Dataset approximating to a normal distribution	31
Figure 2.1	Phase 1 habitat map in the UK	39
Figure 2.2	Portable weather station	43
Figure 2.3	Maximum/minimum thermometer	44
Figure 2.4	Soil thermometers	45
Figure 2.5	Whirling hygrometer	46
Figure 2.6	Anemometers	47
Figure 2.7	Environmental multimeter	47
Figure 2.8	Penetrometer	48
Figure 2.9	Soil gouge auger, mallet and core-removing tool	49
Figure 2.10	Bulb planters	50
Figure 2.11	Aquatic multimeters	52
Figure 2.12	Secchi disc	53
Figure 2.13	Dynamometer to measure wave action	54
Figure 2.14	Light meter	56
Figure 2.15	Using ranging poles to measure the inclination of a slope	57
Figure 2.16	Using a cross-staff to survey a shoreline	57
Figure 2.17	Using a global positioning system (GPS) receiver	58
Figure 2.18	Lichen zone scale for mean winter sulphur dioxide estimation on trees with moderately acidic bark in England and Wales	61
Figure 2.19	Estimating canopy cover	62
Figure 3.1	Quadrats	75
Figure 3.2	Recording positions on a subdivided quadrat	77
Figure 3.3	JNCC guideline usage of SACFOR scales	78
Figure 3.4	Two nested quadrat designs	81
Figure 3.5	Using random numbers to identify a position in a sampling grid	82
Figure 3.6	Comparison of the perimeter to area ratios of circular, square and oblong quadrats	83
Figure 3.7	Pin-frame	84
Figure 3.8	Comparison of transect sampling techniques	85

Figure 3.9	Kite diagram to indicate the abundance of different species along a transect	86
Figure 3.10	Using a clinometer	93
Figure 4.1	Use of ink or paint spots to identify individual invertebrates	108
Figure 4.2	'W'-shaped transect walk	113
Figure 4.3	Parabolic reflector concentrating sound onto the central microphone	115
Figure 4.4	Flat-bottomed pond nets suitable for catching surface, pelagic and bottom active invertebrates	123
Figure 4.5	Belleville mosquito larvae sampler	124
Figure 4.6	Using a kick net and sorting the sample	125
Figure 4.7	Kick screen	125
Figure 4.8	Surber sampler	126
Figure 4.9	Hess sampler	126
Figure 4.10	Drift net	127
Figure 4.11	Plankton net	127
Figure 4.12	Suction sampler for animals in burrows	128
Figure 4.13	Naturalist's dredge	129
Figure 4.14	Grabs for collecting benthic animals	129
Figure 4.15	The Baermann funnel	130
Figure 4.16	Bidlingmayer sand extractor	130
Figure 4.17	Hester–Dendy multiplate samplers	131
Figure 4.18	Mesh bags containing leaf litter used to collect aquatic invertebrates	131
Figure 4.19	Crayfish traps	132
Figure 4.20	Soil sieves	133
Figure 4.21	Tullgren funnels	135
Figure 4.22	Kempson bowl extractor	136
Figure 4.23	Winkler sampler	136
Figure 4.24	Simple inclined tray light-based separator	137
Figure 4.25	Setting pitfall traps	143
Figure 4.26	Barriers used with pitfall traps	145
Figure 4.27	Bird's-eye view of an H trap	148
Figure 4.28	Suction samplers	149
Figure 4.29	Emergence trap	150
Figure 4.30	Owen emergence trap	151
Figure 4.31	Pooter used to suck up small invertebrates	154
Figure 4.32	Sweep net	155
Figure 4.33	Beating trays	156
Figure 4.34	Fogging in rainforest	157
Figure 4.35	Nets for catching airborne insects	160
Figure 4.36	Rothamsted suction traps	161
Figure 4.37	Positioning of sticky traps	162
Figure 4.38	Bottle trap for flies and other flying insects	163

Figure 4.39	Attractant-based traps	164
Figure 4.40	Assembly trap	164
Figure 4.41	Trap-nests for bees and wasps	165
Figure 4.42	Window trap	166
Figure 4.43	Malaise trap	167
Figure 4.44	Simple light traps for insects	167
Figure 4.45	Examples of moth traps	169
Figure 4.46	Different types of light used for moth traps	170
Figure 4.47	Rotary trap	172
Figure 4.48	Water traps	173
Figure 4.49	Slurp gun	175
Figure 4.50	Using snorkel gear to observe fish	176
Figure 4.51	Sport fishing techniques	178
Figure 4.52	Examples of nets and traps	179
Figure 4.53	Bottle trap for newts	189
Figure 4.54	Drift fence with side-flap bucket trap	190
Figure 4.55	Funnel traps for amphibians	190
Figure 4.56	Examples of layouts for drift nets	191
Figure 4.57	Artificial cover trap for amphibians	192
Figure 4.58	Concrete housing for a camera trap	194
Figure 4.59	Equipment for catching reptiles at a distance	198
Figure 4.60	Pipe trap	198
Figure 4.61	Measuring captured birds	200
Figure 4.62	Permanent bird hide	202
Figure 4.63	Bird observation tower	203
Figure 4.64	Transect layout for Breeding Bird Survey	207
Figure 4.65	Goose droppings surveyed using a quadrat	209
Figure 4.66	Mist netting	212
Figure 4.67	Propelled nets	214
Figure 4.68	Marking birds	215
Figure 4.69	Use of colour rings	215
Figure 4.70	Deer becoming aware of the observer's presence	217
Figure 4.71	Animals caught using camera traps in tropical forest	218
Figure 4.72	Small mammal tracking tunnel	220
Figure 4.73	Badger dung pit with bait-marked dung	221
Figure 4.74	Sampling mammal hair	222
Figure 4.75	Bat detector (heterodyne system)	223
Figure 4.76	Triangle bat walks with frequency settings appropriate for UK bats	223
Figure 4.77	Small mammal traps	226
Figure 4.78	Longworth trap for small to medium-sized mammals	226
Figure 4.79	Poison bait dispenser	228
Figure 4.80	Mole traps	229
Figure 4.81	Harp trap to capture bats	231

Figure 4.82	Cage trap	232
Figure 4.83	Badger trap	233
Figure 5.1	Transformations for skewed distributions	245
Figure 5.2	Truncation of percentage data	246
Figure 5.3	Bimodal distribution	247
Figure 5.4	Pie diagram of the numbers of invertebrates of common orders found in clean ponds.	248
Figure 5.5	Stacked bar graph of the number of invertebrates of common orders found in clean ponds.	248
Figure 5.6	Clustered bar graph of the number of invertebrates of common orders found in clean ponds.	249
Figure 5.7	The mean and standard deviation plotted on a dataset that approximates to a normal distribution	250
Figure 5.8	Comparison of different ways of displaying the variation around the mean using point charts.	251
Figure 5.9	Box and whisker plots indicating different ways of displaying median and quartile data	251
Figure 5.10	Using capture-removal to estimate population sizes	256
Figure 5.11	Comparison of the central tendency of two samples.	262
Figure 5.12	Summary of stages in using inferential statistics.	264
Figure 5.13	Example of a scatterplot.	270
Figure 5.14	Trends of invertebrate numbers with organic pollution	271
Figure 5.15	Regression line between the number of aphids found at different levels of pirimicarb (pesticide) application	272
Figure 5.16	Examples of common non-linear graph types in ecology	273
Figure 5.17	A canonical variates analysis (CVA) of spiders across three management treatments.	289
Figure 5.18	Types of cluster analysis.	291
Figure 5.19	Dendrogram following cluster analysis of different habitat types	291
Figure 5.20	TWINSPAN of quarry sites on the basis of their component plant species	293
Figure 5.21	Ordination of a number of quarry sites on the basis of their component plant species	298
Figure 6.1	Two formats for research report presentation	307
Figure 6.2	Pollution in the Forth and Clyde Canal in Glasgow	313
Figure 6.3	Presenting graphs.	315

Boxes

Box 1.1	Some sources of ecology projects	3
Box 1.2	Suggested minimum equipment required for field work	13
Box 1.3	Keeping a field notebook	15
Box 1.4	Some tips on time management	18
Box 1.5	Differences between interval and ratio data	21
Box 1.6	Terms used in sampling theory	23
Box 1.7	Species accumulation curves for two sites	24
Box 1.8	Checklist for field research planning	34
Box 2.1	Notes on the resources available for the National Vegetation Classification	40
Box 2.2	Examples of vegetation classification systems	40
Box 2.3	Measurements of aquatic invertebrates used in habitat quality and pollution monitoring	59
Box 2.4	Examples of identification guides for British insects	64
Box 3.1	Calculating population and density estimates from counts of static organisms	69
Box 3.2	Techniques used to identify and count microbial diversity	71
Box 3.3	Commonly used plotless sampling methods	87
Box 3.4	Describing the distribution of static organisms using quadrat-based methods	89
Box 3.5	Describing the distribution of static organisms using T-square sampling methods	89
Box 4.1	Avoiding problems in behavioural studies	102
Box 4.2	Butterfly census method	113
Box 4.3	Calculating the density of flying insects from census walks	114
Box 4.4	Taking account of missing traps	144
Box 4.5	Common Birds Census for territory mapping	206
Box 4.6	Restrictions on handling birds	211
Box 5.1	A note of caution about the examples used in this chapter	236
Box 5.2	Some commonly used statistical software	236
Box 5.3	Important terms used in the keys	238
Box 5.4	The Peterson (Lincoln index) method of population estimation	253
Box 5.5	Testing for the significance of multiple tests	265
Box 5.6	Multiple comparison tests	268
Box 5.7	Using a contingency table in frequency analysis	275
Box 5.8	Analysis of covariance	278

Box 5.9	Using classification tables in predictive discriminant function analysis	279
Box 5.10	Generalized linear model: a worked example using a binomial regression	282
Box 5.11	Generalized additive model (GAM)	285
Box 5.12	Distance measurements	287
Box 5.13	Use of analysis of similarity (ANOSIM)	290
Box 5.14	Examples of agglomerative clustering methods	292
Box 5.15	Using principal components analysis for data compression	295
Box 5.16	Using principal components analysis to produce biplots	297
Box 5.17	Example of distance placement using multidimensional scaling (MDS)	300
Box 5.18	Techniques for comparing ordinations and matrix data	301
Box 5.19	Example of use of canonical correspondence analysis	302
Box 6.1	Citing works using the Harvard system	318
Box 6.2	Reference lists using the Harvard system	320
Box 6.3	Commonly misused words	329

Case Studies

Case Study 3.1	The Park Grass Experiment	74
Case Study 3.2	Studying tree growth and condition	91
Case Study 4.1	Cracking the chemical code in mandrills	99
Case Study 4.2	Barnacle larva trap	105
Case Study 4.3	Tarantula distribution and behaviour	118
Case Study 4.4	Stream invertebrates	121
Case Study 4.5	Collecting insects in Costa Rica	141
Case Study 4.6	Butterfly life cycles	152
Case Study 4.7	The birds and the bees	158
Case Study 4.8	Lake fish populations	182
Case Study 4.9	Breeding behaviour of neotropical tree frogs	185
Case Study 4.10	Reptile diet	195
Case Study 4.11	Counting parrots	204
Case Study 4.12	Bat conservation ecology	230

Preface

This handbook is designed as a guide to planning and executing an ecological research project and is intended as a companion to preparing a dissertation, report, thesis or research paper. The idea for the book arose from many years spent in the field sampling animals and plants, as students ourselves or later when leading groups of undergraduate or postgraduate students. In so doing, it was clear that there was a need for a book to cover all aspects of planning, implementing and presenting an ecological research project. Much of the content of this text has been developed from teaching materials we have used over the years in one form or another, refined following discussions with colleagues and the students who used them. We have included those methods that should be accessible to an undergraduate or taught postgraduate student at a university or college. We have purposely tended to avoid devoting too much space to highly technical methods or those techniques that require the user to have a licence. However, we have mentioned some such techniques that generate data sets that may be made available for student projects.

Our experience is that many students develop an interest in a particular group of organisms, sometimes describing themselves as a birder, entomologist or badger watcher. Rarely, one finds a student principally interested in a particular habitat; this is normally secondary and is often defined by the group under study. Consequently, although we have ordered the sampling chapters by the mobility of the organisms, within the chapter on sampling mobile animals, we have dealt separately with each group of organism under study. We have attempted to take the reader through all the stages of conducting a research project starting from finding a topic on which to do a research project; turning the idea into a provisional title and research question (i.e. the aims); thinking about how to achieve the aims (these are the objectives); and then deciding on the methods to be used. The book then summarises key methodological approaches used by ecologists in the field. The intention has been to cover core, well-tested and robust methods relevant to sampling animals and plants in terrestrial and most aquatic habitats including sandy and rocky shores. Owing to additional health and safety requirements and the highly technical nature of off-shore sampling methods, we stopped short of including these techniques in the book.

This book is not just about the activities associated with field sampling. We felt that it was important to take researchers right through to the end of their project. Many of the more technical hurdles occur once the data have been collected. Ecological research frequently generates complex datasets that require statistical analysis to aid interpretation. There is a need for students to understand the range of methods available to explore and analyse their data and to understand what types of data they need to collect in order to use particular techniques. Frequently, students ask us how they should go about finding and using key references, and how to interpret their own data in the light of current research. Consequently, we also give tips on presentation and writing style. Most research projects are completed in a fairly restricted time-scale, therefore we include guidelines for time management during the project. Several experts have generously provided insights into how they approach their own research. These case studies illustrate some of the diversity and complexity of even quite simple research questions. We hope that this text will both encourage and support students in engaging in the fascinating world of ecological research.

Acknowledgements

It would be difficult to find any field biologist who had enough experience to write about sampling animals and plants without contributions from fellow academics. We would like to thank all those who were generous with both their time and expertise:

Joanna Bagniewska, Sandra Baker, Hannah Dugdale and Stephen Ellwood (WildCRU, University of Oxford) gave advice on survey techniques for mammals;
Philip Briggs (Bat Conservation Trust) provided helpful discussions on bats;
Dave Brooks (Rothamsted Research) provided material on CCA;
Paul Chipman (Manchester Metropolitan University - MMU) contributed to mammal sampling and statistics;
Rod Cullen (MMU) contributed to invertebrate sampling;
John Cussans and Sue Welham (Rothamsted Research) advised on generalized linear models;
Mike Dobson (Freshwater Biological Association) advised on aquatic invertebrates;
Mark Elgar (University of Melbourne) commented on the proposal for the book;
Alan Fielding (MMU) advised on Chapter 5;
Chris Goldspink (MMU) advised on fish;
Mark Grantham, David Leech, Rob Robinson (BTO) supplied information on birds;
Ed Harris (MMU) advised on amphibians and reptiles;
Paul Hart (Leicester University) supplied information on electrofishing;
Øyvind Hammer (Paleontological Museum, University of Oslo) discussed applications within the PAST software package;
Alison Haughton (Rothamsted Research) provided internet information and alerted us to the less obvious information sources and advised on Chapter 6;
Mike Hounsome (University of Manchester) advised on bird sampling and statistics;
Martin Jones and Stuart Marsden (MMU) advised on birds;
Mark Langan (MMU) provided information on aquatic invertebrate sampling and statistics;
Les May (MMU) provided guidelines on field notebooks and advised on sampling using animal sounds;
Ed Mountford (JNCC) contributed towards the forest techniques section, especially mensuration;
Richard Preziosi (University of Manchester) commented on the proposal for the book and discussed various sampling and statistical methods;

Liz Price (MMU) helped with plant sampling;
Helen Read (Corporation of London: Burnham Beeches) contributed to Chapters 1, 2 and 3;
Ian Rotherham (Sheffield Hallam University) commented on the proposal for the book;
Robin Sen (MMU) advised on microbial techniques;
Emma Shaw (MMU) advised on sampling spiders including tarantulas;
Rob Sheldon (RSPB) advised on the bird section;
Dave Shuker (University of Edinburgh) commented on the proposal for the book;
Richard Small (Liverpool John Moores University) commented on the proposal for the book and on Chapters 1, 2 and 3;
Graham Smith (MMU) advised on GIS and remote sensing;
Nigel Stork (University of Melbourne) provided information on fogging;
Rob Strachan (Environment Agency for Wales) gave insights into less well-known survey techniques for mammals;
Michelle Tobin (University of Hull) commented on the proposal for the book;
David Wilkinson (Liverpool John Moores University) commented on the proposal for the book and on Chapters 1 to 6;
Derek Yalden (University of Manchester) advised on mammals.

Our appreciation goes to all those who wrote case studies:
Amanda Arnold (University of Edinburgh) – aquatic invertebrates;
Chris Bennett (Rothamsted Research) – plants;
David Brown (University of Cardiff) – snakes;
Robin Curtis (Zoological Society of London) – butterflies;
Jenny Jacobs (Rothamsted Research) – bees;
Erica McAlister (Natural History Museum, London) – insects;
Stuart Marsden (MMU) – parrots;
Vicky Oglivy (University of Manchester) – tree frogs;
Helen Read (Burnham Beeches) – trees;
Jo Setchell (University of Durham) – mandrills;
Emma Shaw (MMU) – tarantulas;
Christopher Todd (University of St Andrews) – barnacles;
Ian Winfield (CEH, Lancaster) – fish;
Matt Zeale (University of Bristol) – bats.

Thanks to the Ordnance Survey for permission to use the map fragment in Figure 2.1. All photographs are used with permission and plates are marked with the appropriate initials (e.g. JRB – James Bell, and CPW – Phil Wheater). Unless otherwise stated, the photographs used within each case study are used with permission of the scientist profiled within the case study. The Park Grass photograph is courtesy of Rothamsted Research (RRES). Thanks to:

Sandra Baker (SB);
Chris Bennett (CB);
David Brown (DSB);
Robin Curtis (RC);
Mike Dobson (MKD);
Hannah Dugdale (HD);
Paul Higginbottom (PH);
Mark Langan (AML);
Erica McAlister (EM);
Mark Mallott (MM);
Stu Marsden (SJM);
Sharon Matola (SM);
Vicky Oglivy (VO);
Helen Read (HJR);
Kelly Reynolds (KR);
Miira Riipinen (MR);
Rob Robinson (RAR);
Jo Setchell (JS);
Emma Shaw (EMS);
Nigel Stork (NES);
Rob Strachan (RS);
Christopher D. Todd (CDT);
Ian Winfield (IJW);
Matt Zeale (MZ).

A huge thanks to all of those generations of students and colleagues on many a field course who commented on the early and developing ideas behind this book, discussing the merits of particular techniques, the ways in which to introduce the information to students, and the intelligibility (or otherwise) of the handouts and other teaching materials from which this text derives. In addition to those mentioned above, Gordon Blower, Glyn Evans and Robin Baker provided much early inspiration. JRB would like to thank Ian Denholm for his support and members of PIE for their varied contributions. Finally, thanks to all of those who have supported us and suffered during the writing of this book. CPW and PAC would particularly like to thank Abhishek Kumar, Charlotte and Henry. JRB is especially grateful to FF, Beanie, Haz and Julia for their understanding.

1
Preparation

For many students, honing their research skills is an important component of their academic development. However, inexperienced researchers can be naïve in their approach, and may even attempt highly complicated studies that have little chance of being completed in the time and with the resources available. This chapter describes the thought and preparation needed to plan your project, particularly how to formulate your ideas into something structured and workable before going out into the field. Your research will search for explanations or relationships; make comparisons, predictions and generalisations; and formulate theories. Research is not simply an exercise in information gathering; rather, research is about asking questions that go beyond description and require analysis. Your research will be highly individual, and there are no set outcomes. You will form your own opinion, even if this disagrees with previous work. This is because progress in science results from the continual testing, review and criticism of other researchers' work. Do not expect your research project to answer all your original questions: it is much more typical to find that research generates more questions than it answers. Research submitted for publication or for examination should show evidence of originality. Even if your research is not wholly original, it can show evidence of original thinking. Although the prospect of carrying out original research may seem rather daunting, providing you do not exactly copy someone else's experimental design, methods, sites, etc., your research is almost certainly going to be original. There are several ways in which your work can be original:

- Executing an entirely new piece of work (e.g. studying a plant or animal for which there is little or no information currently available).

- Adding knowledge in a way that has not been previously done before; many empirical studies do not develop new topics to study but instead angle their work with the use of original experimental designs, new statistical methods, etc. (for example, new insights might be generated from exploring the ecology of an otherwise well-studied animal at different sites to see whether food preferences differ between locations).

- Showing originality by testing somebody else's idea, or by carrying out an established idea in a new area, new experimental subject, etc., or by using existing data to develop new interpretations.

- Continuing an existing piece of work usually at your university or with a partner institution; for example, there are many long-term experiments that invite students to participate in summer work. These opportunities can be symbiotic and provide both you and the scientist running the project with more data that could elucidate a mechanism or generate new hypotheses.

- Originality may only be apparent in the breath of the study. Increasingly popular is 'cross-disciplinary' science, where, for example, soil scientists, botanists and entomologists converge on a subject matter or site and work together to test an overarching hypothesis.

All research needs careful planning (whether in the field or not). It is perhaps self evident that such planning should involve the correct use of equipment and choice of appropriate sampling methods and collection sites. In addition, a wide range of associated logistic, legal, and health and safety implications are also very important. Although many of these issues are equally important in field or laboratory-based investigations, field research may be more limited by time and other factors (access to sites, time of year, weather conditions) than research based entirely in the laboratory. Thus field research may need more careful consideration prior to implementation. Chapter 1 details some of the issues involved in planning and designing field research, and culminates in a checklist that may help to prevent problems once research is implemented. Chapter 2 deals with the techniques required for monitoring sampling sites and measuring physicochemical factors. Chapter 3 covers the methods used to sample static or relatively immobile organisms, and Chapter 4 extends this to studying mobile animals. Chapter 4 includes a consideration of monitoring behaviour and of dealing with both direct and indirect observations, as well as covering trapping and marking individual animals. In Chapter 5 we summarise a large number of different approaches to the statistical analysis of ecological data. Finally, in Chapter 6 we cover how to present your results and write appropriate reports.

Choosing a topic for study

The first stage of a research project is choosing a subject area on which to research (see Box 1.1 for a list of some texts that include ecological project ideas). As you will be devoting substantial time to your project, it is important to choose a topic that interests you. You may also wish to make your research relevant to your current or future employment. Pick a topic of the right size: neither too big nor too small. Looking at successful previous projects may assist you in judging how much can be done in the available time (ask those more experienced for examples of good projects to look at). Finally, your proposed project has to be feasible, for example in terms of equipment, access to sites and time-scale. Once you have selected your subject and provisional title, be prepared to be flexible and, if necessary, to change direction. This may happen for a variety of reasons, for example if a pilot study reveals a more interesting avenue for research or if your original ideas turn out to be unfeasible. You should note that the planning process should involve a consideration of the whole project to enable you to identify and deal with any potential problems before they become major issues (Figure 1.1). In all aspects, reading around the subject will allow you to use appropriate techniques, build on existing knowledge and avoid reinventing the wheel. Inevitably there can be logistical problems that influence your choice of site, or species, or otherwise prevent you from proceeding exactly as you would have wished. Although you can avoid such problems by careful planning, there are some aspects that you will not think about until you implement the research. A pilot study will help to identify such issues and may allow you to refine the study in advance of full implementation.

Box 1.1 Some sources of ecology projects

There are many resources that give examples of feasible ecological research projects. The series listed below are examples of some of those that cover either a wide range of habitat types or a range of organisms.

The Practical Ecology Series provide project ideas associated with grasslands (Brodie 1985), freshwaters (Gee 1986), the seashore (Jenkins 1983) and urban areas (Smith 1984).

Routledge Habitat Guides each include a section (Section 5) giving project ideas for the habitats associated with grasslands (Price 2003), uplands (Fielding and Howarth 1999), urban habitats (Wheater 1999) and woodlands (Read and Frater 1999).

The Naturalists' Handbooks (Richmond Publishing Company, Slough) contain many ideas related to studying a group of species (e.g. Gilbert 1993 on hoverflies or Majerus and Kearns 1989 on ladybirds), or different habitats (e.g. Hayward 1994 on sandy shores or Wheater and Read 1996 on animals under logs and stones), or implementing different techniques (e.g. Unwin and Corbet 1991 look at insects and microclimate, whilst Richardson 1992 examines pollution monitoring using lichens).

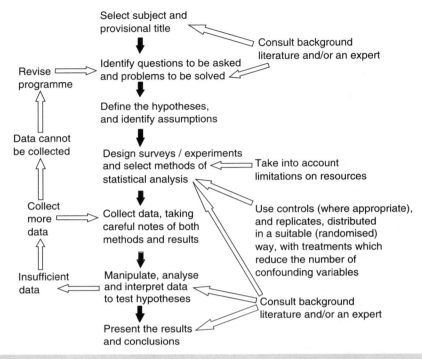

Figure 1.1 Flowchart of the planning considerations for research projects

Ecological research questions

Having decided on a provisional topic, the next step in the successful planning of any research project is to identify those questions you wish to ask and then to formulate the aims and objectives. There are various reasons for researching particular plants, animals or environments and this section provides a quick overview of the scope of ecological projects.

Many studies involve monitoring the number of species, number of individuals (relative abundance), estimates of population size or density (absolute abundance) or community structure (diversity, evenness and richness). Additionally, studies on animals may require observations of the behaviour of individuals or groups and their interactions with each other and their environment.

Monitoring individual species and groups of species

Sometimes, projects may be targeted at a single species. For example, where an important species is present because of its positive interactions (including any conservation or commercial value) you might require information about its distribution, population size and dynamics, age structure, behaviour, etc. Where you have a more negative view of a species (e.g. because it spreads disease, competes with native fauna and flora or is an invasive species that dominates a habitat to the exclusion of other species), you may need information about its distribution, dispersal ability, vulnerability to disturbance and predation, etc. An interesting aspect of biogeographical study is the examination of species distributions where species are expanding or contracting their ranges, perhaps as a result of climate change or other factors (either natural or human-influenced, e.g. habitat disturbance and fragmentation). Conversely, you may be interested in groups of organisms, examining the diversity of communities, the interrelationships between plants and animals in protected areas, or in establishing ecosystem function in relation to environmental legislation (e.g. the EU Water Framework Directive). Studies spanning a wide range of different taxa can be particularly valuable in understanding complex environmental systems, although they may be difficult to implement and the subsequent analysis and interpretation of the results can be complicated.

Monitoring species richness

In studies examining species richness, you might be interested in the presence or absence of one or more species (or other taxonomic group) in order to investigate the links between such species and aspects of the environment (perhaps in terms of the ecology of the species concerned, or in studies of pollution where the species may be useful as a biological indicator of certain toxins). Here, simply listing the plants and animals present may suffice. Although this may appear to be a quite simple approach, care needs to be taken to ensure that sampling techniques are used that are appropriate to both the organisms under consideration and the habitats in which they are found. For example, studies on the richness of bird species in urban parks may be complicated if some parks are dominated by relatively open habitats of amenity grassland and formal flower gardens, whereas others feature dense shrubberies and even woodland. Observations of the species present may be easier in the open habitats than under dense canopy. Care will thus be needed to ensure that all species are counted in an appropriate way at all sites. For these reasons, issues around surveying habitats and sampling organisms are considered in the next three chapters.

Monitoring population sizes and density

In population and density studies, it is the number of plants or animals of particular species that is of importance. Such studies may look at the number per unit area (i.e. the density) or calculate estimates of population sizes. Densities are taken from the estimated population divided by the area sampled. However, for mobile organisms it may be difficult to identify the spatial limits of the population (e.g. in studies of butterflies in agricultural sites some species may be highly mobile with individuals not being restricted to defined small sites). Under such circumstances, densities may be less useful than population estimations of the animals using particular sites. If populations of several species are being studied then it is important to ensure that the sampling methods used are appropriate to all the species being monitored. For example, in rainforests, some species of butterflies are found mainly within the canopy and are only occasionally caught at ground level and, conversely, some are predominant at ground level. Clearly, any survey comparing such study sites should incorporate sampling at both levels.

Monitoring community structure

Other studies might involve establishing the structure of the community of a specified area or habitat type (e.g. the community of fish in a lake, or the community of insects inhabiting a certain species of tree). Such studies may involve sampling a large range of quite different organisms. These may differ in size, distribution (both spatially and temporally), use of micro-habitats and, in the case of many animals, mobility. As such care needs to be taken to ensure that the methods are as comprehensive as possible and are not biased towards or against any particular species or groups of species. For example, sieving soil to examine the communities of animals living within different layers (leaf litter, humus layer, A horizon of the soil, etc.) may underestimate larger animals that are found at low densities (e.g. large ground beetles), and may overestimate species that are found in large aggregations if sampling happens to coincide with these groupings (e.g. some woodlice). Several different techniques may need to be used together during a single study in order to obtain a broad understanding of the community structure of such habitats.

Monitoring behaviour

Studies on animals may involve monitoring the behaviour of individuals, even if this is not the primary purpose of the study. Knowing whether rabbits are feeding, being vigilant for predators, etc., may be useful if numbers are being counted in particular sites. Of course, other research projects will focus primarily on animal behaviour. Such behavioural studies may involve the observation of a number of individual animals in a variety of settings, or the interactions that animals have with others of the same, and/or different, species. It is essential that the location and methods used by the observer do not influence the behaviours being monitored. For example, working too close to large mammals with young may be dangerous to the researcher and may mean that the major behaviour monitored is vigilance directed against the observer.

A note of caution

Although focusing in on the main aim of the research will help to formulate the procedure to be followed, you will also need to understand the limitations of the approach that you take. Census methods (e.g. simple species counts) can be quick to implement and provide substantial amounts of data in a short time. In contrast, techniques to assess population sizes or community structure tend to be much more time-consuming and may produce complex datasets. However, you should be aware that although it is usually possible to extract census information from population or community study datasets (albeit with a loss of detailed information), it is not possible to use census methods to assess community structure or population levels. In general, it is important to have at least some knowledge about the ecology and behaviour of the species or group of species under investigation when designing the research project, irrespective of the type of study being undertaken.

Creating aims, objectives and hypotheses

Once a topic for research has been chosen, you can work out the aims of the study. These are important since tightly defining the aims helps to focus more clearly on the work in hand and can avoid problems in implementation. Woolly aims such as 'to investigate invertebrates under logs' may be a starting point for a more focused aim such as 'to determine whether the number of invertebrates found under logs is related to the size of the log'. This then leads to further questions, including:

- Which invertebrates are to be examined, that is should they be identified to species, or merely counted *en masse* or allocated to ecological groups (e.g. predators, herbivores, etc.).

- What is a log (i.e. when is a fallen piece of wood a log rather than a twig?) and how many logs should be investigated.

- How should we standardise or otherwise account for the condition and type of the logs (degree of decomposition, species of tree, etc.).

- Which measurements of size should be incorporated (e.g. length, width, surface area touching the ground, volume).

- Where should we sample the logs.

- Which statistical method(s) should we use to analyse the data?

Once these questions have been answered, they become objectives that can be used to determine the methods. The aims and objectives lead us to the setting up of working hypotheses. For example in our study of possible relationships between log size and the numbers of invertebrates found beneath them, we would set up a hypothesis to be tested. It is common practice that the hypothesis to be tested is a null hypothesis; in this context that 'there is no relationship between log size and the number of invertebrate animals found underneath them'. Most univariate statistical tests examine the likelihood of the null hypothesis being true (see Chapter 5). A null hypothesis should meet the following criteria:

- Be a single, clear and testable statement – where more complex research questions are asked, you should break these hypotheses down into individual statements that are treated separately and tested in turn.

- Have an outcome, typically either 'accept' or 'reject' the null hypothesis.

- Be readily understandable to someone who is not a scientist.

Reviewing the literature

You should always review the planning and implementation of each stage of your research project by using current information, either from others who have been involved in similar research, or using texts, papers in journals or other information sources (e.g. the Internet), or a combination of these. Be aware of possible biases in the information used, especially where this is obtained from websites belonging to individuals (rather than respected organisations) that have not been independently validated. Most papers in reputable journals and many text books have been examined by independent referees, although even these may contain factual inaccuracies and personal opinions that may not conform to current opinion. Although considered the gold standard of information sources, even peer reviewed journals are subject to bias against the publication of negative results. It is important to start your review of the literature as early as possible, since it is an ongoing process throughout your research and should inform each stage of your project. At the very least you should begin by reading the literature to establish that your proposed idea has not been already published and to define the gaps in knowledge that you will attempt to fill. It is likely that as you read one paper, you will find references to other work that may be important.

If you are new to a subject matter, you should first try and locate seminal piece(s) of work in the field. Typically, this will be close to the top of a search list of highly cited papers and can be found by ordering a search by 'times cited'. Take a detailed look at the seminal paper(s), the reference list and who is citing that paper. In journal databases (e.g. Web of Knowledge), citation networks can be viewed to examine the connectedness between a seminal paper and all those papers that cite it. This is useful because it can elucidate key papers in the field and reduce the search effort dramatically. Typically, your first search should include seminal works and a collection of the most recent papers in the field (i.e. last few years). It might be helpful to order these by journal impact factor (if available), since parochial journals may not contain as high-quality science, although sometimes smaller research papers with less apparent impact can provide valuable information in the form of species lists, new methods and negative findings that are often not reported in more mainstream journals. An additional word of warning: highly cited papers can also be poor papers in the field since other authors might simply be referencing them to make an example of that piece of work (e.g. 'Black and White's (2000) experimental design has been shown here and by others to be flawed'). Knowledge of the literature can assist in avoiding 'blind alleys' and unfruitful lines of enquiry or techniques. There are two main types of literature, primary and secondary.

Primary literature

This is first-hand information, for example, articles in specialist journals, reports, MSc and PhD theses. Journals that publish only refereed papers (i.e. those that have been through a peer review process) are the most important sources of primary, up-to-date information, and where possible your literature review should focus on this type of source. Other primary sources include technical reports, management plans, consultancy reports and species lists (e.g. from annual recorder reports). All of these can be useful sources of information for ecological projects, but you should be aware that they may not have been edited or their quality controlled.

Secondary literature

This is prepared from other sources of information, including textbooks, review articles, etc. If you are lucky there may be specialised books covering your subject area. These may provide a good starting point; however, since books are secondary sources of information, journals are a preferred source of reference for most research projects. The coverage in student textbooks is rather more superficial than that in specialist texts. If there are review articles on your subject, these may be useful to obtain an overview and as a source of new references. Reviews are found in edited book sections, journals that specialise in reviews (e.g. *Trends in Ecology and Evolution*[1]) and journals that have occasional review papers. Review papers in established scientific journals have usually been subject to peer review.

Other sources of information

Maps, personal notes, museum collections and archives may all contain information that can be useful in supporting your research. There are several software systems used within professional ecology that can be useful for research projects. These can help to set your work into a context grounded in practical conservation issues as well as supplying data on either a wider spatial or temporal basis. One such package, the Conservation Management System (CMS) software[2] is used by many countryside

---

[1] http://www.cell.com/trends/ecology-evolution/home
[2] http://www.cmsconsortium.org/software.html

Search terms

managers to produce management plans, and Recorder[3] is the software used by the UK National Biodiversity Network to record, manipulate and map biological records.

Search terms

The use of targeted search terms can identify appropriate works and avoid too many articles or irrelevant papers from being selected. First, identify a list of key words or phrases that could be included in the title, abstract or key words of an article. Begin by using simple combinations of terms (e.g. woodland beetles, hedgerow birds). If you are finding too many papers, then either restrict your search to more recent publications or use more complicated combinations (e.g. by adding 'predation' to the key terms 'woodland beetles' you can significantly reduce the number of articles returned by a search). Note that most databases allow wildcard entries (e.g. an asterisk) to truncate key works (so that space, spaces and spatial can all be covered by spa*), although such terms can increase irrelevant returns (e.g. spa* will also cover any word beginning with spa: e.g. Spain, Spanish, etc.).

Reading papers

You will find more source papers than you have time to read. It is easy to get bogged down in the wealth of published material. Keep your subject area in mind and do not read everything indiscriminately. Skim-read a new reference to decide how much attention it deserves. Start by reading the abstract, skimming the subheadings and then the first paragraph or so of the introduction and the last paragraph of the discussion. Only read in detail those papers that are particularly relevant.

Keep a copy (photocopy, scanned image or electronic copy) of key references and make notes of (or highlight on photocopies) any useful information. Write down the full reference, since all material cited in a research report must be listed in full in the reference list; there are few things as annoying as having to refind the details of a reference that you read much earlier. For journals, the full reference includes the authors' names and initials, year, journal title, volume number and page numbers. For the other types of references see the guidelines for the reference list in Chapter 6. For books, in addition to noting the authors, publishers etc., take down the library classification number for your own reference in case you need to return to it at a later date. Increasingly, many papers are available on-line. The level of access depends on the services subscribed to by your organisation. All libraries offer an interlibrary loan service to provide access to works that are difficult to get. When using such a service, note the date on which you request any interlibrary loans to help to keep track of your requests.

---

[3] http://www.nbn.org.uk/getdoc/4bd40fa0-3692-4cea-9d81-701838bc8e87/Recording-software.aspx

Be critical as you read; do not accept everything as true just because it is published. Look at the evidence and decide whether the conclusions are justified, or whether the results could be interpreted differently. It is, unfortunately, not uncommon for assertions to be made with no supporting evidence. You will find that different authors in the same field may disagree. It is particularly important to distinguish opinions (and speculation) from evidence. You can make your own interpretations and conclusions from the work of others and cite them using expressions such as 'an alternative explanation for the results of Green and Brown (2010) is that...'. Read critically and keep your use of the information relevant for constructing your own account.

Practical considerations

Research (and especially field research) can be an unpredictable business. However, with careful thought it should be possible to ensure that most eventualities are covered. One of the major issues is the legal aspect (including access rights to land and the impacts on protected species or habitats). Health and safety is another obvious concern and it is essential that you ensure that there is no danger either to you or to those around you. In addition, practical approaches like effective time management, efficient data recording and security, and the appropriate use of equipment and techniques will also help to deliver a research project successfully.

Legal aspects

When planning fieldwork it is important to take into account your responsibilities and any legal implications of the work. At an early stage in the planning of the project, always seek permission to work on a site from the landowner and any other interested parties (e.g. relevant statutory bodies if the sites have some form of special protection, such as Sites of Special Scientific Interest or National Nature Reserves). Keep disturbance to a minimum, and remove as few plants or animals as possible. Identify specimens *in situ* (if you can) so that they need not be removed from the habitat. Whole plants should not be taken without the express permission of the landowner. It is good practice in the field not to pick plants of any kind, unless absolutely necessary. In many countries there are a number of protected species (e.g. orchids) that should not be uprooted, picked or harmed in any way. You should check with the appropriate governing organisation for the country involved for details on protected species. In some countries there is specific legislation covering protected species (e.g. the Endangered Species Act 1971 in the USA, the EU Birds Directive 1979, and the Wildlife and Countryside Act 1981 in the UK).

Some animals may not be disturbed or handled without a permit (e.g. birds and bats, among others, in the UK), nor may some microhabitats (e.g. badger setts in the UK).

Rare animals and plants are often protected by law and a licence may be required to handle or disturb them, without which you could face a large fine. You should also consider the ethical aspects of your study. This is particularly important where animals (especially vertebrates) may be harmed. Under these circumstances appropriate licensing authorities should be consulted. Even where this is not the case, the benefits of the study must be considered against the removal of individual animals and plants (sometimes in large quantities), possible damage to the environment, and other impacts including the spread of disease. Ethical aspects of work on animals are covered in work by Reed and Jennings (2007). Where relevant, your work may need to be examined through a local ethical review process. Most universities and similar organisations operate these, and further details can be found in RSPCA and LASA (2010). It is also worth reading the discussions by Minteer and Collins (2005a, b; 2008) and Parris et al. (2010) on this subject with reference to fieldwork. **Note:** this handbook is a guide that does not definitively outline the legal position or interpretation of any act or regulation. In all cases of protected species it is the responsibility of the researcher to check with the relevant bodies to understand what guidelines and regulations are in force.

Health and safety issues

Look after your own health and safety and that of those around you and try to avoid adversely influencing the environment. All investigations should be assessed for any risks, including those caused by the terrain, the techniques and any sudden changes in weather (see Barrow 2004). Any chemicals being used should be checked against appropriate regulations, and risk assessment should be produced to identify safe use, disposal and how to deal with spillage and accidents. In the UK such regulations (COSHH – Control of Substances Hazardous to Health) are covered by the Health and Safety Executive (HSE[4]). Many organisations have their own health and safety guidelines; in the absence of these, advice is available in Nichols (1999), Winser (2004) and Aldiss (2007). Pay particular attention when working at the coast (especially with regard to tides and hazards, including quicksand), rivers (above all with regard to potential flash floods), the uplands and mountain areas (especially with regard to sudden changes in weather conditions and the risk of exposure) and in situations where there is a risk from disease transmission, poisonous and venomous animals, or antisocial or violent behaviour from other people.

Try to avoid working alone in the field. If you must work alone, always carry a mobile (cell) phone and check out and back in with someone who knows your planned routine. Clothing and footwear should be suitable for the terrain and climatic conditions (warm and waterproof). Safety glasses and gloves should be worn to handle chemicals, and suitable gloves to protect you against thorns and infection from soil and

---

[4] http://www.hse.gov.uk/coshh/

water-borne disease. Keep your tetanus injections up to date, and take particular care where there is risk of Weil's disease (near to rivers and canals) and Lyme's disease (transmitted by ticks). Be aware of any other risks, including bites from venomous creatures (e.g. snakes), other toxic species (e.g. poisonous plants, scorpions, spiders and stonefish, which have poisonous spines), and the possibility of rabies from mammals. In general, ensure you are properly equipped (see Box 1.2) and avoid risks to help to ensure problem-free project work.

Box 1.2   Suggested minimum equipment required for field work

Always recommended

First aid kit
Map(s) of the area
Paper for recording (preferably in notebook form and waterproof if possible)
Pencils and sharpener (avoid ink if possible; even waterproof inks run when wet)
Mobile (cell) phone (fully charged and with spare batteries)
Whistle
Compass or global positioning system (GPS) receiver
Watch
Appropriate clothing and footwear
Appropriate safety equipment (e.g. gloves, safety glasses, etc.)
Appropriate sampling equipment (nets, traps, plastic tubes, plastic bags. NB: put any samples in a double plastic bag and label each bag so that if one label does come off, the other is there for reference)

Recommended depending on terrain, weather and timing and extent of work

Survival bag
Emergency food
Torch (fully charged and with spare batteries)

If working outside of your usual comfort zone, for example in locations overseas, check with researchers or organisations (e.g. The Royal Geographical Society of London) who have experience of working in such countries or habitat types. Even within your sphere of experience, avoid complacency since conditions may change and even small risks can be hazardous unless planning is comprehensive.

Implementation

A thorough literature search and any knowledge you have gained from other scientists experienced in the specific field of your proposed research is vital to define and refine the appropriate methods for your study. There will usually be a balance between the ideal solution in terms of the methods used, and logistical restrictions of time, and availability of equipment and expertise. Provided that your project has been well designed, and pilot studies have enabled you to refine your techniques, the implementation of the project should be straightforward. Here we emphasise the importance of careful note taking and time management during your project.

Equipment and technical support

Ensure the availability of equipment before starting, and obtain essential items well in advance of beginning your research project. You may need to allow adequate time to order specialist equipment or materials. If your project requires technical support, arrange this as far in advance as possible.

You need to be as familiar with your equipment as possible. This includes knowing how reliable it is likely to be under the conditions in which you are working and whether you need to have access to spare components or extra full items of equipment. For example, small mammals will eventually chew through the sides of aluminium small mammal traps (they rather more quickly get through the sides of equivalent plastic traps). Although it is possible to patch these up, this is tricky in the field and therefore spares should be taken. Anything that runs on batteries (e.g. data loggers or light traps) need to be recharged on a regular basis and spare batteries, bulbs, etc., should be available while in the field. If you are using multiple pieces of equipment, then you should ensure they are comparable. Different makes of bulb may provide different light levels in light traps, and monitoring equipment from different companies may have different levels of accuracy and resolution. Wherever possible, ensure that as similar as possible equipment is used for an individual project. Instrumentation errors may occur if users are unaware of the limits of the equipment (where attempts are made to estimate between gradations on an analogue scale). Make sure that you are familiar with all aspects of your equipment before engaging in field work. Calibration may also be important with equipment that requires regular calibration against standards of approximately similar values to the variables being measured (e.g. calibrating pH meters at pH 7 for neutral soil and water pH measurements). It is also important to take care of equipment, including protecting it against vandalism, theft and animal damage (many a moth trap has been trampled by inquisitive cattle when placed in their pasture).

Field/Laboratory notebook

Keep all your data and notes in an organised format, preferably in a hard-backed notebook (Box 1.3). Have a standardised way of recording your data, including everything that might be relevant: the date, weather conditions and notes of any important points that occur to you while carrying out the project. It is useful to record data in the same layout as you will on a computer spreadsheet for analysis (see p. 22). If you do use sheets of paper (similar to the one illustrated later in Figure 1.3), make sure that they all go into a ring binder as soon as possible. It is very easy for single data sheets to get lost. It is worth checking to see whether there is a standardised recording sheet available for use with the technique that you are employing. Biological Records Centre species recording sheets,[5] Breeding Bird Survey recording sheets[6] and freshwater invertebrate recording sheets[7] are some examples. Make photocopies of data at frequent intervals and scan them into a computer if possible. Where data loggers are being used either to note climatic variables (see Chapter 2) or to log behaviour (see Chapter 4), then make sure that you take backups of your files as soon as possible. Enter data and comments in electronic form whenever possible and create backup copies on a regular basis, including copies lodged on a networked drive or Internet hub.

Box 1.3 Keeping a field notebook

Use a field notebook to write down data, ideas, observations, tentative conclusions and hypotheses as you do your fieldwork to create an immediate and faithful history of your research. Produce comprehensive, clearly organised notes as a reference and so that you can reconstruct the research time-line and follow the development of your thoughts and ideas. Although you may use other collection sheets (e.g. pre-printed data collection forms to ensure data are collected consistently in different locations and at different times), your field notebook should provide the context for data collection and help resolve ambiguities or inconsistencies when preparing for analysis. After data analysis, reference to your notebook may generate further hypotheses and suggest further lines of enquiry.

Select an A5 or A6 hardback notebook with a spiral binding and wide-ruled lines, ideally on waterproof paper. Use a clutch-type propelling pencil with a moderately soft lead (HB or B). If you do not use waterproof paper, then encase your notebook in a plastic bag large enough to cover your hand and the notebook when writing. In very wet conditions write on an A4 sheet of white plastic with a thick soft pencil (use kitchen cleaner to erase your notes after transcription).

What should be recorded?

The first page should include contact details in case of loss, the subject of your research and the start and end dates of the period covered by that notebook. Include any conventions used, for example 'All times are recorded as local time'. Number the pages and ideally add a contents table to make searching for information easier. Write on the right hand page only so the left hand page can be used for ideas generated by reading about similar observations or relevant research papers. Leave a few lines between observations for comments to be inserted later (e.g. 'No bark damage here 23 June, see p39'). Add a 2 cm margin to write the time, location (e.g. from a GPS reading) or other identifying labels. Create lists of codes, acronyms, specialist terminology, etc., at the back and include any emergency numbers (e.g. those of field buddies). Other useful notes about equipment (how to use, limitations of instruments, etc.) and any numerical information you might require in the field (simple formulae for calculations, random numbers, etc.) can also be added here.

Before starting work each day, write down the date, weather, general location, nature of the habitat and purpose of the day's work. Write down any changes in weather or habitat that occur during the day, for example 'At 15.00 hours snow began to fall and visibility was reduced to 20 m'. When observing behaviour, note the sampling method, how animals were chosen for observation and the recording method (e.g. whether you noted all occurrences or used a time-sampled method). If animals or start times are chosen at random, note how this was done.

---

[5] http://www.brc.ac.uk/record_cards.asp
[6] http://www.bto.org/bbs/take_part/download-forms.htm
[7] http://www.fba.org.uk/recorders/publications_resources/sampling-protocols.html

> Note the type and model number of any equipment (e.g. GPS receiver type Garmin 12). Some instruments need calibrating at intervals, so record the time of calibration and any raw data and subsequent calculations so that any arithmetic errors can be identified and corrected later. Use your notebook to create rough species accumulation curves, etc., so you can tell when you should stop collecting data (see Chapter 2). Along with observations, note the time and, if possible, the location from a GPS receiver. Although notes should be made at the time observations are made, it may be difficult to observe and write at the same time, but if you do rely on memory, you should note this. Write exactly what you see or hear, for example when describing behaviour do not ascribe a function to it in the guise of a description (i.e. do not write that a goose was 'vigilant' when you mean that the bird was in a standing posture with an elongated neck and raised head).
>
> Sketches enhance any photographs you take of your study sites and you will have a sketch available in your notebook the next time you visit the area. Sketches can be added to subsequently (annotating any changes with the date of the amendment). The value of sketches can be increased by explanatory labels. A careful sketch can aid species identification and will help to jog your memory when you encounter a species in the future; such sketches are more valuable if labelled with the diagnostic feature(s) *you* use (e.g. 'two spots on forewing' or 'sepals reflexed'). Landscapes change over time and maps may not reflect this. In some cases no map of a suitable scale may be available and a sketch map can be made using compass and tape, or by pacing out distances using a pedometer. This may be adequate to note the locations of those animals or plants of interest.
>
> It is also useful to record any notes and actions from supervisory or team meetings both as a reminder and to ensure that any designated actions have been completed as planned.
>
> See also http://www.geos.ed.ac.uk/undergraduate/field/fnb

Time management

Conducting a piece of research within a relatively short time span is a demanding process, and will require careful time management. You will need to allow time to check the feasibility of the research project and should also add sufficient time for method training or familiarisation. Ensure that you leave time spare to allow for the almost inevitable problems associated with both field work and laboratory analysis. Check out your methods by first running a pilot to help identify any pitfalls and inadequacies. It is easy to underestimate how long it takes to analyse, interpret and write up a research report. Figure 1.2 illustrates a timetable (in the form of a Gantt chart) for a research project expected to last 33 weeks – around a full academic year: many research projects are substantially shorter than this. Note how the time allocated to actually collecting the data is relatively short and overlaps with the ongoing literature review and writing process. Also, you are strongly urged to begin writing the report before completing your data collection; it should be possible at this stage for you to write the methods and introduction since the former will describe what you have been doing, whereas the latter reflects your background reading of the literature. Table 1.1 shows an example time-scale for a research project lasting a week (e.g. on a field course). Note that even with such a short time-scale, sufficient time needs to be devoted to planning the project to ensure the best chance of a successful outcome (Box 1.4).

PRACTICAL CONSIDERATIONS

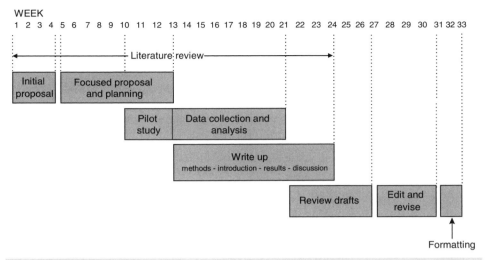

Figure 1.2 Example time-scales for a medium-term research project

Table 1.1 Example time-scales for a short research project

Day	Morning	Afternoon	Evening
1	Select topic, identify aims and hypotheses, create programme of study, select sites and techniques, identify major resources required, complete risk and ethical assessments	Test techniques, train on equipment, implement a pilot study	Evaluate pilot study, amend programme of study
2	Implement amended programme and collect data		Enter data into spreadsheet, write methods
3	Collect data		Enter data into spreadsheet, write introduction
4	Collect data		Enter data into spreadsheet, edit introduction and methods
5	Collect data		Enter data into spreadsheet, plan results tables and figures
6	Collect data		Analyse data
7	Write results section	Write discussion	Complete report

> **Box 1.4 Some tips on time management**
>
> - Be realistic and work within your strengths and weaknesses.
>
> - Plan your long-term goals.
>
> - Have a weekly plan, with realistic and achievable targets, and update this on a regular basis to reflect your progress.
>
> - Identify not only the key phases of your research project but also other areas that will take up your time (both in terms of study and general living) to ensure that your research time-scales are realistic and achievable.
>
> - Prioritise your work into that you have got to do, that you ought to do and that you would like to do but may not have time.
>
> - Use odd snatches of time: a trip by train may be an ideal opportunity to read references or edit your manuscript.

Project design and data management

Since research is about asking questions, you need to design your project so that it will answer the question effectively, without allowing your design to introduce ambiguous results or results that are open to other interpretations. This is where the planning phase starts to define what you are going to measure and how. If we look at an example where we investigate the types of birds found inhabiting a woodland patch, then we have a choice of ways in which we record the data. We might note how many individual birds there are, or the numbers of each feeding type (insect feeders, seed feeders, etc.) or how many individuals there are in each species. These measurements enable us to obtain a picture of the birds found in a woodland patch. If we only monitor birds in a single woodland patch, we could worry that our chosen woodland is unusual in some way and therefore not representative of woodland patches in general. We could therefore examine a series of patches and obtain data for 10 or more. Now if we wish to describe how many birds were found in all of these woodlands, we require some sort of descriptive statistic to summarise the information across 10 or more patches. Descriptive techniques include estimates of the average values per sampling unit (e.g. per site), population estimates and densities, methods of describing distributions (i.e. whether organisms are distributed randomly, evenly or in aggregations) and measures of community richness, including diversity and evenness indices. These techniques are discussed in more detail in Chapter 5.

Most projects go beyond a simple description of particular species and sites in an attempt to make comparisons or generalisations that can hopefully have wider applicability. For example, if we decide to investigate whether the number of animals found under decaying logs on a woodland floor is influenced by the size of the log, we might approach this in one of three ways:

1. by looking at possible differences between samples, for example if the logs were easily divided into two classes (large and small), we could compare the numbers of animals found under each size class.

2. by looking at possible relationships between variables, for example we might have a wide range of sizes of logs and decide to examine whether the number of animals varies in some systematic way (either increasing or decreasing) as log size increases.

3. by looking at possible associations between frequency distributions, for example, we could compare the frequency of obtaining predators, herbivores, decomposers, etc., under each of two size classes of logs.

We will examine each type of question in a little more detail later in this chapter (p. 29), while the analytical techniques needed to answer these questions are described in Chapter 5. Other questions that might be asked include looking at the similarity of sites based on their species composition (i.e. are the animals found under logs from different tree species similar in species composition to each other) or predicting the presence or numbers of a species from a knowledge of the environmental conditions (i.e. is the presence of wood ants nests predictable if we know the woodland type, topography, microclimate, etc.).

We need to ensure that our study does not produce ambiguous results. For example, in a comparison of the invertebrate diversity between urban ponds and rural ponds we could aim to include the size of each pond studied into the survey design. If we did not manage this, and found that the rural ponds surveyed happened to be both larger and contain more invertebrates, it would not be clear whether the results were due to rural ponds being more diverse or whether it was simply an effect of pond size. The correct experimental design would be to either standardise on a given pond size for both environments, or to make sure that the full range of pond sizes were included in both environments (and measured, recorded and built into the subsequent analysis). Other factors that would have to be standardized, or at least recognised as covariates, in this particular study would be the quality of the water, the pH, age of pond, and so on.

The goal of the study may be to get a deeper understanding of the system by gathering a wide range of variables. In the pond survey example, this might mean that in addition to pond size, various measures of water quality and chemistry (nutrient status, oxygen content, pH, etc.) and the numbers of each species of plant and animal we observe. For a large number of ponds (each usually recorded as a single row in a data spreadsheet) there may be a large number of variables (usually recorded as columns) leading to a large data matrix. In order to examine and make sense of such a complex dataset, we would need to move into the realm of multivariate analysis (see Chapter 5).

Designing and setting up experiments and surveys

There are two main approaches to collecting data: experiments and surveys. An experiment involves the manipulation of a system, whereas a survey depends on observations being taken without manipulation. For example, if we were interested in how many invertebrates could be found under logs of varying size, we could either survey a woodland floor finding as many logs as possible and recording both the number of invertebrates and the size of the log, or we could devise an experiment where we placed logs of differing sizes on a woodland floor and after a period of time examined the number of invertebrates underneath them. The advantage of the experimental approach would be that we could standardise all aspects of the logs except for size, for example age, the degree of decay, the type of wood and the distance between logs. All of these factors may influence the invertebrates found and confuse any relationship with log size. However, with a survey we would get an impression of what was happening in a real life situation (i.e. under logs that had been naturally deposited). Moreover, we may decide that the experimental approach is damaging to the environment, here adding unnaturally deposited logs to a natural system. In addition, for practical reasons we might decide that the colonisation of newly introduced logs by invertebrates would take longer than the time available for the project to be completed. In most environmental research programmes, surveys are useful for generating ideas about important factors, but because of the additional complexity in real situations, cannot identify cause and effect. Because experiments strip away the additional complexity, they are more useful in identifying cause and effect, but less likely to be applicable to real life situations. When designing experiments it is important that as many factors as possible are kept constant. So, for example, if we are interested in identifying whether an increase in pesticide concentration will lead to a decrease in aphid infestation of a crop, then the same amount of water (assuming this is the solvent or carrier for the pesticide) should be used for each application (irrespective of the concentration applied) so that we are testing the amount of pesticide added, rather than the amount of water added. In addition, it is important where possible to include a control treatment. In this example we would use a water-only treatment to see if the addition of any water had an impact. If we did not do this and found a reduction in aphid numbers with any application of pesticide, we would be unable to tell whether this was due to the pesticide or the fluid added.

Types of data

In order to design an appropriate experiment or a survey, you need to think about the type of data you wish to collect. The pieces of information that are recorded (e.g. height of tree, number of birds, density of plants per unit area) are termed variables and may be in the form of one of three types of data. The simplest type is categorical or nominal data where each value is identified as one of several distinct categories (e.g. male or female animals; purple, red or yellow flowers; grasses, ferns, herbaceous plants, shrubs, trees). Where we can place the categories in some kind of logical order, so that the data are able to be ranked, this is called ordinal data (e.g. large, medium-sized or small ponds; above the high-tide line, mid-shore and below the low-tide line on a rocky shore). The most detailed types of data are those measurements that not only can be placed in a logical order, but where there is a known interval between adjacent items in the sequence (e.g. the number of deer in a herd; the temperature in the centre of patches of plants of differing sizes; the depths of a series of ponds). There are two types of measurement data: interval data and ratio data (Box 1.5). In most cases the analysis of interval and ratio data uses the same techniques and so in this text we will tend to combine them and refer to them as interval/ratio data or measurement data.

Box 1.5 Differences between interval and ratio data

Interval data have no true zero so that negative values are possible (as in temperature measured on the Celsius scale where 0 °C refers to the freezing point of water rather than the lowest possible temperature) and where measurements cannot be multiplied or divided to give meaningful answers (as in dates).

Ratio data are measurements that have an absolute zero point that is the lowest possible value (as in temperature measured on the Kelvin scale where 0 Kelvin is absolute zero) and so negative values are not possible (e.g. you cannot have –6 foxes). With ratio data all basic mathematical operations can be performed to give meaningful answers; e.g. you can derive a ratio of water lost from soil following drying out as follows (where the original mass = 20 g, and dried mass = 16.5 g):

the proportion lost on drying = $(20 - 16.5)/20 = 0.175$, i.e. 17.5%

Note that we can readily reduce measurement data to ordinal or categorical, but not the other way around. Thus, if we count the numbers of invertebrates of different species on a particular type of plant, we could subsequently express this in order of dominance from abundant through to rare (an ordinal scale) or indicate the presence or absence of different species (categories). However, if we originally merely record presence and absence of species, we cannot subsequently calculate the numbers of individuals. Thus, if in doubt, it is safest to collect the information at the highest resolution possible.

It is good practice to use a standardised data recording sheet (ideally in your field notebook) that is as similar as possible to the way in which data will be entered into a computer for analysis to avoid data transcription errors in moving from paper to a computer spreadsheet. In our example (Figure 1.3) we have two types of variables: fixed and measured. It is easier to deal with these in order so that fixed variables come first, followed by measured variables. Fixed variables are those determined by the research design and do not vary during the investigation (record number, site, day and time). Hence these can be added to the recording sheet early in its production. Measured variables on the other hand are those factors recorded during the investigation, values of which will vary depending on the site, day, time, etc. (numbers of wrens, blackbirds, etc.).

Sometimes derived variables are also required (i.e. variables produced from measured data, e.g. the proportions that each species forms of the whole catch). Such derived variables can be added to the right of the measured data once the latter have been entered on a computer spreadsheet, since the required computations are easily carried out using spreadsheet functions. In most cases, data will be recorded as numerical values. Where categories (e.g. site) occur, codes or names can be used, although some computer programs will not accept letter codes, so you may need to allocate numeric codes to such variables. You should make sure that any paper copies of results sheets are photocopied or scanned as soon as possible after completion, and that electronic copies are properly backed up.

Data recording sheet: 12						
Comments and notes: Weather mild, sunny at first then clouded over a little around 9 am Ramblers walked by at about 8:30 am – didn't make much noise (no dogs!)						
Record number	Site	Date	Time	Number of wrens	Number of blackbirds	Etc.
0101	Black Wood	02/05/09	08:00	4	0	
0102	Black Wood	02/05/09	08:15	3	2	
0103	Black Wood	02/05/09	08:30	2	2	

Figure 1.3 Example of a data-recording sheet for an investigation into the distribution of woodland birds

Sampling designs

When implementing a project, it is rarely possible to collect information on all the animals or plants present. Usually we need to use a sample that we hope to be representative of the situation as a whole. The total number of data points that could theoretically be gathered is known as the population (this is a statistical population rather than the population of animals or plants – Box 1.6); the actual number of data points is termed the sample size. Larger samples are usually more representative of populations, although this depends on the variability of the system being studied (small samples may be reliable representations of populations with low variability). Those elements of a system that are calculated (e.g. the mean number of plants per square metre in a meadow) are termed statistics and are estimates of the true attributes of a statistical population (called parameters – Box 1.6). So if we counted all the plants in the entire meadow, we would be able to calculate the actual mean value per square metre (a parameter). Since it is usually impractical to count all individual plants, in reality we usually count plants in a subset of the meadow (i.e. take a sample) and calculate the mean numbers per square metre using this sample on the expectation that it will be representative of the whole site (a statistic). This sort of situation occurs in many types of survey. For example, market researchers obtain opinions from large groups (samples) of people and use these to indicate the attitudes of the population as a whole.

> **Box 1.6   Terms used in sampling theory**
>
> A **population** is a collection of individuals, normally defined by a given area at a given time. For example, scientists refer to the decline in the world population of Atlantic cod in the last century or the annual harvest of Northeast Atlantic cod. These are both true populations.
>
> A **sample** is a term that can be used ambiguously but is a subset drawn from a population, which usually includes a quantity (e.g. 100 individual fish taken from the Northeast Atlantic cod population).
>
> A **parameter** is a population metric that is estimated from a variable (e.g. the mean body size of Northeast Atlantic cod) and can be used to summarise data. Importantly, statistical tests aim to derive parameters from a population in order to test for differences, relationships, associations, etc.
>
> A **variable** is a measurement that may change from sampling unit to sampling unit (e.g. the body size of Northeast Atlantic cod taken from a sample) and can be used to summarise collected data (e.g. by taking the mean).

Choosing the sample to be taken needs some care and at this point it is worth discussing why replication is important. Since environmental systems are usually intrinsically variable (i.e. physical, chemical and biological factors differ spatially and temporally), the larger the sample, then the more representative it will be of the population (i.e. the more of the natural variation will be covered). However, the larger the sample, the more time and effort it will take to collect it. There are methods to calculate the optimum sample size; however, these rely on knowledge of the variability of the system. This is rarely known in advance, although a small pilot study may give some indication. If it is known or suspected that there is substantial variability, then a large sample should be taken. In most ecological surveys a large sample would include over 50 data points. However, where the population is likely to be very large and variation is expected to be great, even larger sample sizes may be required. Otherwise it is best to aim for as large a sample as possible after taking into account constraints including the size of the workforce, the time available and how much material is present in the system under investigation. Where several levels of a number of variables are to be analysed (e.g. male and female animals of each of three different age groups: young, mature and old), then it is important to take sufficient replicates of each subgroup (e.g. young males, mature males, etc.) to be able to account for within-group variability. This will inevitably have an impact on the required sample size and is another reason why the intended statistical analyses should be considered at an early stage of project planning. See Krebs (1999), van Belle (2002), and various on-line calculators[8] for further details of the different calculations that can be used to estimate sample sizes depending on the intended statistical analysis technique to be used.

---

[8] For example, Rollin Brant at the University of British Columbia (http://www.stat.ubc.ca/~rollin/stats/ssize/index.html) and The Australian National Statistical Service (http://www.nss.gov.au/nss/home.nsf/pages/Sample + Size + Calculator + Description?OpenDocument)

In surveys of community structure, it may be important to know that the majority of species in an area have been recorded. In this case, species accumulation curves may help. At its simplest, this involves plotting the accumulated number of species against increasing sampling effort. Sampling effort is the number of sampling units (quadrats, pitfall traps, animals handled, hours of observations, sites surveyed, etc.). Box 1.7 illustrates the use of species accumulation curves in quadrat sampling (see Chapter 3). There are a variety of methods of modelling species accumulation curves (see Colwell, Mao and Chang 2004; Magurran 2004 for further information) and many standard software packages include routines for this (e.g. those obtained from Pisces Conservation[9]).

Box 1.7   Species accumulation curves for two sites

By plotting the cumulative number of species found against the number of quadrats examined, it can be seen that as the number of quadrats used increases, the number of species also increases. At the point at which the curve levels off towards horizontal (the asymptote), we may assume that we have obtained the maximum number of species and can stop sampling. For site A (dashed line, diamonds), we may not yet have reached the total number of species, even after 30 quadrats and should consider increasing the sampling effort. For site B (dotted line, squares), it appears that we have reached about the maximum number of species that we can expect to get. In fact, we probably reached this number at round about 16 or so quadrats. This difference between sites A and B might reflect not only a difference in the number of species found there, but also a difference in heterogeneity of the site, with site A being less homogeneous than site B. Note that had we looked at the data for site A after 12 quadrats (solid line), we might have assumed that we had reached the maximum number of species as the curve levels off. This highlights the importance of collecting past the initial point of curve levelling to check that it truly does reflect the asymptote.

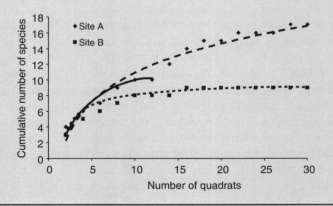

---

[9] http://www.pisces-conservation.com/

Since you are generally taking a sample in order to make a valid estimate of a parameter of the population (e.g. the number of species, the mean temperature, the proportion of predators), a central requirement is that the individuals sampled are independent of each other. It is important to recognise (and avoid or account for) situations where the individuals sampled are linked in some way as a result of the sampling design. For example, we might compare the number of spangle galls found on leaves chosen at random on oak trees growing in clumps with those on isolated oak trees. If we found over 20 trees in separate clumps but only 10 isolated trees, we might be tempted to take double the measurements from the isolated trees. However, this would mean that individual data points from isolated trees were linked by virtue of the tree on which they were growing, and shared many different attributes with each other. Such data would not be independent of each other and hence may cause problems in interpretation since we would be unsure whether any differences between clumped and isolated trees were due to the multiple measurements from some trees. It would be better to use unbalanced sample sizes (i.e. 20 clumped and 10 isolated trees) than use non-independent data. Similarly we should not take data from more than one tree in any clump in case these are linked in some way. From a statistical analysis point of view, few tests require equal sample sizes, and even where this is a problem it would be preferable to reduce the number of trees from clumps that were measured.

If we survey a pond in order to look at the animals and their relationships with several physical, chemical and/or biological factors, then no matter how many replicates we take, we are merely describing what happens in a single entity (i.e. this one pond). Such a study does not tell us anything about pond ecology in general, and the use of such replicates is termed pseudoreplication and should be avoided (Hurlbert 1984, van Belle 2002). In order to broaden our approach and gain more of an understanding of ponds in general, we would need to study a large number of separate ponds. Thus, studies of single sites or small parts of sites may not reveal information applicable to the wider ecological context.

In some situations the data collected are linked to each other by design. For example, we might be interested in comparisons of matched data (e.g. examining the animals found on cabbages before and after the application of fertiliser or pesticide, or the numbers of mayfly larvae found above and below storm drain outflows into a series of streams). These designs can be perfectly sound, but because the data are matched (by cabbage or by stream) we require a slightly different approach to the resulting analysis (see Chapter 5).

When designing your sampling strategy, it is important to consider the variability and whether the timing or order of sampling might bias the result by measuring only part of the potential variation. For example, sampling the insects present on thistle flower heads will be biased if all the data are collected in the early morning since this will miss any animals that are active later in the day. If two areas are being compared, sampling one site early and one site later will introduce another variable into the comparison: we would not just be looking at the two sites, but also at two times of day. Since it would be impossible to separate the two variables, it would be difficult to draw conclusions from such a survey design. It is in managing some of this variability that experiments come into their own, because they standardise as far as possible the conditions under which the subjects are examined. It is much easier to design an experiment where only one factor (also known as the treatment) is manipulated, while all others remain constant. However, if we wished to survey a real life situation (as opposed to examining a rather more artificial experimental design) then we would take into account the time of day. We could do this by designing our survey so that we alternated the measurements or observations that we took from our two sites, sampling first one then the other, then back to the first, and so on, to get a spread of measurements for each site over the day. Alternatively, we could sample on successive days, reversing the order in which we sampled the sites on each day.

There are several sampling layouts that help us to avoid bias. One commonly used approach is random sampling. Here a random sequence is used to determine the order in which to sample plants, or the coordinates to sample experimental plots or survey sites. Hence, if we wanted to randomly sample 1 m × 1 m quadrats in a field, random coordinates can be used to position the sampling sites (Figure 1.4a) using pairs of random numbers generated using a calculator or computer, or obtained from a table (Table 1.2). This works by using pairs of numbers as sampling coordinates, so if we have coordinates of 23 and 85 in a sampling grid that is 10 m by 10 m, we would place our quadrats 2.3 m along the base and 8.5 m up the vertical axis. Random sampling may also be used to determine which site is visited first if sites are allocated number codes that are then selected randomly from the table.

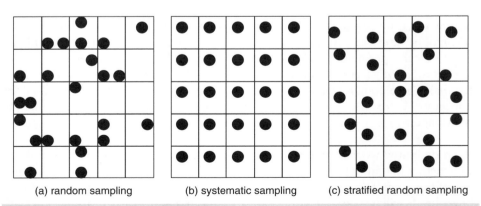

Figure 1.4 Examples of sampling designs

Table 1.2 Random numbers
Coordinates can be extracted simply by taking pairs of random numbers in sequence from the table (e.g. 23, 85 – shaded values – provides the position within a sampling area where we would take the first measurement of a series).

23	85	56	84	92	4
62	51	27	74	83	84
56	32	87	75	95	5
87	7	20	30	25	12
99	86	29	41	29	39
31	73	30	73	27	97
24	38	91	16	17	66
94	59	12	17	37	39
41	67	25	42	2	84
32	67	48	99	74	3
68	1	59	20	25	7

Although random sampling is often appropriate for selecting sampling points, where there is a great deal of variation across a sampling unit, such as a site, by chance the coverage may not include all of the heterogeneity present. For example, in Figure 1.4a, the two squares in the lower right of the sampling site have no sampling points. If the site was reasonably homogeneous, then this would not be a problem. However, if these small squares represented the only damp area within the site (covering around 8% of the total area), then this particular variation would have been missed altogether. An alternative strategy would be to use systematic sampling (Figure 1.4b). This is an objective method of spreading the sampling points across the entire area, thus dealing with any spatial heterogeneity. So, to systematically sample the insects on trees, we might collect from every tenth tree in a plantation.

Usually systematic sampling would provide us with random individuals unless for some reason every tenth individual is more likely to share certain characteristics. Suppose we used systematic sampling to examine the distribution of ants' nests in a grassland. We could place 2 m by 2 m quadrats evenly 10 m apart across the site and then count the number of nests within each quadrat. However, if ants' nests are in competition with each other, they are likely to be spaced out. If this spacing happens to be at about 10 m distances, we would either overestimate the number of nests if our sequence of samples included the nests, or underestimate if we just missed including nests in each quadrat. It would be better in this situation to use a mixture of random and systematic sampling (called stratified random sampling – Figure 1.4c) where the area was divided into blocks (say of 10 m × 10 m) and then the 2 m by 2 m quadrats were placed randomly within each of these. This type of sampling design can also be applied to temporal situations by, for example dividing the day into blocks of 4 hours and allocating the order of the sites to be sampled within each block using different random numbers.

More sophisticated methods of laying out sampling plots (or allocating sampling periods) may be useful for experiments. One example is the use of a Latin square design (Figure 1.5) that ensures that experimental treatments are equally distributed across the rows and columns of a sampling design.

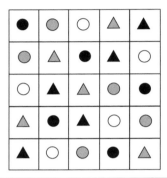

Figure 1.5   Latin square design for five different treatments
Each treatment is represented by a different symbol.

Planning statistical analysis

Although at this stage we will not discuss in detail the ways in which data are analysed, it is important to at least have sight of the likely methods that may be used. This is because different statistical methods are required to deal with different research questions. Most techniques require specific types of data to be gathered, and lack of care at this stage could result in data being collected that cannot answer the question posed. In addition, some statistical analyses have minimum numbers of data points that are needed to obtain meaningful results and a few need balanced designs (i.e. the same number of data points for each factor measured). In this section we will discuss some of the major types of analyses that you could employ to answer certain commonly asked types of questions. Further details are given in Chapter 5. As always it is worth looking at the literature to see what types of analysis have been used in similar studies to the one you propose to do. There are several major groups of analysis based on the broad types of approach required.

Describing data

We need a variety of techniques to describe the data that we collect. This might be as a data exploratory technique (to check the data to see how variable a dataset is, or what sort of distribution we get, etc.), to understand some aspects of the data (e.g. how diverse communities are) and for communication purposes (to be able to discuss the results, orally and in writing, with other people).

Here simple plotting of measured variables on frequency histograms (or tables), cross-tabulation of one (nominal) variable against another, and examining the range of the data (from the minimum to the maximum) may help to check for errors and ensure that we can chose the correct type of test for subsequent analysis. Extracting statistics, such as diversity indices (to describe species richness) or evenness (to describe how equal the proportion of species is within a community) can be important in being able to assess what sort of community we have. Likewise the estimation of population size or density can be important in some studies. Similarly, we can use the average value of a variable to describe the magnitude of the majority of the data points (usually in conjunction with some measure of how variable the values are and how many data points were collected). Chapter 5 reviews these descriptive statistical techniques and more detail can be found in Wheater and Cook (2000; 2003).

Asking questions about data

If we wish to ask specific questions of the data, then we are in the realm of inferential statistics. These usually involve the testing of hypotheses. It is standard practice to set up a null hypothesis alongside the questions to be asked. The null hypothesis tests the chance of there being no significant difference between samples (or relationship between variables, or association between categories of variables). So if we wish to know whether there is a difference between two samples (e.g. comparing the number of birds found in deciduous woodlands with the number found in coniferous woodlands), then we actually test the null hypothesis that *there is no significant difference between the number of birds in deciduous and coniferous woodlands*. Note that we are looking at 'significant' differences. These are differences that are unlikely to have resulted from random variation in the individual woodlands sampled. For this we need a method that tests the null hypothesis that there is no significant difference in the sample averages (see Chapter 5 for more details). In addition to difference tests between samples, there are also relationship tests between variables, and tests designed to examine associations between categories of variables. Table 1.3 summarises some commonly used statistical approaches to these research questions.

There are various questions that we might ask as part of an investigation, and it is important to be clear about possible analysis methods in advance of any sampling. The choice of test depends not only on the question being asked, but also on the data types being used. Where data are ranked, but not measured (i.e. ordinal data – p. 21), then a suite of tests called non-parametric tests may be used. The alternative (using parametric tests) is more robust and generally preferred, but requires data to be on a measurement scale (i.e. interval/ratio data). Therefore it is usually an advantage to obtain

Table 1.3 Common statistical tests
Note that in each case, there are possible questions (and analyses) dealing with more than 2 samples and/or variables – see Chapter 5 for further details

Example question	Null hypothesis	Type of test	Data required
Is there a difference between the number of birds found in deciduous woodlands and coniferous woodlands?	There is no significant difference between the number of birds in deciduous and coniferous woodlands	Difference tests e.g. a t test or a Mann–Whitney U test (see Chapter 5 p. 265)	Two variables: one nominal describing the woodland type and one based on either measurements (i.e. actual numbers) or on a ranked scale that describes the number of birds
Is there a relationship between the number of birds and the size of the woodland?	There is no significant relationship between the number of birds and the size of the woodland	Relationship tests, e.g. correlation analysis (see Chapter 5 p. 269)	Two variables: one (either measured or ranked) that describes the number of birds and one (either measured or ranked) that describes the size of the woodland
Is there an association between whether birds are resident or not and whether the woodlands are deciduous or coniferous?	There is no significant association between the frequency of residency and the frequency of woodland type	Frequency analysis e.g. a chi-square test (see Chapter 5 p. 274)	Two variables: one nominal describing the residency status of the birds and one nominal describing the woodland type

measurement data rather than rankable data wherever possible. Even where measurements are taken, parametric tests may not be the most appropriate. This is because most parametric tests require the data to conform to a type of distribution called a normal distribution. Briefly, this is determined by examining histograms of the data (with the variable of interest plotted on the x axis and the frequency of its occurrence on the y axis) to see whether they have a symmetrical pattern rather than a skewed distribution where the mode (the value with the maximum frequency) lies towards the left or right rather than the centre of the histogram (Figure 1.6). For further details of which test to use, see Chapter 5. There are also different tests depending whether the data are matched or unmatched (see p. 266).

To illustrate some of the considerations in project design and data collection, we start with a research question that sounds relatively simple on the face of it: is there a

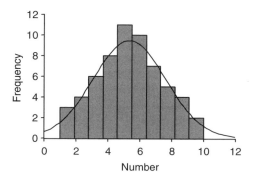

Figure 1.6 Dataset approximating to a normal distribution

relationship between the size of trees and the number of squirrels' dreys in the canopy of the trees? Ideally we would want to measure the canopy height with some degree of accuracy. This would enable us to work out whether the relationship exists using a parametric statistical technique called Pearson's product moment correlation analysis (see Chapter 5). However, it may be difficult even to see the tops of very tall trees and those obscured by other trees. Thus, we may estimate tree height, perhaps into several groupings. We can of course rank these data but this means that we need an alternative approach for analysis that is suitable for ordinal data. This is Spearman's rank correlation coefficient analysis, which is not quite as powerful as the Pearson method. The power of the test is its ability to detect a true relationship (or difference, or association) if one exists. If we knew that any such relationship was likely to be fairly weak, then the less powerful technique might not reveal it and we could be wasting our time in not measuring the trees relatively accurately to obtain measurement data and thus employ the more powerful test. Alternatively, if we are only interested in revealing strong relationships, then using ranked size classes to indicate tree height may be acceptable. We will review these techniques in more detail in Chapter 5. The other complexities in this apparently simple question include ensuring that all other aspects are as constant as possible (e.g. species of tree, surrounding landscape, density of the squirrel colony, etc.).

Predictive analysis

We may wish to collect data to set up a model that enables us to predict the outcome in a hypothetical situation, one of the simplest of which is known as a linear regression model. Thus, if we are interested in looking at a possible relationship between woodland size and the number of birds and knew that this was likely to produce a significant linear relationship, then we may wish to use this fact to calculate the expected number of birds found in any particular woodland. This could be used theoretically or in conservation management to check that we have the sort of bird biodiversity that we expect from other data. Here it is important to note that any such prediction can only be made if the woodland area in which we are interested lies between the minimum and maximum value of the dataset we used to establish the model. We first need to establish which variable is the dependent and which is the independent variable: that is, which is likely to be affected (the dependent variable – plotted on the y axis of a scatterplot) by the other (independent variable – plotted on the x axis of a scatterplot). Here obviously the number of birds (the dependent variable) is more likely to be dependent on the size of the woodland (the independent variable) than vice versa. To develop such predictive models, data should be measured values. We can extend the technique to cover the case where there are a number of independent variables (e.g. woodland area, habitat diversity, area of associated green space, distance to nearest water body) that might influence the number of birds (see Chapter 5 for further details).

Multivariate analysis

Where the question to be asked is a complicated one involving a number of dependent and/or independent variables then multivariate analyses may be appropriate. The choice of analysis depends on whether the dependent variable is a category, or a ranked or measured variable, and on whether the independent variables are categories, ranked or measured (or even a mixture). Although most (but not all) such analyses only have one dependent variable, there may be multiple independent variables. For example, we may want to know whether the number of birds differs in different types of woodland when we take into account the woodland size (measured variable), woodland type (nominal variable), distance to the nearest neighbouring woodland (measured variable), age of woodland (measured variable) and the land use type surrounding the woodland (nominal variable). Here, we could enter all of the data into one analysis that would take into account the interrelationships between each variable and produce a model describing the relative importance of each variable on the number of birds (this particular example could be analysed using a generalized linear model – see Chapter 5 for further details). Such techniques are powerful but require a full understanding of the data and their attributes and may be quite complex to interpret (see Chapter 5 for further details).

Examining patterns and structure in communities

Ecological datasets can be very complex and difficult to visualise. For example, a dataset might include many variables collected as measurements (including counts), as ranks (e.g. scores of abundance) or in a binary form (e.g. presence or absence data). Chapter 5 introduces a number of techniques for visualizing complex datasets to enable the use of a range of different types of data, although variables with large numbers of zeros (as can occur when surveying relatively rare species), cases where data are heavily skewed, or situations where variables are measured on scales of greatly differing magnitude may require data transformation before using these techniques (see Chapter 5 for further details).

As an example, we might collect information about woodlands on the basis of their size, age, distance to the nearest neighbouring woodland, etc. Since some of these variables will be related to each other, we might wish to find out the underlying pattern of interrelationships within the data and hence identify a number of unrelated factors that can be used instead of our large number of variables. This is a data reduction exercise, reducing the number of variables we have measured into a smaller number of factors that take into account the interrelationships between the variables.

Alternatively, we might wish to look at a range of species found in each of several woodlands and see which woodlands have similar species types. This is a similarity or clustering analysis and depending on the technique used to calculate the similarities, data are normally recorded as a matrix that contains either measurements (e.g. counts), ranks (e.g. ranked abundance) or binary data (e.g. species presence or absence). A similar technique to clustering enables us to visualise patterns in either the individuals (in this example, the woodlands) and/or the variables (here, the types of species). This is known as ordination and there are a number of different methods available depending on the algorithm (i.e. statistical formula). Such methods can utilise data comprising measurements, ranks or binary information.

Choosing sampling methods

The choice of sampling method will usually be dependent upon the habitat type and organisms being studied (see Chapters 2–4). However, all sampling techniques have limitations, and there are some general principles that are applicable to most sampling methods, for example:

- Some techniques may be suitable for a limited range of habitats, or be biased in favour of active rather than sedentary animals, or collect only a subset of the population being examined (e.g. males rather than females, or those migrating rather than those resident). It is therefore very important that limitations are known and dealt with during the design of the research to avoid later problems in interpretation.

- Usually we wish to collect as many data as is feasible, bearing in mind any restrictions in terms of time and personnel. However, different techniques require different skills or time frames, and may collect differing amounts of data, thus our choice may be restricted by logistical considerations.

- Many techniques are not directly comparable with each other, and even using the same technique but under different conditions (e.g. between habitats with very different vegetation layers, between night time and daylight collections, at different times of the year) may not produce comparable data.

- Limitations of the equipment being used may mean that monitoring environmental variables is restricted if, for example, differences between areas are smaller than the accuracy of the equipment allows.

- Resource issues may determine the methods available for use: the cost of equipment, necessity for training, ease of relocation of apparatus between sites, and health and safety issues could all limit the choice of methodology.

Summary

Ensure that you take sufficient time in the planning phase of your research project to cover all of the component parts. This includes health and safety and legal issues as well as making sure that your aims and objectives are focused and that any methods employed are appropriate to gather and analyse data. At each stage, consider the details of the implementation, whether this is in the practicalities of sampling or data management. Box 1.8 gives some general guidelines that should be ticked off in advance of implementing your project.

Box 1.8 Checklist for field research planning

☑

- ☐ Determine the question and formulate the aims, objectives and hypotheses
- ☐ Determine whether a manipulative experiment or observational survey would be the most appropriate method to use
- ☐ Decide whether you are looking primarily at the presence or absence of a species, relative abundance (e.g. counts of organisms), absolute abundance (e.g. population sizes or densities), community structures, behavioural responses, etc.
- ☐ Determine the statistical analyses that are likely to be employed
- ☐ Select the appropriate sampling technique, taking into account the intended statistical analysis
- ☐ Decide on the taxonomic level for identification, or the appropriate ecological grouping you will use (see Chapter 2)
- ☐ Assess what types of data will be collected and produce a standardised recording sheet
- ☐ Work out the sampling design including sample sizes
- ☐ Select appropriate sample sites
- ☐ Determine the site characteristics that will be monitored and choose appropriate techniques for doing this
- ☐ Obtain permission to use the sites and check any legal restrictions
- ☐ Risk assess all the work to be carried out
- ☐ Employ a pilot study and amend your protocol if necessary

2
Monitoring Site Characteristics

When studying an ecological system, it is often useful to monitor aspects of the environment that are either fundamental to the study or may provide background information for your subsequent interpretation of your results. For example, knowledge of the tree management history in a woodland may be important in a study of bird breeding success, whereas an appreciation of the water quality in a series of ponds will be useful in interpreting the numbers of different species found. When designing your research project, it is important to consider which features of the local and wider environment you should measure and take into account. It is also important to have clear reasons for the selection of study sites to avoid possible bias. This section introduces some considerations for choosing and profiling study sites.

Site selection

The selection of sites depends on the study to be undertaken. Where the emphasis is on obtaining species lists for a particular locality, then obviously site selection is predetermined. However, if a broader investigation is being considered, then it is important to carefully consider which sites are to be examined in order to maximise the number to be used, while bearing in mind logistical aspects, including the ease of travelling between them. Site proximity is also important if sites need to be matched in some way. For example, if the bird fauna of coniferous and deciduous woodlands is being compared, then the sites should all be at similar altitudes, be of similar sizes and shapes, have similar canopy cover, etc. Access may also be a problem and could influence site selection. It is also likely that in some habitats disturbance may be more problematic than in others. For example, working on farmland may involve disruption because of spraying (both for pests and adding fertilisers), weeding and harvesting. It is therefore important to build any known or likely disruption into the research design (or at the very least note any disturbance as it occurs).

Practical Field Ecology: A Project Guide, First Edition. C. Philip Wheater, James R. Bell and Penny A. Cook.
© 2011 John Wiley & Sons, Ltd. Published 2011 by John Wiley & Sons, Ltd.

The number of sites to choose is usually based on a balance between the ideal scenario of as many sites as possible to take into account site variability and the logistics of the study in terms of workforce, time and equipment availability. Previous discussions (p. 23) have considered sample size and this should be a prime consideration, even if it means re-evaluating the aims of the study if adequate replicates cannot logistically be taken. Broad-based habitat surveys (e.g. Phase 1 habitat surveys – see p. 39) record semi-natural vegetation and other wildlife habitats, and can be useful in helping you to decide geographically which sites to consider.

Site characterisation

Living organisms are affected by the physical, chemical and biological characteristics of the environment (Table 2.1), which can influence survival and growth, as well as behaviours including dispersal and reproduction. Physical and chemical factors include microclimate, soil characteristics, attributes of water and pollutant levels (e.g. sewage in freshwater and toxins, including heavy metals and pesticides). Monitoring pollution may require sophisticated equipment, although the presence and absence of tolerant and intolerant species can be used to identify whether habitats (especially freshwaters) are heavily polluted (see Hellawell 1986; Mason 1996; Wright, Sutcliffe and Furse 2000). Watts and Halliwell (1996) and Jones and Reynolds (1996) discuss techniques for measuring a wide variety of environmental variables.

Even factors that are not major aspects of the study may need to be recorded. For example, it is important to record if there are any unusual or extreme weather conditions, or if sites and equipment are disturbed or damaged by people or other animals. Knowing when and to what extent this has occurred may assist in the subsequent interpretation of the data. Without such knowledge, discussion of your results may be difficult and could lead to incorrect conclusions being drawn. In some study designs you can control for habitat effects such as altitude and aspect by including sites with a range of attributes in these variables within the sample set and using the statistical analysis to take into account the variability in the habitat variable concerned (e.g. see Chapter 5, Box 5.8).

Habitat mapping

Habitat mapping is a valuable tool for ecologists enabling estimations of land coverage for different vegetation types. Since habitats are usually identified through their dominant vegetation types (grassland, heathland, woodland), it is perhaps no surprise that habitat mapping involves a closer look at the vegetation. Surveys can be quite general, as with the Phase 1 habitat surveys used in the UK (Figure 2.1 and JNCC 2003), or much more detailed descriptions of habitat types. Sometimes the survey can be tackled using remotely sensed data or aerial photographs as long as these are recent, clear images that are at an appropriate scale to transpose onto the map. The finished

Table 2.1 Common factors influencing living organisms

Physical factors	Chemical factors	Biological factors
Microclimate/weather	Substrate characteristics	Food for heterotrophs
Temperature	pH	Seasonal availability
Humidity	Oxygen content	Quantity
Rainfall	Salinity	Quality
Wind speed	Organic content	Accessibility
Altitude	Chemical elements	Competition for resources
Latitude	Water characteristics	Space
Substrate characteristics	pH	Light
Type	Oxygen content	Food
Texture	Biological oxygen demand	Mates
Compaction	Ammonia	Death and disease
Temperature	Salinity	Predation
Water	Conductivity	Parasitism
Water characteristics	Chemical elements	Disease
Temperature	Nutrient levels	Exploitation
Depth	Nitrates	Removal of pests
Inundation	Phosphates	Vegetation features
Flow	Potassium	Sward height
Suspended solids	Sulphates	Diversity
Turbidity	Calcium	Structure
Tides	Magnesium	Patchiness
Inundation	Silicon	Habitat features
Force	Micronutrients	Type
Habitat features	Pollutant levels	Diversity
Size	Organic	Structure
Shape	Toxins	Species features
Age	Pesticides	Invasive species
Shade	Heavy metals	Algal blooms
Shelter	Acidity	Disturbance
Aspect	Alkalinity	By humans
Insolation	Atmospheric	By other animals
Temperature extremes	Aquatic	Management
Inclination	Rainfall	Species removal
Instability	pH	Species (re-)introduction
Run-off	Sulphur dioxide levels	Habitat restoration
Natural disturbance	Nitrogen oxide levels	Habitat reclamation
Landslip	Management	Habitat rehabilitation
Flood	Liming	Habitat stabilisation
Fire	Fertilising	
Management	Detoxification	
Landscaping		

survey can be used to give estimates of county woodland coverage, for example, or to identify habitats of local or regional importance (e.g. oligotrophic mires). Land managers or reserve wardens often use Phase 1 maps for habitat management. As fairly broad surveys, Phase 1 maps are of limited use in ecological studies, although they can help to visualise a study site.

In the UK, Phase 2 surveys normally follow Phase 1 surveys. Phase 2 is much more powerful as it maps the vegetation stand types. This involves identifying particular vegetation community types. In Britain these have been defined through the National Vegetation Classification (NVC) system, which details over 250 community types (Box 2.1; Rodwell 2006). Hence, specific types of woodland that in Phase 1 would be identified as semi-natural broadleaved woodland (Phase 1 code – A1.1.1) are subdivided into communities (e.g. W4b: W indicates woodland; 4 states that it is *Betula pubsecens–Molinia caerulea* stand type and, b indicates that the subcommunity is a *Juncus effusus* type). The method involves surveying a number of quadrats (see Chapter 3 for details of quadrat sampling) of certain sizes (depending on the habitat involved) and recording the plants found and their frequency (measured on the Domin scale – see Chapter 3). Although NVC maps give a clearer picture of the type of communities present in an area than do Phase 1 maps, they do not describe small variations within a stand type, which is often only covered by one code. Some UK ecologists have criticised the NVC system as being too general, preferring to use established systems like Peterken's method for woodlands (Peterken 1993). However, the NVC has become the standard in Britain.

Across Europe, there are a number of different vegetation classification systems in use (Box 2.2), which, although they do not map perfectly one on another, are in the process of being developed through the European Vegetation Survey (EVS) network to enable common survey methods and analytical software (Rodwell 2006 – NVC user handbook).

The distribution, extent, shape and boundaries of habitat types can be extracted from remotely sensed data, including aerial photographs or satellite imagery. Aerial photographs may need to be interpreted and integrated with other spatial data in order to use them within maps. This involves the creation of orthophotographs, which correct for perspective and variation in topography. Similarly satellite remote sensing requires geometric correction (georeferencing). Images from satellite photography are also useful in identifying habitat features. Some techniques use multispectral imagery (exploiting the spectral properties and characteristics of different earth surface phenomena), together with *in situ* verification, to identify different land classes. False coloured images involve the allocation of different colours to different wavelengths of radiation in multispectral images and can be used to provide strong contrasts between different land use classes or types of land cover (including vegetation classes), allowing them to be visually separated more readily. Both aerial photographs and satellite images can be used in geographic information systems (GIS) to derive a range of

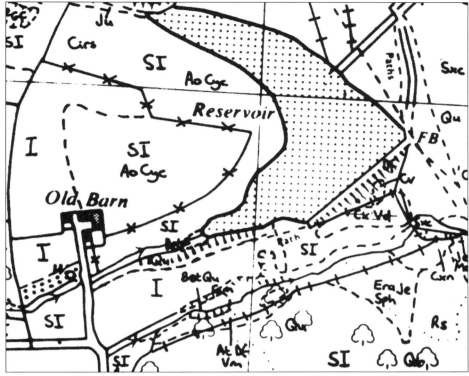

Ordnance Survey Crown Copyright. All rights reserved.

Figure 2.1 Phase 1 habitat map in the UK
Phase 1 habitat surveys involve mapping broad habitat types onto Ordnance Survey maps (1 : 25000 or, more usually, 1 : 10000 scale for rural habitats and 1 : 10000 or even up to 1 : 2500 for urban areas). This involves marking the boundaries of woodland, grassland, heath, fen, etc., ideally using a global positioning system (GPS) to establish exact positions. In the example above, the habitats are labelled with capital letters (I – improved grassland; SI – semi-improved grassland) and the dominant vegetation types are identified using species codes (e.g. Bet – *Betula* sp., Cirs – *Cirsium* sp., Qu – *Quercus* sp.). See JNCC (2003) for further details.

variables (e.g. boundary length, size, distance to the nearest similar habitat, etc.) and to create digital elevation (digital terrain) models that represent the topography of an area. It is often useful to ground truth aerial or satellite images by surveying some sites to check that the features apparent on the images are indeed what you expect. For example, ponds can be hidden from view by woodland canopies and it may not be possible to tell whether ponds seen on maps and aerial or satellite images are always water-filled. These remote imaging techniques can be useful in landscape ecology, for example in examining spatial and temporal changes in ecological process and patterns over a range of scales. Similarly, such methods can be used to examine interactions

Box 2.1 Notes on the resources available for the National Vegetation Classification

Rodwell's (2006) Handbook gives background to, and details of, the method. The primary reference describing the communities is Rodwell (ed.) British Plant Communities, published in five volumes:

Volume 1: Woodlands and scrub (1991a)
Volume 2: Mires and heaths (1991b)
Volume 3: Grasslands and montane communities (1992)
Volume 4: Aquatic communities, swamps and tall-herb fens (1995)
Volume 5: Maritime and weed communities and vegetation of open habitats (2000)

Note: Some summary descriptions are also available published by JNCC, Peterborough (see http://www.jncc.gov.uk/page-4265)

It is worth noting that some NVCs (e.g. Volumes 1 and 3) are relatively easy to use compared with others such as those for mires (Volume 2) and some montane communities (Volume 3), which rely heavily on the moss flora and it can be hard to key out stand types without extensive knowledge of bryophytes. To help with this particular issue, see the paper by Hodgetts (1995) who describes how to identify *Sphagnum* species, the commonest moss in mires, and rather usefully describes the NVC stand types in which different *Sphagnum* are likely to occur.

There are several different pieces of software that can be used to derive NVC types from species data. These include several that are available from the Centre for Ecology and Hydrology (http://www.ceh.ac.uk/products/software/land.html):

MAVIS (Modular Analysis of Vegetation Information System) includes several classifications including NVC, Ellenberg scores (Hill et al. 1999), and Grime's CSR model (Grime, Hodgson and Hunt 1988).

Two older programs are MATCH (a windows-based program that classifies vegetation data against the NVC system) and TABLEFIT (a DOS-based program to identify vegetation types including the NVC and CORINE systems – see Box 2.2 and http://www.eea.europa.eu/publications/COR0-landcover for further details of the latter system).

Box 2.2 Examples of vegetation classification systems

Systems that are used outside Britain include the examination of phytosociological associations (e.g. Braun-Blanquet Associations – see Kent and Coker 1995 for further details), and the use of the Coordinated Information Network on the Environment (CORINE) Biotypes Classification (http://www.eea.europa.eu/publications/COR0-landcover) and the European Nature Information System (EUNIS) habitat classification (http://eunis.eea.europa.eu/introduction.jsp). The CORINE system is aimed at a larger scale than are the other methods. Some methods (e.g. classical phytosociological and EUNIS systems) involve a similar approach to the NVC in that they utilise lists of plants collected in specified ways to identify agreed communities. Rodwell (2006) compares some of these systems to the NVC. Other vegetation classifications have been, or are being, developed in other areas of the world (e.g. in Australia, Canada and the United States). The United States National Vegetation Classification (USNVC) and the Canadian National Vegetation Classification (CNVC) are currently being integrated into the International Vegetation Classification (IVC) to enable the identification of vegetation associations, which is itself is being extended to cover the Caribbean and Latin America (see http://www.natureserve.org/explorer/classeco.htm).

across landscapes between the ecological and physical and/or anthropogenic characteristics. See Heywood, Cornelius and Carver (2006) and Campbell (2007) for introductions to GIS and remote sensing respectively. Fortin and Dale (2005) discuss the relevance of spatial analysis in ecology.

Examination of landscape scale

An ecological research project usually focuses on a particular scale, for example, very small-scale variations can influence the availability of micronutrients and water for plants, larger scale influences on microclimate alter the temperature and humidity, and thus some physiological processes such as transpiration. All of these can impact on the development of the vegetation and hence the nutritional content of leaves. Herbivores will interact at this scale and also at those larger scales that affect dispersal, such as habitat fragmentation and ecotone boundary type. For many animals, hard boundaries can restrict dispersal between adjacent habitat patches, with animals often showing more movement along the boundary interface than across it.

In some projects the effect of scale is the topic of interest. For example, the size of a habitat patch can influence the viability of organisms that inhabit it, with small patches supporting smaller numbers of organisms and being particularly vulnerable to disturbance and fragmentation. The examination of the shape of habitat patches can have an important influence on the ecology of the system via edge effects. Here, round habitat patches have smaller edge-to-central habitat ratios than do long thin patches.

Where animals move relatively rarely between habitat patches, they can be viewed as forming a metapopulation; that is a series of populations that, while being separated in space, do interact in some way. The connectivity of habitats can influence the dynamics of metapopulations with those with relatively free dispersal capability being potentially more viable than those without. The degree of connectivity including the presence of corridors and the distances between habitat patches (acting like stepping stones) can be an important line of research. Some animals require particular mosaics of habitat, for example badgers (*Meles meles*) often build setts in woodland (taking advantage of the root structure to support tunnels), but will forage more often in grassland to find the earthworms that constitute an important source of prey. Red kites (*Milvus milvus*) similarly nest in large trees but hunt for food in open areas.

GIS and remote sensing are particularly important techniques in landscape ecology where the digital mapping of habitat boundaries and elucidation of habitat mosaics can be of major interest. Historical documentation can also assist in understanding the context in which you are working. For example, old species lists, land management information and knowledge of pollution events can all help to understand the dynamics of the landscape in which you are working.

Measuring microclimatic variables

The microclimate (i.e. the local climatic conditions) is crucial for the development of organisms and the maintenance of particular communities, and even small changes in structural features (e.g. shading due to vegetation height and density) may result in large differences in factors such as temperature and humidity. High temperatures can cause desiccation in seedlings and even established plants can lose more moisture than is available in the substrate. This can be exacerbated when associated with dry winds. Small invertebrates and vertebrates (including amphibians and reptiles) may depend on external temperatures to enable them to move around to find food and mates. Insect flight is energetically expensive and may be curtailed at low temperatures. Larger animals lose heat relatively slowly and may either bask or use internal production of heat during activity to raise their body temperatures. Many animals avoid climatic extremes by hiding in burrows, and some undergo hibernation or aestivation to cope with seasonal decreases or increases in temperature and food availability. Temperature can also interact with other variables; warm water contains less oxygen than cold water and so increased temperature in ponds and streams can have an impact on the organisms that live there. Although instruments for monitoring microclimate are often expensive, it is possible to buy (or even make) equipment with a reasonable degree of accuracy quite cheaply (see Unwin 1978, 1980; Unwin and Corbet 1991 for further details). Quite sophisticated weather stations can be obtained that will record a range of climatic variables (pressure, temperature, humidity, rainfall, wind speed and direction, etc.) for study sites (Figure 2.2).

It is often useful to record the daily fluctuations in temperature of sampling sites by measuring the maximum and minimum temperatures (using maximum/minimum thermometers; Figure 2.3). At a smaller scale, more sophisticated and targeted techniques are required. Alcohol or mercury thermometers can be used to measure air or water temperatures, but should be placed within in brass tubes before being pushed gently into the ground to measure soil temperatures (Figure 2.4). For air measurements, it is important to avoid the bulb of the thermometer being directly influenced by sunlight or wind. If this is a problem, then a shield of aluminium foil will help prevent inaccurate readings. Small thermometers or thermocouples can be used to measure the temperatures under leaves, in flowers or under debris like logs and stones. Thermisters that make continuous temperature measurements can be used to obtain continual records for relatively small scale sites such as animal burrows.

Another important element of microclimate is humidity. Plants growing in areas of low humidity may have (xerophytic) adaptations to prevent the loss of water from stomata when transpiring. Invertebrates without waterproof cuticles tend to lose water faster

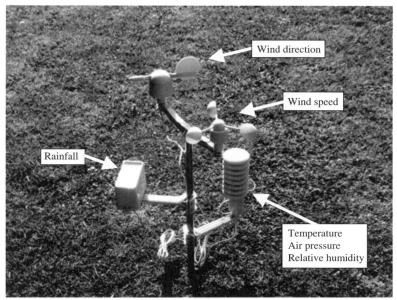

Figure 2.2 Portable weather station
Many automated weather stations will automatically store data and some may transmit it directly to a nearby computer. These machines should be sited away from any object (such as a building) that is likely to shade it from the sun or rain, and ideally at a distance that is over 10 times the height of anything that might obstruct the wind flow.

than those that have some protection. This may result in differences in behaviour, for example vulnerable animals may aggregate together under stones and other debris when the risk of drying out is the highest (e.g. during the day or in the height of summer). Positions on plants even a few millimetres apart may differ in the degree to which they expose animals to drying out and many insects avoid the upper surface of leaves during the day, especially in direct sunlight (unless basking). Conversely, the underside of the same leaf, or adjacent leaves in the shade, may be covered with animals such as aphids. Air humidity may be measured using simple hygrometers (Figure 2.5), although these are slow acting and are not suitable for smaller scale measurements. More precise instruments are often expensive. See Unwin and Corbet (1991) for more details about the measurement of humidity.

Wind speed is an important environmental variable that interacts with the microclimate (disturbing moist air and providing cooling in sunlit areas). Wind speed can influence the rate of transpiration in plants and can have a direct effect on flying insects; even slight breezes can prevent small insects like mosquitoes from flying, whereas higher wind speeds may ground larger animals including some butterflies. Some

Photo CPW

Figure 2.3 Maximum/minimum thermometer
As the temperature rises, the expansion of alcohol in the left-hand tube pushes a mercury column around the bend of the thermometer up the right-hand arm. This in turn pushes a small steel marker up the right-hand tube to mark the maximum temperature. As the temperature falls, alcohol contracts and the mercury retreats to the left. This leaves the maximum temperature marker in place on the right-hand arm and pushes the minimum temperature marker up the left-hand arm. The thermometer is reset by using a small magnet to drag each marker down its tube to make contact with the mercury column. Sometimes the mercury column can become separated. If this happens, try swinging the thermometer in an arc with your arm straight to rejoin the column. Increasingly this type of thermometer is being replaced as the use of mercury thermometers is reviewed and rejected on environmental grounds.

invertebrates that lack wings (e.g. spiders) make use of the wind to carry them large distances. This is most common in small spiders (under 1.6 mg) that crawl up vegetation or other upright objects, exude a length of silk from their spinnerets and are dragged by the wind (a type of behaviour known as ballooning).

Wind speed can be measured using an anemometer, a simple version of which is a cup anemometer, which has three cups that rotate in the wind at a speed proportional to the wind speed (Figure 2.6a). Alternative anemometers utilise the speed at which the wind moves a small fan blade (Figure 2.6b). Where the space in which wind speed is measured is too small for such anemometers, a hot wire anemometer can be used, which exploits the way in which the wind cools a heated wire to measure wind speed

Photos CPW

Figure 2.4 Soil thermometers
Old style soil thermometers like the one shown in (a) are simply glass thermometers containing alcohol (or sometimes mercury) that are housed in steel, brass or aluminium cases to prevent breakage when they are pushed into the ground. You should be careful using glass thermometers in this way since (especially in stony ground) breakages are not uncommon. Modern equivalents include dial thermometers (b) or probe-based digital thermometers (c). It is important to wipe soil from any metal probe after use to prevent rusting and prolong the life of the equipment.

(Unwin and Corbet 1991). The most advanced method of measuring wind speed is by sonic anemometers, which use ultrasonic sound waves to detect wind speed and direction in three dimensions (Figure 2.6c). Sonic anemometers are rather expensive and hence not always widely available.

In addition to specific instruments that measure individual attributes of the environment, multimeters are available that can monitor a wide range of environmental factors (Figure 2.7). Care should be taken to ensure that you are aware of any limitations of such equipment: multimeters can be less discriminating for some factors than can instruments designed specifically to measure that factor. In addition, multimeters may have different levels of precision for the different factors being measured.

Monitoring substrates

For most organisms, the soil, mud or sand in or on which they live has a direct or indirect influence, and several important components of this may usefully be measured. For substrates (including soil), the chemical and physical properties will

Photo CPW

Figure 2.5 Whirling hygrometer
This is also called a psychrometer and consists of two thermometers: a normal dry bulb thermometer and a wet bulb thermometer that is kept wet by water wicking from a container through a muslin sheath that surrounds the bulb. When the device is whirled in the air, the dry bulb records the temperature of the air. The water surrounding the wet bulb evaporates at a rate that depends on the relative humidity of the air and cools the wet bulb down. In lower humidity, there will be more evaporation and faster cooling of the wet bulb. Reference to a wet bulb depression chart will enable the relative humidity to be determined.

directly influence the plants growing there and may affect animals directly or indirectly (e.g. via influences on the vegetation). Most of the biologically available water in soils is held by capillary forces associated with the soil particles. Where there are more small particles, there are more small channels that increase the water-holding capacity of the soil. Sandy soils comprise large particles with rather few, large channels having weak capillary forces that increase drainage, leaving the soil quite dry. Clay soils comprise very small particles with many tiny channels that hold water well and so are much wetter. Heavy clays may become waterlogged, although when they dry out they may become hard and form surface cracks. The water content can be measured by taking the mass of a sample, then heating it at about 105 °C to drive off the water until the mass does not change. The difference between the original mass and that after the water has been driven off is the moisture content of the soil and is usually expressed as a percentage of the original mass.

Soil water content can also be estimated indirectly using a penetrometer (Figure 2.8), since the force used to penetrate the soil is generally regarded as inversely proportional to soil moisture: the harder the soil the less water is held. This technique can also be used to measure the compaction of the soil.

Figure 2.6   Anemometers
(a) Cup anemometer – the device is held in the wind stream, which catches the cups and rotates them with the speed being calculated automatically from the speed of rotation.
(b) Fan blade anemometer – the device is held in the wind stream and rotates the fan blade with the speed being calculated from the speed of rotation.
(c) Sonic anemometer – the device is sited in an appropriate area and records the wind speed in three dimensions based on the way in which the wind affects the speed of travel of sonic waves between adjacent pairs of transducers. Sonic anemometry is very accurate and useful for understanding turbulence. Wind is typically sampled every second and the data are then automatically saved to a computer for later manipulation.

Figure 2.7   Environmental multimeter

Photos MM

Figure 2.8 Penetrometer
(a) Comprises a gauge and a small cone connected to a rod. (b) The cone is pushed through the soil profile in order to measure the force (kgf – kilogram force) needed to penetrate the profile at several depths within the soil profile which is taken as being proportional to the water-holding capacity of the soil. Note that the soil type should not vary between samples and stony soils should be avoided as they potentially cause soil hardness to be overestimated.

Sandy soils tend to be warmer than clay ones, but can have more variable temperatures since they warm up faster in the sun and cool down faster at night. There are many other soil types (e.g. loams that show intermediate characteristics), which can be identified using keys (including those by Brodie 1985 or Trudgill 1989). Soil samples should be taken randomly or using a systematic approach to ensure that heterogeneity is covered (see Chapter 1). Plugs of soil are usually removed using soil augers (Figure 2.9), although bulb planters can provide reasonably standardised samples for surface soils (Figure 2.10). The downside of using bulb planters is that (unlike the cores from soil augers) the soil profile cannot easily be viewed once extracted. Soil samples should be transported carefully to maintain the integrity of their structure, stored at low temperatures (in a fridge or cool box) and analysed as soon as possible after removal from the field.

The nutrient status of the substrate is important, especially for plants and those animals feeding on the plants. Sophisticated instruments can be used to measure the concentrations of major nutrients (e.g. nitrogen, phosphate and potassium). Although simple soil testing kits (of the type used by gardeners) are much cheaper and easier to use, especially in the field, these should be used with care since they can be rather inaccurate. Some soil measurement kits can also estimate soil pH. The measurement of soil pH helps to understand the chemistry of the soil. For example, in alkaline soils some nutrients (e.g. phosphate) may be biologically unavailable since they are converted to insoluble compounds, whereas in acid soils some chemicals (e.g. aluminium) may be released as soluble salts, sometimes reaching toxic concentration levels. It is common to dry soils out before measuring the pH in order to remove the buffering effect of the soil water. Equipment used *in situ* measures the overall pH

Photo CPW

Figure 2.9 Soil gouge auger, mallet and core-removing tool
Used particularly in peaty soils, the gouge auger is pushed (and if required hammered) into the soil, then twisted and pulled out to collect a reasonably undisturbed core of soil: there may have been some compaction of the core when pushing the auger into the soil. The core can be removed using the supplied tool to show a profile of the soil collected. Different lengths of auger and extension rods are available if you wish to collect from depths further below the extraction of the initial cores.

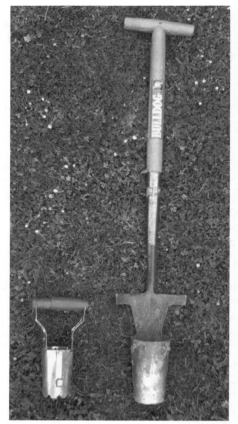

Photo CPW

Figure 2.10 Bulb planters
Gardeners' bulb planters can be used to take reasonably standardised cores of soil for analysis of water and/or organic content. These tools can also be used to produce the holes in which pitfall traps can be placed (see p. 139).

(including of the soil water) and hence may give different values to that of soil tested back in the laboratory.

The organic content influences not only the nutrient status but also the water-holding capacity of the soil. Organic content can be measured by comparing the mass before and after carbon is burnt off at high temperatures in a special furnace. Approximations may be made by floating the organic material off in a bowl of water. The two fractions (one that sinks – mainly the inorganic component – and one that floats – mainly the organic component) can then be dried (at 105 °C until they reach a constant mass) and their proportions compared (usually expressed as percentages of the original soil mass).

Pollutants and other contaminants of substrates are usually measured using expensive analytical equipment (often requiring sophisticated laboratory facilities), although some soil testing kits include facilities to estimate some of these, for example salinity. The salinity of the soil can be especially important where roadside de-icers may have been used, or on transects across salinity gradients (e.g. in dune or salt marsh systems). Conductivity measurements may be made using special meters that measure the electrical resistance. This is related to the total concentration of dissolved substances (particularly inorganic compounds) within the soil. If salt is the major dissolved substance, then the conductivity may provide a reasonable comparison between soil samples. However, if other similar substances are also present in large quantities (e.g. in agricultural land where agrochemical applications have been used), the conductivity will also represent their levels and assumptions about the salinity of the soil may be more difficult.

Monitoring water

A range of chemical and physical factors influence the organisms living in aquatic habitats. The depth can be measured using a metre rule if shallow or a plumb line if deep (taking especial care when working near to deep waters). Mackereth, Heron and Talling (1989) summarise many chemical analysis techniques for freshwaters. Modern sophisticated multimeters can measure a range of factors including temperature, pH, and oxygen concentration (Figure 2.11).

The oxygen content of the water determines the organisms that can survive. It is possible to use an oxygen meter for this or a chemical titration technique such as the Winkler method (see Stednick 1991 for further details). Where the water is polluted with organic matter (from sewage or farm waste), then decomposition may use oxygen faster than it can be replaced. The resulting oxygen deficit is the biochemical oxygen demand (BOD), which can be measured by establishing the oxygen concentration of a sample of the water and then incubating the sample in the dark (to prevent the generation of oxygen by any photosynthetic organisms in the water) for 5 days at 20 °C and remeasuring the oxygen concentration. The difference between these two measurements is the amount of oxygen used in the biological oxidation of organic matter.

The nutrient status of freshwaters may also be associated with organic pollution and can lead to nutrient enrichment, often by nitrates and phosphates (eutrophication). Eutrophication can greatly increase the growth of algal communities (resulting in algal blooms) that, following death and decomposition, can seriously deplete oxygen levels, killing fish and invertebrates. Nitrate and phosphate concentrations can be estimated using colour comparator kits, testing sticks or measured by spectrophotometry (see Stednick 1991 for further details of chemical analysis on water samples). Other contaminants, including heavy metals or pollutants from agrochemical or urban runoff, may require more sophisticated analysis (see, for example, Allen 1989).

Photo CPW

Photo AML

Figure 2.11 Aquatic multimeters
These can be used to measure the physicochemical attributes of a stream. They can be used in shallow streams as illustrated, or lowered from boats or bridges. Calibration of the sensor may be required, especially for pH measurements.

Thermal pollution occurs when water is extracted from rivers to be used as a coolant in factories and then reintroduced later several degrees warmer. A rise in water temperature reduces the level of oxygen that can be held in the water, although this might be complicated by any turbulence associated with extraction and reintroduction that increases the diffusion of oxygen between the air and the water. Water temperature can be measured using simple spirit thermometers or thermocouples.

High levels of turbidity can have a serious impact on animal and plant communities. Suspended solids from storm water, or erosion following heavy rains or from industrial processes may increase the cloudiness of the water and can prevent plants much below the surface of the water from photosynthesising, reducing oxygenation of the water. Suspended solids may also clog the feeding apparatus of filter-feeding invertebrates (e.g. freshwater mussels and fairy shrimps) and interfere with the gills and vision of fish. Although spectrophotometry can be used to measure the amount of suspended material in water samples, more commonly a Secchi disc is used as a simple comparative method of estimating turbidity (Figure 2.12).

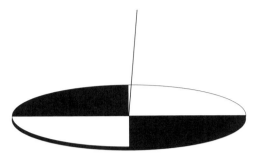

Figure 2.12 Secchi disc
These are circular disks of 0.2–0.3 m diameter, coloured white (for marine studies) or with alternate quadrants coloured white and black (for freshwaters – as illustrated). The disk is lowered slowly into the water column and the depth at which it disappears from view is recorded. It is then raised slowly and the depth at which it reappears is recorded. The mean of these two depths (called the Secchi disc depth) is a useful measure of the turbidity of the water since in clear waters the disc will remain visible for longer than in cloudy water.

Salinity levels determine the organisms that can survive following, for example, tidal surges in estuaries and saline pollution from road runoff following de-icing in winter. The distribution of species along a salinity gradient within an estuary or other brackish water habitat can be an interesting avenue for investigation. Salinity meters can be used, as can chemical techniques (e.g. the argentometric method of chloride determination – see Stednick 1991 for further details). As with measuring salinity in soils (p. 51), using a conductivity meter provides a measure that is associated with the total concentration of dissolved substances within the water. This will only provide a reasonable estimate of salinity if salt is the major dissolved substance.

Water samples brought back to a laboratory for analysis should be placed in inert clean containers that must be filled completely to reduce the amount of air in the container. Water samples should be carried in cool boxes (at about 4 °C) in the dark and should be processed as soon as possible following removal from the field. The maximum period for which water samples can be kept before analysis depends on the variable being measured: only a few hours for the initial measurement when estimating biochemical oxygen demand; up to a day for pH; under a week for nitrate, phosphate and chloride; and a maximum of 6 months for heavy metal concentrations. Even within these time-scales, water samples being kept before analysis may require some sort of fixation process to avoid deterioration (see Stednick 1991 for further details).

Where the environment is not permanently under water, it can be important to understand the level of inundation by water. This can involve monitoring the extent of flood events, the position (and frequency) of water levels on the bank of a river (or flood plain) and the tidal range. The action of waves or other water flow can be measured indirectly as water velocity or erosion. Velocity in freshwaters can be measured as the flow, using flow meters that can be used at several depths within a river.

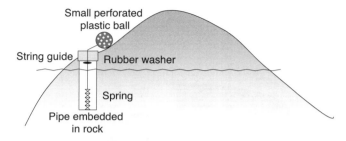

Figure 2.13 Dynamometer to measure wave action
As waves drag the ball, the maximum extension of the spring (measured by the displacement of a rubber washer) is proportional to the wave velocity. The apparatus can be removed from the rock to view the displacement of the washer.

A simple method of estimating surface flow in rivers with long straight sections is to time a suitable floating object as it moves between two points a known distance apart. It is best to use an object (e.g. an orange) with a specific gravity similar to water so that it neither sinks nor floats too high in water.

In tidal zones the velocity of breaking waves can be measured using a dynamometer,[1] which can be made from a small ball attached to a rock using a spring (Figure 2.13). Erosion is another characteristic of the tidal zone that can be monitored. This can be established by examining the percentage loss of mass from a ball of plaster of Paris anchored in place over a set time period (e.g. the tidal cycle). The position on the shoreline in relation to the tidal range can be used as a descriptive measure of tidal inundation and impact, with organisms near the top of the high-tide mark being assumed to be exposed to less inundation than those lower down the shore. In coastal systems it is commonplace to assume that shores on open coastlines are exposed, and imply that they are less protected from the impacts of tidal waters. However, local channelling of tides can funnel water and produce major impacts of drag and erosion, even on some so called sheltered shores. Thus, it is important to measure local conditions rather than rely on assumptions that may be misleading.

Other physical attributes

Other physical attributes include the site's dimensions, aspect and topography. The size and degree of isolation of the habitat being studied influences colonisation rates and the size of the populations that can be sustained: larger habitats are easier to find and will support larger populations, especially of large animals and plants (e.g. deer and trees). Habitat size can be measured in the field (e.g. measuring the width of a stream or river, or the length of a hedgerow) or from maps (e.g. the area of a lake or

---

[1] http://www.stanford.edu/group/denny/dynam_intro.htm

woodland). Distances between habitat blocks are usually easier to measure from maps (including using GIS with digital maps), although, where appropriate, GPS readings in different blocks can help. If a number of sites are to be compared, the degree of connectivity can be estimated on a ranked scale, or measured using an objective value (e.g. the distance to the nearest similar habitat of a defined minimum size).

Aspect influences a number of the factors already considered in this section, including microclimate and moisture availability, and is easy to measure (using maps or a magnetic compass). In the northern hemisphere, south-facing habitats gain more sunlight and are warmer and dryer than north-facing habitats. Animals can also be influenced by the aspect, either through the impacts on vegetation or more directly. For example, some species of ants build nests where the southern slope is shallower than the other slopes, so that the nest gets more insolation at noon (Skinner and Allen 1996). Some animals have shorter developmental periods at higher temperatures, for example in grasshoppers and solitary wasps. Similarly, many plants have preferences for warmer, south-facing slopes. Interesting comparisons can be made between the populations on warmer south-facing slopes and those on cooler north-facing slopes, or even on different sides of the same wall. Light levels associated with aspect or in areas of shade (e.g. under woodland canopies) can be measured using light meters (Figure 2.14), although these do not directly represent the wavelengths utilised by plants. A quantum flux meter can be used to measure the level of photosynthetically active radiation (PAR).

Topography also influences the distribution of plants and animals. Steep slopes can be unstable, especially when they lack vegetation. Even small changes in slope may create differential drainage resulting in heterogeneity in plant cover. Slope angle can be measured using simple surveying techniques. Sighting horizontally from a known position on one of a pair of calibrated posts (ranging poles) to another a known distance away and reading the difference in height provides the height increase (or decrease) over the distance. Alternatively, the angle of the slope between ranging poles can be measured directly using a clinometer (Figure 2.15). Depending on the steepness of the slope, the gradient can be measured every metre and the change in angle of slope plotted on graph paper with a protractor to give a profile of the landscape. On shorelines, it is common practice to use a cross staff to survey the profile (Figure 2.16). Here positions are identified at set heights from each other and the distances between the positions measured using a tape measure, enabling a profile to be drawn.

In field projects it is always necessary to record the location of the site(s) being studied. Maps or global positioning systems (GPSs) (Figure 2.17) can be used to obtain grid references. Many GPSs can also be used to estimate the altitude, which can have an influence on the exposure and microclimate of the area and therefore the animals and plants living there.

Photo CPW

Figure 2.14 Light meter
Light, or lux, meters such as this can give an indication of the light levels at particular zones within vegetation. However, they usually measure within the range perceived by the human eye. If you intend to measure the wavelengths that are used by plants for photosynthesis (photosynthetically active radiation – PAR), then a more sophisticated quantum sensor should be used.

Measuring biological attributes

The living components of a system are usually the factors of primary interest to the ecologist studying them, and the methods and rationale for these will be covered in more detail in Chapters 3 and 4 on sampling methods. Monitoring the non-living (abiotic) factors at a sampling site may be quite complex, and there is increasing interest in using living (biotic) indicators of some environmental factors. This may include species that are present or absent as a result of the tolerance (or intolerance) they have to specific environmental factors (e.g. pollution). Commonly used indicator groups include freshwater invertebrate indicators for organic pollution (Box 2.3) and lichens for sulphur dioxide levels (Figure 2.18). A number of insect groups (e.g. ground beetles and hoverflies) have been identified as having the potential to be indicators of terrestrial habitat quality (Speight 1986). Such groups must be fairly ubiquitous, relatively easily identified and have been well researched, so we have a wealth of previous information (e.g. for ground beetles see Forsythe (2000) and Luff (2007), and for hoverflies see Gilbert (1993) and Stubbs and Falk (1996)). Similarly, plants

SITE CHARACTERISATION

Figure 2.15 Using ranging poles to measure the inclination of a slope
(a) Ranging poles have height markings at set intervals and are placed a known distance apart. The angle of the slope can be measured by sighting from a given position on one pole to the equivalent position on the other using a clinometer. (b) Clinometer used to measure the angle of a slope by sighting from one ranging pole to the same position on another at a known distance apart.

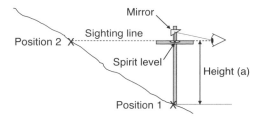

Figure 2.16 Using a cross-staff to survey a shoreline
The observer sights along the arm of the cross-staff (using the spirit level to ensure it is horizontal) to identify position 2 at a set height (a) above position 1.

including aquatic macrophytes and particular woodland ground flora have been used to monitor freshwater nutrient levels and ancient woodland status respectively.

Other features of the biotic component of the habitat being studied may be important in determining why plants and animals are distributed or behave in the ways that they do. This can be as a result of predation, competition, disturbance or interference with resource availability. One example we shall cover here is monitoring light levels under woodland canopies. The percentage canopy closure can be measured by lying prostrate on the ground holding a Perspex grid divided into 100 parts and counting each square in which sky is clearly visible. This can also be done using a photograph of the canopy

Photo CPW

Figure 2.17 Using a global positioning system (GPS) receiver
A GPS can be used to estimate location and/or altitude, as well as to record the path of a transect. When using a GPS receiver it is important to record the accuracy measurements given by the instrument (e.g. an altitude of 247 m ± 12 m).

that is analysed later by overlaying an appropriate grid or series of dots printed on clear acetate (Figure 2.19). Alternatively, image analysis software can be used to identify the amount of light space within a digital photograph. Other attributes of trees can be used to indicate some aspects of the historical ecology of a site. For example, the branches of trees growing in the open begin low down on the trunk, whereas those growing in dense forest are often only present near the canopy. A change from open to dense cover can be inferred from branch scars (i.e. the patterns in the bark that result from a branch being lost). If trees in forests have branch scars low down, this suggests that they originally grew in an open area (perhaps a large gap in the canopy), but then branches were lost as a result of over-shading and hence lack of photosynthetic activity. Where the branch scars are asymmetrically distributed on the trunk, this may indicate that there was a gap to one side of the tree at an earlier stage in its life.

Identification

In order to be sure that the organisms under study are the ones you think you are studying, it is important to be able to identify them to a particular taxonomic level. Some groups of plants and animals are easier to identify than others with some being better served by identification guide books than others. In some cases, relatively small differences between species may necessitate detailed examination, whereas in others

SITE CHARACTERISATION

Box 2.3 Measurements of aquatic invertebrates used in habitat quality and pollution monitoring

This table lists several different measurements of aquatic invertebrates that are commonly used to describe the community in terms of habitat quality and levels of pollution including organic pollution (see also Hellawell 1986; Mason 1996; Wright, Sutcliffe and Furse 2000; Chadd 2010). Those near to the top of the table (including taxon richness, taxon counts and EPT-based measures are simple to use and commonly applied. Those in the middle of the table (including the Trent and Chandler indices and the BMWP score) are more sophisticated. Although less frequently used these days, you may find published work on specific rivers that have used these and so may find them useful in projects comparing developments over time. The most commonly used methods at present are those at the bottom of the table (e.g. ASPT and LIFE scores and the RIVPACS technique).

Measurement	Comments
Taxon richness	Total number of species found (generally larger numbers of taxa indicate better stream conditions)
Specific taxon counts	Some groups are intolerant to poor water conditions (e.g. organic pollution), thus their abundance can be a measure of water quality – Stoneflies (Plecoptera) are particularly intolerant, followed by mayflies (Ephemeroptera) and caddisflies (Trichoptera)
EPT index	The intolerance of Ephemeroptera, Plecoptera and Trichoptera (EPT) to poor water quality can be further utilised by combining the counts of each group to form an index (low values indicate poorer conditions than high values) – alternatively Coleoptera can be added to make the EPTC index.
EPT to total ratio	Dividing the EPT by the total number of individuals found (multiplied by 100 to express this as a percentage) produces a ratio of intolerant groups to total captures (larger values indicating cleaner waters)
Percentage of Diptera	Some animals (e.g. Diptera larvae) are tolerant of poor water conditions and the percentage of these found can be used as an indicator of poor conditions
EPT : Diptera ratio	The number of EPT divided by the number of Diptera gives a ratio of intolerant to tolerant groups (larger values indicating cleaner waters)
Hilsenhoff Biotic Index (HBI)	The number of individuals of each taxon found is multiplied by their HBI – from 0 (sensitive) to 10 (tolerant) – and then each product is summed and the sum divided by the total number of individuals in the sample giving a combined HBI from 0 (clean water) to 10 (polluted water): $$\mathrm{HBI} = \frac{\sum(\text{number in taxon} \times \text{HBI score for taxon})}{(\text{total count in sample})}$$
Trent Biotic Index (TBI)	A score based on the number of specified taxonomic groups found, balanced by the presence of particular pollutant tolerant (or intolerant) taxa – from 0 (low quality) to 10 (clean water). This has been described as a relatively insensitive index

Measurement	Comments
Extended Trent Biotic Index	This takes account of a wider range of organisms than the TBI, resulting in a score range from 0 (low quality) to 15 (clean water)
Chandler's Biotic Score System	Each group of invertebrates is given a score based on its abundance and these are summed to give the index (higher values indicate cleaner water)
Biological Monitoring Working Party (BMWP) Score	Invertebrates are identified to family and each family is scored between 1 (tolerant) and 10 (intolerant). See http://www.fba.org.uk/recorders/publications_resources/sampling-protocols/contentParagraph/03/document/BMWPLIFEtaxa_Modified.pdf. The BMWP score is obtained by summing the family scores, with higher values indicating cleaner water. BMWP scores have also been used to enhance conservation scoring (e.g. the Community Conservation Index (CCI) of Chadd and Extence 2004). BMWP is used less frequently these days; the ASPT being preferred. A revised version of the BMWP (the WHPT scores – Walley Hawes Paisley and Trigg) incorporates abundance data (see Davy-Bowker et al. 2008 for further information)
Average Score Per Taxon (ASPT)	This is calculated by dividing the BMWP score by the number of families found. BMWP scores above 100 combined with ASPT values of 5 or more indicate clean water
Lotic-invertebrate Index for Flow Evaluation (LIFE)	This is similar to the BMWP score except that scoring is by flow rate tolerance (see Extence, Balbi and Chadd 1999)
Acid Waters Indicator Community Index (AWIC)	This is similar to the BMWP score except that scoring is by acid tolerance (see Davy-Bowker et al. 2005)
River InVertebrate Prediction And Classification Scheme (RIVPACS)	A computer-based comparative system based on knowledge of high-quality reference sites that are short river sections covering all the types of sites (e.g. covering all geological attributes) within a region (including headwaters and downstream sites). Macroinvertebrate species are used to classify sites and are combined with physiochemical data from the same sites to produce a predictive model that is used as a reference set against which other sites can be compared (http://www.ceh.ac.uk/products/software/RIVPACS.html gives further information). This produces reference values that can be used to create Ecological Quality Indices (EQIs – see below)
Ecological Quality Index (EQI)	This is used to compare sites between regions in a standardised way and is calculated by dividing the BMWP and ASPT scores by the RIVPACS value and taking the lower of the two values (higher values indicate higher water quality and a value of 1 or above indicates that the site achieves or exceeds the value predicted by RIVPACS)

Lichen desert	Inner transition	Outer transition	Clean zone
Zones 0 – 1	Zones 2 – 3	Zones 4 – 5	Zones 6 – 10
Ranges from no epiphytes on trees to trees bare except for green algae at base	Green algae higher up. Encrusting lichens at base, each species found higher up as others move in	Green algae and encrusting lichens higher up tree. Foliose lichens at base, each species found higher up as others move in	Wide range of species all over tree, leading to crustose and fruticose lichen flora being well developed in very clean zones
200 190 180 170	160 150 140 130	120 110 100 90 80 70 60	50 40 30 20 10

Mean winter sulphur dioxide levels (μg m^{-3})

Figure 2.18 Lichen zone scale for mean winter sulphur dioxide estimation on trees with moderately acidic bark in England and Wales
Sources: Richardson (1992), Hawksworth and Rose (1976). Wheater (1999) gives a more detailed summary of this scale.

the species are reasonably distinct. Unfortunately, these extremes can be found with the same group as well as between groups. In many areas of the world, the fauna and flora have been relatively poorly studied and although there may be particular experts who have made major contributions to certain groups from some regions of the world, there may be little support for identification either in terms of guides or keys, or expertise or even museum collections. In many heavily populated areas of the world, especially those with a long history of interest in natural history, there is more likely to be information on the native species. For example, across Western Europe and North America there are many field guides and much expertise on a wide range of species. In more diverse habitats, especially those that are relatively remote and have not had a sustained long history of investigation, there may be many species that are unknown, even among larger and better studied groups. In general, the broader the appeal to the public at large (usually on the basis of economic and aesthetic factors), the more is known about the fauna and flora. Groups such as birds, mammals (especially larger ones), flowering plants (including trees) and diurnal showy invertebrates (e.g. butterflies) are more likely to have been relatively well studied and therefore to have guides or keys available for their identification. There is a wide range of general and specialist guides that can be used to identify live organisms in the field and/or preserved animals in the laboratory. One major distinction is between the mainly descriptive, illustrative field guides and those containing keys (stepwise pathways that assist in identifying organisms to produce a more objective identification of individual specimens). Box 2.4 gives examples of some of the types of identification guides available for one group of organisms for one region of the world. Wherever your studies take you, it is worth finding out at an early stage what identification support is available. In addition to books and papers, there may be experts in the field, societies

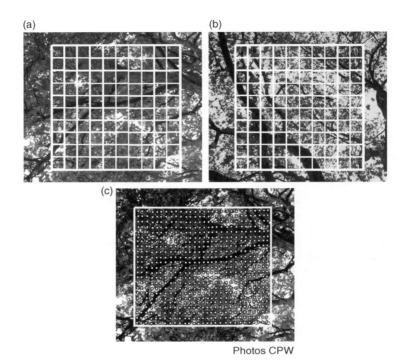

Photos CPW

Figure 2.19 Estimating canopy cover
Using a 10 × 10 grid on top of pictures of the canopy to estimate cover in a: (a) relatively closed canopy; (b) relatively open canopy. Estimates of the canopy cover can be made by averaging the percentage cover in each of the 100 small squares or more quickly by using 25 larger squares comprising groups of four smaller squares. Such grids of squares can be printed onto clear acetate and either overlaid on photographs of the canopy (as here) or used directly in the field. In either case, any comparisons between sites will be more meaningful if they are based on woodlands with similar heights of canopy. If used in the field, then make sure that the grid is held at about the same distance from your eyes in each area surveyed.
(c) Using a grid of dots overlaid on a picture of a canopy to estimate cover.
The cover is proportional to the ratio of dots that lie on branches and leaves compared to those falling in gaps. Yates and McKennan (1988) suggest covering the area with around 1000 dots and repeating the count six times. See Wilkinson (1991) for further details of this technique.

targeted at either scholars or amateurs or both, and collections in museums either in the region concerned or other countries from where previous interested collectors came. The Victorians were very good at accumulating exotic species from around the world and their collections may be found in many regional as well as national museums.

When deciding on the taxonomic level (Table 2.2) to which to identify organisms, bear in mind that the most useful is often that of species. This is because scientists over

Table 2.2 Major taxonomic groups

Major taxonomic groups	Plant example	Animal Example
Domain	Eukarya	Eukarya
Kingdom	Plantae	Animalia
Phylum/Division	Tracheophyta	Arthropoda
Class	Angiospermae	Insecta
Order	Asterales	Coleoptera
Family	Asteraceae	Carabidae
Genus	Taraxacum	Pterostichus
Species	*T. officinale* agg. Wigg	*P. madidus* (Fabricius)
	Photo CPW	Photo JRB
Common name	Dandelion	Strawberry ground beetle

a wide range of taxonomic groups tend agree on the status of a species. Unfortunately, the status of other levels (including genus, family, order and class) tends to differ between groups of organisms with one group having a large number of genera compared to another, or taxonomists working with one group of organisms splitting families into subgroups (e.g. subfamilies and tribes), whereas those working with a different group may not. Even the level of species is not always immune from this. For example, the ground beetle *Pterostichus madidus* (Table 2.2) is sometimes referred to by the name *Steropus madidus* (using the subgenus name *Steropus* instead of the genus name) or (more rarely) as *Feronia madidus* (a old name for some of the genus *Pterostichus*). Note that in Table 2.2 species names are indicated as a two-word scientific name (known as the binomial) in italics (the convention for identifying scientific names). If the genus has already been referred to, it can be abbreviated (e.g. *T. officinale*). The words after the scientific names (Wigg in the case of *T. officinale* and Fabricius for *P. madidus*) are the authorities – that is the taxonomists who originally gave the species their scientific names. Where the authority is enclosed by brackets, as with *Pterostichus madidus* (Fabricius), this indicates that the current name is different from that originally given to the species. In the case of *T. officinale*, the word *agg.* has been added to the scientific name. Extra words following the binomial are used for a variety of reasons. Sometimes a subspecies name is used, or (as here) the word is

Box 2.4 Examples of identification guides for British insects

This is not intended to be an exclusive list but to illustrate the range of types of guides available to assist with the identification of particular types of organisms (in this case insects) from particular regions of the world (in this case Britain). Similar lists could be drawn up for other taxa (e.g. birds, mammals, flowering plants, lichens) and areas of the world (e.g. North America, Australia, Mediterranean Europe, the Falkland Islands).

Collection of key works
One starting point to find identification guides is to look at articles or books that cover key works:

- Barnard P. C. (1999) *Identifying British insects and arachnids: an annotated bibliography of key works*. Cambridge University Press, Cambridge. *Lists appropriate identification texts for British insects.*

Field guides
For field use there are many field guides that cover multiple groups, are focussed on specific habitat types, or the animals of a particular group found in a specific habitat, or that concentrate on particular groups of insects. Although many of these have good-quality illustrations or photographs, it is not usually possible to identify insects to species using the pictures alone unless the book is highly targeted and there are few species that can be easily confused with each other. Some of these guides also contain keys to particular groups of insects. A selection are listed below:

General insects
- Chinery M. (2007) *Insects of Britain and western Europe*. A and C Black, London. *Illustrated picture guide.*
- Chinery M. (1997) *A field guide to the insects of Britain and northern Europe*. Harper Collins, London. *Includes some keys and information about different groups of insects.*

Habitat guides
- Fitter R. and Manual R. (1994) *Photoguide to lakes, rivers, streams and ponds of Britain and North West Europe*. Harper Collins, Hong Kong. *Photographic guide includes some keys.*
- Olsen L-H., Sunesen J. and Pedersen B. V. (2001) *Small woodland creatures*. OUP, Oxford. *Illustrated guide.*

Guides for specific groups
- Smallshire D. and Swash A. (2010) *Britain's Dragonflies*. WILDGuides, Berkshire. *Photographic guide.*
- Waring P. and Townsend M. (2009) *Field guide to the moths of Great Britain and Ireland*. British Wildlife Publishing, Dorset. *Illustrated guide.*
- Zahradnik J. (1998) *A field guide in colour to bees and wasps*. Blitz Editions (Bookmart), Leicester. *Illustrated guide.*

Keys
These are the major tools for identification and can cover general identification for major groups of insects or be focused on particular orders or families of insects, or that concentrate on the insects found in particular habitat types. There are also some guides that act as a halfway house between field guides and more formal keys. These include the Naturalists' Handbooks that cover either particular groups of insects, or common insects living in specified habitats and aim to be reasonably

comprehensive in that they will key out common species and tell you when the species you have found does not fit the key:

- Forsythe T. G. (2000) *Common ground beetles.* Naturalists' Handbooks 8. The Richmond Publishing Co. Ltd., Slough.
- Wheater C. P. and Read H. J. (1996) *Animals under logs and stones.* Naturalists' Handbooks 22. The Richmond Publishing Co. Ltd., Slough.

Keys available on digital media

- Lawrence J. F., Hastings A. M., Dallwitz M. J., Paine T. A. and Zurcher E. J. (1999) *Beetles of the world.* CSIRO Publishing, Victoria. *Interactive key to families and subfamilies available on CD-ROM.*
- Amateur Entomologists' Society (1997–2009) *Key to adult insects.* London. http://www.amentsoc.org/insects/what-bug-is-this/adult-key.html *Key to orders available on-line.*
- Wheater C. P., Read H. J., Wright J., and Cook P. A. (2010) *Key to common animals found on, or just below, the ground surface.* Manchester Metropolitan University, Manchester. http://www.sste.mmu.ac.uk/teachers_zone/ *Key to orders found in Britain available on-line.*

AIDGAP keys

These are scholarly (yet accessible) guides that have been tested by scientists before publication. They cover discrete habitat types or the families (or even genera) of particular groups:

- Skidmore P. (1991) *Insects of the cow-dung community.* (Occasional Publication No. 21) Field Studies Council, Shropshire. *Key to various taxonomic levels including some groups to species.*
- Tilling, S. M. (1987) *A key to the major groups of British terrestrial invertebrate.* (FSC publication 187). Field Studies Council, Shropshire. *Key to major orders (and some suborders) of invertebrates including insects.*
- Unwin D. (2001) *A key to the families of British bugs (Hemiptera).* (FSC Publication 269). Field Studies **10**: 1–35. *Key to families.*

Monographs and specialist works from learned societies

Where available, these should be used to identify insects to species:

- Eddington J. M. and Hildrew A. G. (1995) *A revised key to the caseless caddis larvae of the British isles, with notes on their ecology.* FBA Scientific Publication 53. Freshwater Biological Association, Ambleside.
- Joy N. H. (1932) *A practical handbook of British beetles.* H. F. and G. Witherby, London. *Although old, this still covers some species groups for which there are no other keys (E. W. Classey of Faringdon, Oxfordshire produced a reduced-sized reprint in 1997 and Pisces Conservation Ltd. produced a digital version on CD-ROM in 2009).*
- Luff M. L. (2007) *The Carabidae (Ground Beetles) of Britain and Ireland.* Royal Entomological Society Handbook Volume 4 Part 2. Field Studies Council, Shropshire.

Note that for non-insect invertebrates, other specialist guides are also available (e.g. the series of books of the Linnaean Society of London Synopsis of the British Fauna published by the Field Studies Council).

used to indicate that a number of very similar species are aggregated together in one species group. Common names should be used with care since different people use different common names for the same species: in the UK the names peewit, green plover and lapwing refer to the same species (*Vanellus vanellus*). Conversely, the same common name can be used for a number of different species: *Aphis fabae* (an aphid which is a major pest of sugar beat and beans) and *Simulium* species (a genus of Diptera which transmits diseases, including river blindness) have both been called black flies. Similarly, *Harpalus rufipes*, *Pterostichus cupreus*, *P. madidus* and *P. melanarius* have all been referred to as strawberry ground beetles. More information on classification and biodiversity can be found at the Tree of Life Project[2] and the Catalogue of Life.[3]

Ecologically functional groups or life-form categories are also useful categories in which to classify organisms. For example, when monitoring birds in investigations of the impact of disturbance due to logging on tropical rainforests, they may be divided into those that are insectivorous, or fruit eaters, or seed-eaters, etc. Organisms may be placed in large groups (e.g. by major feeding type – predators, herbivores, detritivores) or more specialist groups (e.g. by dividing leaf herbivores into leaf miners, chewers, gall formers, etc.). An example for plants (the Raunkiær system for classifying plants by their life-form) is given in Table 2.3.

Table 2.3 Major divisions of the Raunkiær plant life-form system

Major divisions	Descriptions
Phanerophytes	Plant stems project into the air with buds more than 25 cm above the ground surface; subdivided on height (often woody perennials e.g. trees and shrubs, also epiphytes, which are sometimes considered a separate division)
Chamaephytes	Buds on shoots within 25 cm of ground surface (dwarf shrubs and creeping woody plants)
Hemicryptophytes	Buds at or near ground surface (subdivided on the types of leaves present, i.e. stem leaves and/or basal rosette leaves)
Cryptophytes	Buds beneath surface of dry ground (Geophytes), marshy ground (Helophytes), or water (Hydrophytes)
Therophytes	Annual species surviving unfavourable conditions as seeds
Aerophytes	Plants that obtain moisture and nutrients from the air and rain

See Gibson (2002) for further details.

---

[2] http://www.tolweb.org/tree/
[3] http://www.catalogueoflife.org/info_about_col.php

3
Sampling Static Organisms

Organisms that are firmly attached to rock surfaces, to dead wood or trees, or within substrates, including sand, mud or soil (e.g. algae, fungi, lichens, mosses, ferns, vascular plants, and encrusting animals such as barnacles) can be monitored using methods suitable for static organisms. Animals' homes (e.g. crab burrows, badgers' setts and ants' nests) and by-products (e.g. dung and leaf litter) can also be sampled using such techniques. Importantly, unlike mobile organisms, because we do not need to take into account movement, we are certain that once a static individual has been monitored in a particular place it will not turn up again in another place. Even relatively mobile groups of animals, including roosting birds and basking seals can be treated as if they are static, as long as the time-frame for sampling is short in relation to individual movement times. In addition, we can use these techniques at larger geographic scales if we census groups of organisms (e.g. woodlands).

Because of their static (or relatively static) nature, these groups are usually quite well studied and are reasonably accessible as potential material for research projects. In many cases, the long history of study has led to the provision of a wealth of identification material both field and laboratory based, especially for vascular plants (i.e. those with a vascular system – xylem and phloem cells – that translocate substances from one part of the plant to another, including flowering plants and ferns) and seashore organisms. Although some others are more difficult to identify, groups, including lichens, pteridophytes (ferns) and bryophytes (mosses), are increasingly covered by identification courses, field guides and keys, and supported by experts based in museums and local and national societies. Where necessary for identification purposes, vascular plants (including flowering plants and ferns) can be preserved by pressing, drying and mounting on herbarium sheets. Larger structures such as seeds and fruits can be dried and kept in envelopes, and regularly checked to ensure that they are not attacked by mould. Bryophytes and lichens can also be dried and kept in envelopes. See BCMF (1996) for further details of appropriate techniques for the preservation of botanical specimens. Samples required for DNA analysis need to be preserved appropriately (see p. 96).

At very large scales (e.g. where a large woodland or grassland is being surveyed), then we may need to subdivide the area into a grid of sampling blocks and adopt an approach that enables us to take samples from each block (e.g. a stratified random strategy – see Chapter 1). The size of the blocks should be determined by the size and distribution of the organisms in question, while the number of blocks and the number of samples within each block will depend not only on the size of the area and the size and distribution of the organisms but also on the heterogeneity of the habitat.

The study of some groups of organisms can present challenges: some can be patchy (e.g. lichens), seasonal (e.g. fungi) or so numerous and small that they present logistical survey problems (including many mosses). If you plan to work on such species, try and limit the study by concentrating either on the ecology of a particular habitat (e.g. in surveying rocky shores for lichens) or a particular genus or family (e.g. *Cladonia* spp. for lichens). Many macro-fungi are particularly problematic, since what we can usually see above ground are just the fruiting bodies; the bulk of the organism (which can be enormous) being hidden underground. In addition, there can be extremely large variations between the numbers of fruiting bodies seen in successive years that may not reflect changes in the amount of fungal material found in the soil. Therefore, density measurements based solely on the fruiting bodies can be meaningless. See Mueller, Bills and Foster (2004) for further details regarding surveying fungi. Although it may be relatively easy to count large and obvious static individuals, static sampling is not necessarily always straightforward. There can be substantial logistic problems in ensuring that all available individuals are counted accurately when:

- organisms are very small;
- surveying taxa in which it is difficult to distinguish individuals (e.g. grasses);
- large geographic areas are being surveyed;
- the individuals that are to be sampled are rare, either seasonally (i.e. occur in a very restricted time period) or geographically (i.e. only locally abundant);
- large surveys are carried out that require many field workers who may have different abilities in terms of identification and recording skills.

Counts are, therefore, easier and more evenly accurate when there are relatively few large organisms in relatively small areas and field work is completed by a single individual. In many other situations, compromises may have to be made (e.g. taking a (sub)sample rather than trying to count the entire sample or population). In general, the techniques for estimating population sizes or densities of static species involve fairly straightforward calculations (see Box 3.1).

Box 3.1 Calculating population and density estimates from counts of static organisms

The population estimate $(\widehat{P}) = \dfrac{\text{total number counted in the samples}}{\text{sampling fraction}}$

or $\widehat{P} = \dfrac{\text{mean number per sample} \times \text{number of samples}}{\text{sampling fraction}} = \dfrac{\bar{x}\, n}{SF}$

Where $\bar{x}$ = the mean number of organisms per sample;
n = the number of samples;
SF = the sampling fraction

The sampling fraction is the proportion of the total area that has been sampled. Where the whole area has been sampled, then the sampling fraction is 1 and the population is the total number counted. If a tenth of the area has been sampled, then the sampling fraction is 0.1. The density is the population estimate divided by the area (expressed in appropriate units of measurement, e.g. per square metre, or per hectare, or per square kilometre).

For example, if the area is 50×10 m and there are five random samples of 0.5×0.5 m ($0.25\,\text{m}^2$) in each of five 10×10 m blocks (placed using a stratified random sampling strategy) then to calculate the sampling fraction:

$SF = \dfrac{\text{area sampled}}{\text{total area}} = \dfrac{5 \times 0.25 \times 5}{50 \times 10} = 0.0125$

So, if we have a mean number of plants per sample of 1.2 from the 25 samples, then:

$\widehat{P} = \dfrac{\bar{x}\, n}{SF} = \dfrac{(1.2 \times 25)}{0.0125} = 2400$

The fact that we can also calculate both the mean number counted per sample and the standard error of the mean (see Chapter 5 for further background on means and standard errors) allows us to also calculate the 95% confidence interval (CI) for the population estimate:

$95\%\, CI = \dfrac{2\, SE\, n}{SF}$

SE = the standard error of $\bar{x} = \dfrac{\text{standard deviation}}{\sqrt{n}}$

Where: standard deviation $(S) = \sqrt{\text{variance}}$

variance $(S^2) = \dfrac{\sum x_i^2 - (\sum x_i)^2 / n}{n-1}$

In our example, if the variance had been 0.48, the standard error is:

$SE = \dfrac{\sqrt{0.48}}{\sqrt{25}} = 0.139$

The 95% confidence intervals are:

$95\%\, CI = \dfrac{(2 \times 0.139 \times 25)}{0.0125} = 556$

$\widehat{P} = 2400 \pm 556$ (i.e. the population probably lies between 1844 and 2956)

> The density is therefore $\dfrac{2400}{500} \pm \dfrac{556}{500} = 4.80\,\text{m}^{-2} \pm 1.12\,\text{m}^{-2}$
>
> If you are monitoring more than one species, the relative (percentage) density of each can be calculated as:
>
> $$\text{Relative density} = \dfrac{\text{number of individuals of a species}}{\text{total individuals of all species}} \times 100$$

Sampling techniques for static organisms

There are a number of techniques available to estimate the numbers of organisms, percentage cover, or number and diversity of species within static (or relatively static) communities. With the exception of some forestry techniques applied specifically to trees, most can be used on a wide range of organisms, be they animals or plants, habitats or other ecological features (e.g. ants' nests). The major determining factors are whether they are small or large and whether they are relatively common or rare. Generally static organisms are grouped into those sampled at different scales – small scale (lichens, liverworts, mosses), medium scale (macro-fungi, ferns, herbaceous vascular plants, small heathers, aquatic macrophytes) and large scale (shrubs, trees). Static animals can be similarly grouped: barnacles and bryozoans at small scale; limpets and other shelled molluscs at medium scale; and aggregations of marine mammals or roosting colonial birds at large scale. Common techniques include quadrat sampling and the use of pin-frames, the use of transects and plotless sampling.

The microbial communities of soil, water and air are arguably some of the richest and most diverse contributors to global biodiversity (see Dance 2008). Microbes are also found living on plants and animals, for example on the surfaces of leaves and in moss cushions. Microbes comprise eubacteria, algae, cyanobacteria, fungi and protozoa. Although some classes of micro-organisms are motile, we include all microbes in this static sampling section because they are usually sampled within the substrate by relatively straightforward sampling methods (taking soil cores, water samples, etc.), identified back in a laboratory, and their densities estimated in a similar way to those of static organisms (i.e. by multiplying the numbers found in a sample to provide an estimate per square centimetre of surface, or per millilitre of substrate or water – see Box 3.1). Owing to their small size, the challenge for environmental microbiologists is to be able to view their subjects or gain some other indication of their presence and abundance. A range of methods has been used depending on the size of the organisms and the level of sophistication of the available laboratory, and the main ones are summarised in Box 3.2.

Box 3.2 Techniques used to identify and count microbial diversity

The range of techniques used depends on the micro-organism involved. Many larger protozoa can be seen under relatively low-magnification microscopy and identified using keys. The diversity of bacteria and fungi means that identification keys are not readily available and their size results in the use of more sophisticated techniques to culture, separate and monitor their presence and abundance. The choice of method depends on the level of resolution required and current scientific knowledge of the micro-organism involved, and should be informed by reference to appropriate texts (e.g. Hurst et al. 2007, and Maier, Pepper and Gerba 2009).

Technique	Comments
Culturing	This enables different groups of microbes to be isolated to estimate their diversity and abundance after extraction, and either dilution or concentration depending on the microbial levels in the substrate being examined. It has been estimated that only 1–5% of micro-organisms are able to be cultured, hence other techniques are often employed. Various methods include: • Plating (often for bacteria) involves adding the sample to a selected growth medium to enhance the growth of colonies that can subsequently be identified and counted. • Tube or flask cultures (often of algae and cyanobacteria) can be made in a similar way to plating by adding the specimen to appropriate growth media and keeping in suitable conditions (sometimes for several weeks). • Most probable number technique involves serially diluting the sample to extinction and incubating replicates of each dilution, the number present in the original sample being indicated by the number of replicates at each dilution that contained active microbes.
Microscopy	This enables the physical structure of whole (or part of the) organisms to be viewed and photographed, organisms to be identified, and for counts of individuals and colonies to be made. Enhanced contrast and increased resolution can help to visualise morphological differences that enable identification. Different types of microscopy can be used to view different ranges of micro-organisms: • Light microscopy (up to 200 nm resolution, which enables the morphology of larger organisms to be seen) includes bright-field, dark-field, phase-contrast, interference-contrast, polarisation and fluorescence microscopy. The choice of technique depends on the organisms involved and aims to maximise the degree of contrast, and depends on whether visualisation of external or internal structure is needed. • Electron microscopy enables a higher degree of resolution than light microscopy so that smaller organisms and more details (external and/or internal) can be seen, and includes scanning electron microscopy (for whole dead specimens at a resolution of around 10 nm) and transmission electron microscopy (for sections of preserved specimens at a resolution of up to 0.5 nm). • Scanning probe microscopy includes atomic force microscopy, which enables the surface structure of live or preserved specimens to be viewed at a resolution of up to 0.5 nm enabling the morphology of very small organisms to be viewed.

Technique	Comments
Physiological techniques	This involves measuring factors associated with microbial activity to establish presence and/or abundance. Factors measured can include: • The rate of loss of (usually carbon-based) substrates in the culture medium. • The removal of terminal electron acceptors such as oxygen and nitrate. • The cell mass measured through turbidity in a culture sample or through the amount of protein present. • The production of carbon dioxide (in both aerobic and anaerobic conditions).
Lipid analysis	The viable microbial biomass can be estimated using chemical analysis of phospholipids of microbial origin (e.g. using colorimetric analysis or gas chromatography-mass spectrometry).
Immunological methods	Immunoassays use antibody–antigen interactions to identify the presence and estimate the abundance of microbes (which act as antigens). Methods include: • Fluorescent immunolabelling to more easily view antibody-bound antigens using microscopy. • Enzyme-linked immunosorbent assays (ELISA), involving an antiglobulin and enzyme substrate to produce a colour reaction to the antibody–antigen interaction that can be quantified using colorimetric analysis. • Immunoaffinity chromatography, which concentrates the antigens by running the sample through a chromatography column within which antibodies have been bound to an inert matrix. • Immunoprecipitation assays, which use the principle that greatest precipitation will occur at optimal levels of antibody–antigen concentrations, so by using a range of concentrations of antigen added to reaction tubes containing the antibody, the ones with most precipitation will contain the appropriate concentration of antigen for the antibodies used.
Nucleic acid methods*	Some of the most commonly employed techniques are those that identify microbes from their nucleic acids often following amplification of the target DNA using polymerase chain reaction (PCR) or cloning. Methods include: • Gene probes, which are labelled sequences of DNA that are complimentary to the target sequence (e.g. an enzyme) that are unique to the microbe or group of microbes of interest. Multiple gene probes can be synthesised onto microarray slides or chips, which can be used to see how micro-organisms respond to environmental conditions. • Fingerprinting using PCR or terminal-restriction fragment length polymorphism (T-RFLP) analysis (followed by gel electrophoresis), which can be used to detect specific target microbes. • Denaturing (and temperature) gradient gel electrophoresis (DGGE and TGGE respectively), which can be used to examine microbial diversity. • Sequencing, which is an increasingly used technique for the identification of specific sequences of DNA and RNA. Commonly used key markers are 16S for bacteria, 18S for fungi, and cytochrome oxidase subunit I more generally.

*See p. 96 for details about preserving, storing and transporting samples for DNA and RNA analysis.

It may be necessary to revisit a site and re-record the organisms' abundance and/or distribution, to monitor changes in the community including seasonal changes, the spread of invasive species, population dynamics or successional change. In this case, permanent (or semi-permanent) recording stations can be relocated by using a GPS, or by reference to permanent landscape features or other markers. Many communities may be surveyed using photographic techniques. This can include satellite or aerial photography (see Chapter 2) for larger features including woodlands, or more localised photography to obtain details of herbaceous flora, lichens or molluscs on rocky shores. In all cases it is important to have an item of known size within each photograph to enable the scale to be readily determined.

The major techniques used for sampling static plants and animals are discussed in the rest of this section. For further details of these and other methods for monitoring plant communities see Kent and Coker (1995), Elzinga, Salzer and Willoughby (1998) and Gibson (2002). For sampling protocols and methods for monitoring marine and coastal environments see Hayward (1988, 1994) and Davies et al. (2001). Matthews and Mackie (2006) and Mackie and Matthews (2008) give details of measuring trees.

Quadrat sampling

Quadrat sampling is a widely used botanical field technique that can be suitable for monitoring static animal populations. Quadrats are an excellent tool for investigating treatment effects in planned experiments (e.g. determining the effect of herbicide on weeds and non-target flora). In one type of design (known as a randomised block design), the quadrats are randomly placed within treatments, and the treatment plots are distributed within blocks. Quadrats are also widely used in 'natural experiments', where, instead of field manipulations, we study zones (e.g. in studies of coral reefs and sand dunes), habitat types (e.g. woodlands, grasslands) or environmental gradients (including tides, altitude, space). In these protocols it is important that the quadrat is randomly placed, unless they are to be used in conjunction with a transect (see p. 84). Their application in natural experiments is wide and ranges from investigating the effect of tidal range on barnacle distribution, to estimating the density of mammals by sampling their dung or, at their most basic, to providing an estimate of vegetation cover between habitats or to survey encrusting species on rocky shores. Fixed quadrats can be useful for long-term monitoring of species-rich communities where the positions of quadrats are specified, mapped and returned to on subsequent sampling occasions. One long-term study, the Environmental Change Network (ECN), has sampled 12 permanent terrestrial sites across the UK (most since 1992) to examine links between environmental factors and species responses. The ECN protocol[1] for vegetation sampling uses 2 m × 2 m and 10 m × 10 m fixed quadrats that were initially allocated at random. See Case Study 3.1 for a discussion of some of the issues associated with long-term field experiments.

---

1   http://www.ecn.ac.uk/aboutecn/measurements/v.htm

Case Study 3.1 The Park Grass Experiment

Chris Bennett is a plant ecologist based at Rothamsted Research, doing his PhD research on Rothamsted's Park Grass Experiment in affiliation with the University of Sheffield. His case study describes the constraints on field manipulations and the problems of statistical analysis that he has had to overcome as a result of using an inherited experimental design.

Photo RRES

Park Grass Experiment

Photo CB

Chris sampling at Park Grass Experiment

Model system and sampling problems faced

The Park Grass Experiment (PGE) is the world's longest running ecological experiment; started in 1856 by Sir John Bennet Lawes and Sir Joseph Henry Gilbert to investigate the improvement of hay meadows using organic and inorganic fertilisers. Different fertiliser treatments have been continually applied to the plots since 1856, with the addition of a permanent pH gradient introduced in 1965 and maintained using liming treatments. The PGE is cut twice a year in accordance with the traditional management of a hay meadow. The yields of the plots have been recorded every year until 1960 by weighing the biomass of the whole plot, and since then by estimating the biomass from two strips cut through the plots. In addition to records of the yields, dried vegetation samples have been kept in an archive along with soil samples. This makes this experiment very well documented with substantial archived material and provides an ideal scenario to test theories in plant ecology. However, the value of the experiment lies in the consistency of its long-term treatment structure that places constraints on any manipulative work which researchers wish to do on the plots. The integrity of the PGE is very well protected by a committee that discusses all the proposals for sampling on Park Grass and for use of the archived samples. The age of the experiment is a hindrance, as the experiment was started before the development of modern statistical methods and therefore there is no replication of the treatments. Inheriting a long-term experimental design comes with a high degree of responsibility because there is a need to ensure that the integrity of the original experiment is maintained.

How the problem was overcome

Necessarily, Chris has had to compromise what he would like to do in order to obtain permission from the committee. Furthermore, the PGE is not replicated so Chris had to consider alternative statistical methods as advised by the statistics department at Rothamsted. He found that by using restricted maximum likelihood (REML) within a mixed effects model environment, he was able account for correlated structures and also estimate the correct degrees of freedom. Chris used 'year' as both a replicate and a random effect in an attempt to induce a replicated structure while accounting for the source of error due to year-to-year variations. Even so, with some data there is no way to overcome the lack of replication and so he has had to acknowledge the lack of power in his statistics.

Advice for students wanting to use planned experiments
Working on planned experiments often requires a specific statistical test to account for the structure of the experiment. Do not be afraid to consult a statistician at the earliest opportunity, particularly if the design is unusual in some way. Chris found that through collaboration with a statistician he was able to get the best advice. Be as flexible with your ideas as you can and keep in touch with someone on the committee overseeing the experiment to find out if there has been sampling of a similar nature before. It is always useful to formalise your ideas within a proposal that should be submitted to the correct body at the earliest opportunity. Obtaining permission may take some time, especially if you plan to do any large scale sampling or manipulation, so be patient. If your proposal is accepted, Chris advises to sample for a number of years to be able to detect statistically significant long-term patterns.

Chris' work would not be possible without the support of Rothamsted Research, The Lawes Trust and the Biotechnology and Biological Sciences Research Council (BBSRC).

The classic quadrat is square (Figure 3.1) and is used to simply define an area within which you can estimate density, frequency, cover or biomass of the organisms being studied. Quadrats can also be used to sample other materials, such as delineating the area of leaf litter for the extraction of arthropod fauna, or used overhead to examine the leaf cover of canopies. Similarly, large quadrats (or blocks of land) can be used to examine the presence of animals' homes (e.g. fox earths), or applied to maps to estimate the extent of certain habitats (e.g. woodlands or lakes) and is a concept that is now widely adopted in GIS. Some ecologists prefer quadrats divided with cord or wire into smaller quadrats at set intervals to help the surveyor to systematically cover the whole quadrat.

Photo CPW

Figure 3.1 Quadrats
From left to right – subdivided wire quadrat (with pin-frame – see p. 83), fixed square wire quadrat, folding wire quadrat (open and closed).

Density estimations using quadrats

Density is simply the count of the number of individuals found inside the quadrat area converted into the number per square metre (expressed as the number m^{-2}) where there are reasonable numbers within small areas, or per hectare or square kilometre where individuals are relatively rare or cover larger areas. When using quadrats to estimate the abundance of plants, it can be useful in the field to calculate the mean number per quadrat after every fifth quadrat. When this does not change greatly following the inclusion of another five quadrats, you may have reached a point when additional effort will not improve your estimate of abundance.

Frequency estimation using quadrats

This is normally derived from presence/absence data; frequency is the chance of finding a species in a given area. To estimate frequency, a species is counted as present or absent in each of a series of quadrats, or in each section of a single subdivided quadrat. For example, if buttercups were found in 12 out of 50 quadrats surveyed, the frequency would be 24%.

Cover estimation using quadrats

Cover is perhaps the most subjective of all measurements when it is based on the visual estimation of the quadrat area occupied by a species and expressed as a percentage. For some species (e.g. lichens) the quadrat can be photographed or laser scanned and the resulting images explored using specialist software that can digitise each species to assess cover (see for example Wasklewicz et al. 2007). Some authors have suggested that defined scales (e.g. the DAFOR, or Braun–Blanquet, or Domin scales) should be used (see Table 3.1 and Kent and Coker 1995) but many published studies nowadays divide cover into 10 × 10% classes. These scales are used for some standard techniques (e.g. the NVC uses the DOMIN scale – see Box 2.1). Cover is of use when the species are not easily identifiable as with algae, moss, lichens and non-flowering grazed grasses, or where individuals are not easy to distinguish (as in many grasses) or when it is used to estimate canopy closure in woodlands. It is usual to take a bird's-eye view when estimating cover and count those individuals that can be seen from directly above. If the sample is composed of layers of vegetation, the sum of the percentage cover for each plant group may exceed 100%. A more quantitative method of estimating cover is to use a quadrat with cords set at equal distances down each side (Figure 3.2). If the plants found at each of the cross-over points are identified and their frequency summed (based on the number of times they are recorded), this will give a measure of their cover, although small, rare plants may be missed. Subdivided quadrats can also be used to assess impacts such as grazing pressure, for example by counting the proportion of small squares where there is evidence of vegetation having been eaten.

Table 3.1 DAFOR, Braun–Blanquet and Domin scales for vegetation cover

DAFOR scale	Braun–Blanquet scale	Domin scale
D Dominant (>75% cover)	+ <1% cover	+ 1–2 individuals with no measurable cover
A Abundant (51–75% cover)	1 1–5% cover	1 <4% cover with few individuals
F Frequent (26–50% cover)	2 6–25% cover	2 <4% cover with several individuals
O Occasional (11–25% cover)	3 26–50% cover	3 <4% cover with many individuals
R Rare (<11% cover)	4 51–75% cover	4 4–10% cover
This is usually used without the guidance % cover, leading to a lack of consistency between surveyors. The lower end of the scale (frequent to rare) may be prefixed 'locally' or 'regionally' if relevant	5 76–100% cover	5 11–25% cover
		6 26–33% cover
		7 34–50% cover
		8 51–75% cover
		9 76–90% cover
		10 91–100% cover

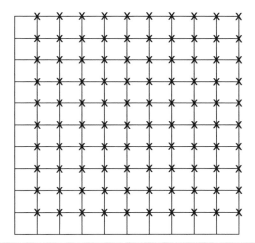

Figure 3.2 Recording positions on a subdivided quadrat
Nine cords are set at equal distances down each side producing 100 squares and (more importantly for cover) 100 positions where the cords cross over (using two sides only, e.g. the top and the right hand side of the quadrat, as points of crossing over) where species can be identified and hence their frequency calculated.

Abundance scales similar to the DAFOR scale can be used for marine algae, encrusting species (e.g. lichens), crust-forming species (e.g. *Lithothamnia*) and static animals (e.g. adult barnacles), or relatively static animals (e.g. marine shells such as limpets). The scales used are termed SACFOR scales (**S**uper abundant, **A**bundant, **C**ommon, **F**requent, **O**ccasional and **R**are). This can be extended to include **E**xtremely abundant, and **N**ot present (ESACFORN scale). In order to standardise these, some guidelines

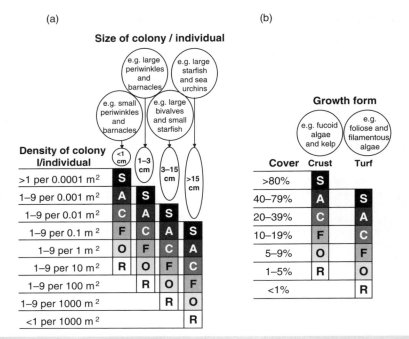

Figure 3.3 JNCC guideline usage of SACFOR scales
(a) Scales used for organisms (mainly animals) that can be measured by the size of the individual/colony; (b) scales for organisms (mainly algae) that can be measured by the type of growth form. See Connor et al. (2004).

have been offered (by for example the UK Joint Nature Conservation Committee – JNCC[2]) depending on the size of the species being surveyed and whether they are found in crusts or turfs or as individuals (see Table 3.2 and Figure 3.3 for examples of how to use such scales in standardised ways). An important consideration is that, if you choose to use any of the abundance scales illustrated in Tables 3.1 and 3.2, they may later limit the range of statistical tests that can be used to analyse them (i.e. they are ordinal/ranked data – see Chapter 1).

Biomass estimation within quadrats

Biomass can be measured as the fresh or dry weight of vegetation expressed in grams (e.g. 15.5 g) harvested from a quadrat. Plants are cut at ground level from the quadrat and bagged and weighed immediately to give a measure of fresh mass. The dry mass can be obtained by heating at 105 °C in an oven until the mass is constant. Fresh mass suffers from various inaccuracies depending on the moisture content of the plants, the time between cutting and weighing, and recent weather events (e.g. heavy dew or rain).

---

2 http://www.jncc.gov.uk/page-2684

Table 3.2 Abundance (ESACFORN) scales for littoral species

Scale	Mussels, piddocks and *Sabellaria*	Dog whelks, topshells, anemones and sea urchins	Limpets and periwinkles	Small periwinkles	Large barnacles	Small barnacles	Lichens and *Lithothamnia*	Algae
Extremely abundant	>80% cover	>100 m^{-2}	>200 m^{-2}	>500 per 10 cm × 10 cm quadrat	>300 per 10 cm × 10 cm quadrat	>500 per 10 cm × 10 cm quadrat	>80% cover	>90% cover
Super abundant	50–80% cover	50–100 m^{-2}	100–200 m^{-2}	300–500 per 10 cm × 10 cm quadrat	100–300 per 10 cm × 10 cm quadrat	300–500 per 10 cm × 10 cm quadrat	50–80% cover	60–90% cover
Abundant	20–49% cover	10–49 m^{-2}	50–99 m^{-2}	100–299 per 10 cm × 10 cm quadrat Extend to mid-littoral zone	10–99 per 10 cm × 10 cm quadrat	100–299 per 10 cm × 10 cm quadrat	20–49% cover	30–59% cover
Common	5–19% cover	1–9 m^{-2}	10–49 m^{-2}	10–99 per 10 cm × 10 cm quadrat Mainly in littoral fringe	1–9 per 10 cm × 10 cm quadrat	10–99 per 10 cm × 10 cm quadrat	1–19% cover Zone well defined	5–29% cover
Frequent	Small patches <5% cover	<1 m^{-2} sometimes locally >1 m^{-2}	1–9 m^{-2}	1–9 per 10 cm × 10 cm quadrat Mainly in crevices	1–9 per 0.1 m^2	1–9 per 10 cm × 10 cm quadrat	Large scattered patches Zone ill defined	<5% cover but distinct zone
Occasional	No patches 1–9 individuals m^{-2}	Always <1 m^{-2}	<1 m^{-2}	A few in most deep crevices	1–9 m^{-2}	1–99 m^{-2}	Widely scattered small patches	Scattered individuals Zone indistinct
Rare	<1 individual m^{-2}	Very few in a 30 minute search	Very few in a 30 minute search	Very few in a 30 minute search	Very few in a 30 minute search	<1 m^{-2}	Very few patches in a 30 minute search	Very few in a 30 minute search
Not present	←——————————————————————— None ———————————————————————→							

See Burrows, Harvey and Robb (2008) for further details of the use of ESACFOR(N) scales.

It is good practice to take the dry mass where possible to avoid these effects. Both fresh and dry measures can be expressed as percentage biomass (i.e. biomass of each species as a percentage of the total). This method is best used in meadows or with similarly sized plants. Making biomass measurements is time-consuming especially if the number of replicates is high. **Note:** before taking plants from a site, ensure that none of the species are protected and seek permission from the landowner or land manager.

Quadrat size

The size of the quadrat should be determined by the diversity of the organisms in the surrounding area and the type of plant or animal under investigation. For example, it is no good using a small quadrat (0.5 m × 0.5 m) to estimate tree diversity in a woodland; instead a 20 m × 20 m would be recommended. Similarly, a 5 m × 5 m quadrat would be too big to estimate the cover of lichens on a rocky shore but a 0.25 m × 0.25 m one may be sufficient. Table 3.3 lists recommended quadrat sizes for various vegetation types. This table should only be taken as a guide: the ideal way to find out the correct size of quadrat is to repeat the survey several times, doubling the size of quadrat each time to work out the gain in the number of species with increasing quadrat size. When very few species are added to the list and upwards of 95% of the total number of species are sampled in the quadrat area, then this should be the appropriate quadrat size to use. However, although this method is recommended by many authors, in practice (because samples are often taken from many different vegetation types) establishing the ideal quadrat size in this way can be problematic (one way round this is to use nested quadrats – see next section). Because of statistical constraints (including variable sampling effort), it is usually best to set one size of quadrat for each of the different zones of vegetation (e.g. moss layer, herbaceous layer and trees) being sampled in a study area.

Table 3.3 Recommended quadrat sizes for various organisms

Vegetation/habitat type	Quadrat size
Algae, mosses and lichens	0.1 m × 0.1 m to 0.5 m × 0.5 m
Seaweeds, grasslands, small heathers and aquatic plants	0.5 m × 0.5 m to 2 m × 2 m
Large heathers or long tall vegetation	1 m × 1 m to 4 m × 4 m
Scrub and woodland vegetation	10 m × 10 m
Woodland canopies	20 m × 20 m to 50 m × 50 m
Habitats (e.g. badger setts, ponds, woodland blocks)	1 km × 1 km to 10 km × 10 km
Static animals	**Quadrat size**
Small animals (e.g. barnacles, Bryozoa) where cover >50%	2 cm × 2 cm to 5 cm × 5 cm
Small animals (e.g. barnacles, Bryozoa) where cover <50%	0.1 m × 0.1 m
Larger animals (e.g. periwinkles, limpets)	0.5 m × 0.5 m
Ants' nests	20 m × 20 m to 50 m × 50 m

Nested quadrats

Some researchers recommend using nested quadrats to examine how species abundance and distribution changes with increasing area. This is also useful when surveying across different sites and/or years in that it allows you to respond to differences in vegetation cover in terms of the optimum quadrat size. Here a large quadrat is subdivided into smaller units such that the area of each subsequent quadrat is larger than the next. There are several ways in which quadrats can be nested (Figure 3.4) and the size of the largest quadrat containing the nests will again depend on the types of species being surveyed.

Note that recording the presence/absence of each species in each size of quadrat enables the percentage frequency to be calculated in the larger quadrat. Counting the number of individuals of each species in each nested quadrat will allow the density and distribution (i.e. clumped, random or evenly distributed – see p. 88) of the species to be calculated.

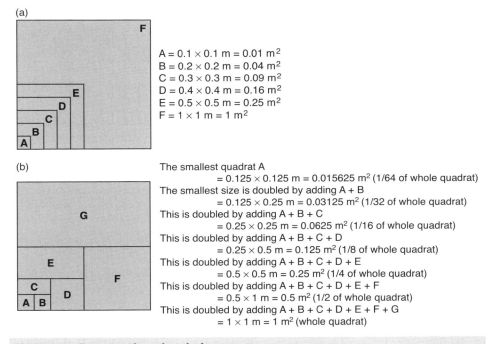

(a)

$A = 0.1 \times 0.1$ m $= 0.01$ m^2
$B = 0.2 \times 0.2$ m $= 0.04$ m^2
$C = 0.3 \times 0.3$ m $= 0.09$ m^2
$D = 0.4 \times 0.4$ m $= 0.16$ m^2
$E = 0.5 \times 0.5$ m $= 0.25$ m^2
$F = 1 \times 1$ m $= 1$ m^2

(b)

The smallest quadrat A
    $= 0.125 \times 0.125$ m $= 0.015625$ m^2 (1/64 of whole quadrat)
The smallest size is doubled by adding A + B
    $= 0.125 \times 0.25$ m $= 0.03125$ m^2 (1/32 of whole quadrat)
This is doubled by adding A + B + C
    $= 0.25 \times 0.25$ m $= 0.0625$ m^2 (1/16 of whole quadrat)
This is doubled by adding A + B + C + D
    $= 0.25 \times 0.5$ m $= 0.125$ m^2 (1/8 of whole quadrat)
This is doubled by adding A + B + C + D + E
    $= 0.5 \times 0.5$ m $= 0.25$ m^2 (1/4 of whole quadrat)
This is doubled by adding A + B + C + D + E + F
    $= 0.5 \times 1$ m $= 0.5$ m^2 (1/2 of whole quadrat)
This is doubled by adding A + B + C + D + E + F + G
    $= 1 \times 1$ m $= 1$ m^2 (whole quadrat)

Figure 3.4 Two nested quadrat designs
In (a) the area sampled and the total perimeter length increase proportionally with each increase in nested quadrat size, whereas in (b) the area sampled doubles with each increase in nested quadrat size, although the perimeter of each subsequent quadrat does not increase proportionally, but varies with the quadrat size depending on whether the quadrat is a square or an oblong. The smallest quadrat in each nest is examined first and the number of species, the presence/absence, number of individuals and/or percentage cover of each species is recorded. Each quadrat size is then surveyed in turn from the smallest up to the largest. If the number of species in each size is plotted, then the point at which no (or relatively few) species are added with an increase in quadrat area will be the optimal quadrat size.

Placement of quadrats

There are a number of issues surrounding whether a quadrat should be placed randomly or systematically. This may be determined by the statistics that you have planned and whether you are using a transect, but you should note that quadrats cannot be 'thrown randomly' (as is implied or even suggested by some texts). Various studies have shown that the arm action used in throwing is not a random event; most people will throw upwards and to the right, few people will throw directly into a patch of nettles or brambles, and in woodlands any quadrat thrown against a tree will drop at its base. If random quadrats are planned, then use a random number generator (see Chapter 1). Either apply the random number to a grid system where each square has its own numeric code, or make a grid reference (i.e. two numbers, one representing the 'easting' and the other the 'northing'). Figure 3.5 illustrates these two systems.

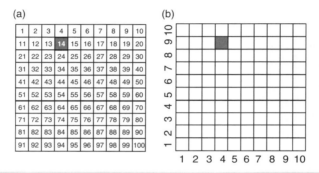

Figure 3.5 Using random numbers to identify a position in a sampling grid
Either (a) each block is allocated a number (the shaded cell is block 14) or (b) each block is identified by coordinates (the shaded block is at 4, 9).

Vegetation can be extremely patchy. When this occurs, you should try to sample about 5% of the habitat using a large number of small quadrats (of a size suitable for the size of the species being examined) rather than using a few large quadrats.

Quadrat shape

Most researchers use square quadrats, although circular or oblong ones may be useful in certain situations. The reason for this is related to two characteristics of differently shaped quadrats:

1. Circular quadrats may be chosen where edge effects are important. The ratio of the length of the edge to the area inside the quadrat changes with the shape so that circular quadrats have a smaller edge effect than square ones of the same area, with oblong shapes having the greatest edge effect (Figure 3.6). Edge effects can lead to

errors in counting since every plant or animal at the edge of a quadrat can be identified as being either in or out of the sampling frame. If all organisms touching the edge are counted as being inside, this may overestimate the cover. Conversely, ignoring all those on the edge may result in an underestimate. One of two strategies are usually employed to avoid such problems:

- Counting only those individuals with more than 50% of their area within the quadrat and ignoring those where more than 50% lies outside.

- Counting all those touching only two of the sides (say the top and right hand side) and ignoring those touching the other two sides.

2. Long thin frames are better than circular or square ones of the same area at covering habitat heterogeneity, since they tend to cross more patches (which is one reason for using transects with quadrats – see p. 85).

Square quadrats are normally used as a reasonable compromise between these two conflicting situations when counts of individuals are being employed. There is less of an issue when using cover estimates since this is not biased by edge effects. However, there are particular situations where round quadrats may be useful. For example when using large quadrats (10 m × 10 m or more) in woodland habitats, it is sometimes difficult to make sure that the quadrat is truly square (deviation from this alters the area being surveyed). It is sometimes easier to set a centre point and use a tape as a radius to define the boundaries.

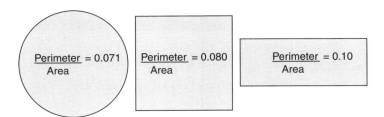

Figure 3.6   Comparison of the perimeter to area ratios of circular, square and oblong quadrats
Note that all the quadrats have an area of 0.25 m^2 and the sizes are based on a 0.5 m × 0.5 m square quadrat and an oblong quadrat with the long sides being twice the length of the short sides.

Pin-Frames

Pin-frames (also known as point quadrats) are simple devices that are commonly used for the estimation of vegetation cover in habitats with short vegetation (<50 cm). These can be made simply and cheaply using a piece of wood or plastic pipe of appropriate length (e.g. 50 cm) with (at least) 10 equally spaced holes drilled through it, large enough for a knitting needle to pass through easily. Two leg supports at each end enable the pin-frame to stand independently (Figure 3.7). As the pins are dropped through the holes, only the vegetation that touches the end of each pin should be recorded. Vascular plants are normally identified to species, but lower plants can be simply listed as moss, lichen, etc. Soil, stone or litter should be noted separately when touched as these can be used to describe the ground zone. Some ecologists prefer to use a pin-frame within a quadrat, moving it 10 times across the whole of the quadrat area (i.e. 100 pins per quadrat). However, pin-frames should be used in the manner that is best suited to the habitat under investigation: if the sward is thick and taking too much time to complete a 100-pin assessment, then fewer pins can be used spaced further apart across the quadrat area. In addition to noting the first plant that is contacted by each pin, a more exhaustive method is to record every touch each pin makes with the vegetation as it is lowered until it reaches the surface of the soil. This more thorough measure records species frequency within the vegetation structure, but is very time-consuming. A third method is to mark the pin every millimetre along its edge and measure the distance (mm) between each hit to give an indication of vertical structural density. This can be used to infer structural differences between stand types, for example whether a thatch or a void below the canopy of vegetation is present. Structural assessment of mature vegetation may take many hours per quadrat.

Photos CPW

Figure 3.7 Pin-frame
(a) Used on its own, (b) used with a subdivided quadrat, (c) pins touching vegetation layers.

Transects

Transects are generally used to sample changes along an environmental gradient, which may include geological, climatic or altitudinal gradients. The length of a transect depends on the gradient under investigation: rocky shore transects may only stretch 1–200 m but montane transects may exceed several kilometres. Two basic types of transect are widely used: line and belt (Figure 3.8).

Line transects are simply a length of tape along which organisms are recorded if they touch the line. Where this is continuously recorded along the line it is called a line intercept transect. Monitoring individuals can be done at intervals (e.g. by recording the plants nearest to each metre along the tape – called a point intercept line transect). These intervals have to be small enough so that changes in the community are not overlooked. This method is good for low growing or sparse vegetation communities (e.g. seaweeds) but is less suitable for very dense, fine vegetation swards. Line intercept transects are also useful for surveying subtidal reefs and intertidal rocky shores.

Belt transects are more widely used and are simply a line transect with quadrats placed adjacent to each other along the length of the tape. Belt transects are more robust than line transects and have a wider application but are much more time-consuming. Ladder transects are belt transects where the quadrats are not placed adjacent to one another but are spaced at equal distances apart. Again, care should be taken to ensure that the distances between quadrats are small enough to reflect changes in the vegetation.

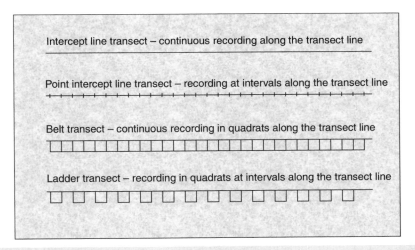

Figure 3.8 Comparison of transect sampling techniques

Used properly, both line and belt transects can identify zones across the gradient, although belt transects provide more data about each position along a gradient. Kite diagrams (Figure 3.9) can be used to display such data. Transects can also be used to place other sampling systems (including pin-frames), or to take other samples (e.g. soil cores, leaf litter, etc.). In addition, transects can be used to sample animals homes (e.g. ants' nests in grassland areas). At larger scales (over several kilometres) transects can be used, either in the field or drawn onto maps, to sample large-scale habitats (e.g. moorland, woodland or hedgerows). When using larger transects in the field, care should be taken to estimate densities accurately because plants and animals that are far away from the surveyor can be under-estimated, particularly if they are small (e.g. mosses) or look similar to other species (as in many grasses).

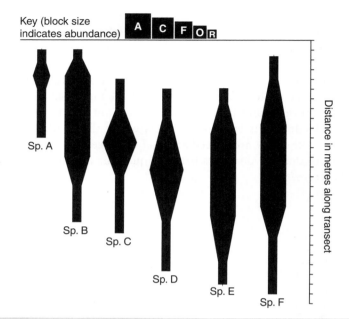

Figure 3.9 Kite diagram to indicate the abundance of different species along a transect
Note that this uses the ACFOR system (Abundant; Common; Frequent; Occasional; Rare) to estimate the abundance of each species.

Data from transects may be unsuitable for analysis using parametric tests, simply because any quadrat placed along a transect is sequentially dependent on its adjacent quadrat. This link between quadrats (i.e. non-randomness) leads to biased estimates of the variance and mean, violating the assumptions of independence and randomness and reducing the degrees of freedom (see Chapter 5). Consequently, the range of tests that can properly be applied to transect data is restricted and may be more complicated than if quadrats were not sequentially dependent. This can be addressed by statistical resampling to take account of the lack of independence and spatial autocorrelation (see Fortin and Dale 2005 for a detailed discussion of spatial autocorrelation). Alternatively, you can be more stringent with your critical probability (using $P<0.01$ rather than $P<0.05$) as a way of being more cautious with the interpretation of your results (see Chapter 5 for a discussion of critical values). More sophisticated statistical models can also be applied (e.g. mixed effects models – see Crawley 2007 for further details).

Plotless sampling

To estimate the density of easily identifiable large organisms that are spaced out (e.g. trees), a method known as plotless sampling can be used. The four most widely used plotless methods are the nearest neighbour method, nearest individual method, point centred quarter method and T-square sampling method (Box 3.3). The nearest neighbour method differs from the other three methods in beginning by identifying random individuals from which to measure. This is problematic in the field since it requires the population to be marked in advance so that random individuals can be chosen. Although some researchers suggest that random positions can be chosen and the nearest individual to that position is then identified as the randomly chosen individual, this is biased towards more isolated individuals. Both the nearest individual method and the point centred quarter method have been suggested by some ecologists as being biased because woodland stands are often clumped or evenly distributed. If species are clumped, the nearest neighbour method overestimates the density whereas the point centred quarter method underestimates it. But if they are evenly distributed, then the opposite is true. The T-square sampling method overcomes some of this bias because it combines attributes of the other two methods. Techniques have been developed that extend this procedure to include extra measurements (e.g. by taking a final nearest neighbour to nearest neighbour measurement). Yet other methods involve measurements to multiple individuals (i.e. to the nth individual), from random starting points (ordered distance method) or within a fixed width strip (variable-area transect method). Plotless methods can be employed on maps, or to examine features such as animals' burrows. Krebs (1999) gives further details regarding plotless sampling methods.

Box 3.3   Commonly used plotless sampling methods

(a) Nearest neighbour method; (b) nearest Individual method; (c) point centred quarter method; (d) T-square sampling method

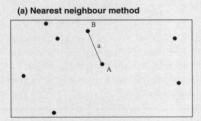

(a) Nearest neighbour method

Here a number of random individuals (usually 20 or more) are located (black circle marked A). The distance between each of these and the nearest individual, e.g. tree (black circle marked B), is then measured as distance (a).

$$\text{Density} = \frac{1}{(1.667 D_1)^2}$$

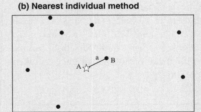

(b) Nearest individual method

Here a number of random points (usually 20 or more) are located (star marked A), each marked with a peg. The distance between each peg and the nearest individual, e.g. tree (black circle marked B), is then measured as distance (a).

$$\text{Density} = \frac{1}{(2 D_1)^2}$$

Where D_1 is the mean distance between all the sampling points (i.e. the sum of all the individual distances divided by the number of points used).

NB: some texts use the same density formula for this method as for the nearest individual method.

Where D_1 is the mean distance between all the sampling points (i.e. the sum of all the individual distances divided by the number of points used).

(c) Point centred quarter method

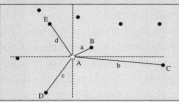

(d) T-square sampling method

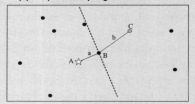

At least 20 random points should be located (e.g. in a woodland), each marked with a peg (star marked A). The area around each peg is divided into quarters using a compass (dotted lines). In each quarter, the nearest tree is identified (black circles marked B, C, D and E) and the distance between each and the peg is measured (distances a, b, c and d).

$$\text{Density} = \frac{1}{(D_2)^2}$$

Where D_2 is calculated by first working out the mean distance to the peg of the four trees from each quarter (i.e. the sum of each set of a, b, c and d distances divided by 4). Once this has been done for all 20 points, then the mean of these 20 averages is taken to obtain D_2.

From a random point (star marked A) the nearest individual (e.g. tree) is located (black circle marked B) and distance (a) is measured. The area is divided into two by drawing a line (dotted line) through the middle of the sample individual (B) at right angles to the line between the random point and the nearest individual (A to B). On the side of the plot opposite the random point, the nearest neighbour is located (grey circle marked C) and distance (b) is measured. Ten random points (n) are recommended as the minimum surveyed.

$$\text{Density} = \frac{n^2}{(2.828 \sum a_i \sum b_i)}$$

That is, add up all 10 distances (a) and do the same with distances (b). Multiply these sums together and then multiply the result by 2.828. Divide this value into ($n \times n$) to give the density estimate.

Distribution of static organisms

Organisms can be distributed evenly, randomly or clumped (aggregated) in space. Both plot-based (quadrat sampling) and plotless sampling can be used to examine this phenomenon. Boxes 3.4 and 3.5 give some examples of the techniques that can be used, both descriptively and statistically to test whether a distribution differs from random either in evenness or in clumping.

Box 3.4 Describing the distribution of static organisms using quadrat-based methods

To test whether a distribution differs from random, count the individuals in each of a series of quadrats (or other sampling blocks) and then calculate:

$$\text{Index of dispersion } (I) = \frac{\text{variance}}{\text{mean}}$$

Where the mean is the sum of all the counts within a series of quadrats, divided by the number of quadrats surveyed (n), and the variance is calculated as:

$$\text{Variance } (S^2) = \frac{\sum x_i^2 - (\sum x_i)^2/n}{n-1}$$

Where x_i is the count of individuals within each quadrat.

To test to see if the distribution differs from a random distribution of organisms, calculate:

$$t = \frac{I-1}{\sqrt{(n-1)}}$$

Where I is the index of dispersion and n is the number of quadrats surveyed.

If the value of t is above $+1.96$ then the distribution is clumped, if the value of t is less than -1.96 then the distribution is even. Values of t between -1.96 and $+1.96$ indicate random distributions.

Box 3.5 Describing the distribution of static organisms using T-square sampling methods

To test whether a distribution differs from random, use the T-square sampling method (Box 3.3) to measure the distances between random points and the nearest individual, distance (a), and the distance between this individual and its nearest neighbour, distance (b), then calculate:

$$t' = \left(\sum (a^2/(a^2+b^2/2)) - n/2\right)\sqrt{(12/n)}$$

Where n is the number of random points surveyed.

If t' is greater than $+1.96$ then the distribution is significantly more even than random, whilst if t' is less than -1.96, the distribution is significantly more clumped than random. Values of t' between -1.96 and $+1.96$ indicate random distributions.

Forestry techniques

Because of the importance of woodlands to humans, a suite of methods have been developed specifically for forestry, including several methods that can be used to describe trees in woodlands in addition to using quadrats and transects (see p. 73 and p. 84).

Tree diameter

One measurement that is often taken to represent the size of trees, is the diameter of the trees (usually the diameter at breast height – dbh – at a height of 1.3 m, or 1.5 m in some surveys, above the ground surface). This is taken for trees of greater than 7 cm dbh (those below being classed as having no volume in forestry terms). Standard tapes can be used to measure the circumference that can then be converted to the diameter by dividing it by pi (π or 3.1416), although it is easier to measure directly using diameter or girthing tapes. These are already calibrated in units of pi centimetres and hence the dbh can be read directly from them. By fastening the loose end of the tape measure to the tree using very sticky tape, it is possible for one person to measure the dbh of even large diameter trees. The mean of the dbh for each species can be used as a measure of average dominance.

The dbh can also be used to estimate the mean diameter of a stand of trees. The standard way of calculating the mean diameter of a stand of trees is to take the quadratic mean of the individual dbh measurements. This is used instead of the arithmetic mean because the quadratic mean gives greater weight to larger trees and correlates better with stand volume than does the arithmetic mean, especially where the tree diameters are highly variable. To find the quadratic mean, each dbh is squared, these are added together and the mean taken (i.e. the sum is divided by the number of trees measured). Finally, the square root of this figure is taken to give the mean dbh.

Measuring old trees can be difficult because of multiple stems and uneven girths. The Specialist Survey Method (SSM) is the standard for recording old and veteran trees. Although there are lots of adaptations across the country for different regional surveys, the *Specialist Survey Method* publication (Fay and de Berker 1997) includes detailed guidelines about measuring the girth on awkward trees that have bulges and are multi-stemmed. It also has guidelines for measuring the amount of dead wood in the crown, the amount of decay, etc. Some of the challenges associated with working with veteran trees are described in Case Study 3.2.

Case Study 3.2 Studying tree growth and condition

Dr Helen Read has worked at Burnham Beeches for over 20 years carrying out and supervising research projects, and monitoring programmes on the nature reserve. Burnham Beeches is a site of European importance for the old pollarded beech trees and the organisms associated with them. Keeping the old trees alive as long as possible while creating a new generation of pollards is the top priority. Evaluating the health of the trees and their response to management work has led Helen to look at similar trees in northern Spain. This in turn has provided information to help with management of the trees at Burnham Beeches.

Photo HJR

Measuring regrowth on an ancient pollard

Model organism and sampling problems faced

Trees are fundamental and structural components in woodlands and other systems. Although they do not move and may appear straightforward to study, trees can present significant problems for researchers. First, they are often tall and it is difficult to reach more than the bottom part of the trunk and the lower branches. Hence, many experimental studies are carried out on saplings and seedlings rather than mature trees. Second, it is tempting to make judgements about trees from the parts that can be seen easily, which may not be a good reflection of what is happening high in the canopy. One option for studying the uppermost branches is to fell the trees, although, aside from the safety issues, it is generally not possible (or allowable) to fell sufficient trees of the type needed for most studies. Some trees are also protected, either in reserves or the wider landscape and felling trees should not be undertaken lightly.

Helen has studied the trees at Burnham Beeches to evaluate the responses of individual branches following pruning for the restoration of old pollards. This pruning was carried out by specialist tree surgeons who climbed high into the branches to make the cuts. Previous work had examined tree growth by measuring the length of extension growth each year on branches: this is the distance between the terminal bud scars formed each winter at the end of the twigs. The problem with this method is that only the lower branches are within reach, which are not be representative of the overall growth. In addition, those branches that could be reached were not necessarily the ones that had been pruned. Even a ladder did not enable the important upper branches to be reached safely and access platforms were too expensive to hire for the period of the study. Felling these trees was not an option because of their high wildlife value.

How the problems were overcome
The simplest method involved recording the tree response using binoculars during winter when the leaves had fallen and the branch structure could be seen more clearly. Because Helen stood on the ground, she did not need complicated safety equipment and details could be seen quite clearly through an average-quality pair of binoculars. The sizes of branches were estimated in a simple way: branch diameter was categorised as finger size, arm size, leg size or waist size; and length was estimated as less than 0.5 m, 0.5–1.0 m or over 1.0 m. Individual shoots growing from cut branches were counted but those arising very close together were recorded as clusters. Tree health was also estimated from the ground. There are standard methods for doing this on beech trees using Roloff scores (Roloff 1985) that evaluate the shape of the canopy and the branching structure of twigs. These measures can be combined with estimates of canopy density by making comparisons with published standards for each tree species (Bosshard 1986). Using such simple estimates, non-parametric tests can be carried out to compare the responses of different groups of trees.

Helen's team used a different technique when researching the responses of trees pruned with a chainsaw or an axe in northern Spain. This time, trained tree surgeons were available to measure growth. This produced more accurate measurements of the responses of the trees, which were complemented by tree health assessments made from the ground. Using tree surgeons to measure the trees was advantageous in that they added value through their expertise.

Advice for students wanting to study trees
Trees are rewarding organisms to study and the range of projects that can be carried out is very wide. However, they are much more three dimensional than ground vegetation and parts of them are out of reach to most researchers. Being long-lived organisms provides opportunities for research but also an extra layer of complication. The number of trees that need to be measured may be large to overcome variation in aspect, exposure, light reaching the branches, and soil type. Helen advises that you stick to simple variables and do not try to make your project too complicated.

Health and safety issues can be problematic with tree studies. You should never use a chainsaw, or work from a ladder, or climb into the tree yourself unless you have been on the relevant training courses. Helen suggests trying to combine your project with work going on at sites or finding a willing tree surgeon who will help by accessing the tops of trees. Always discuss your project with the site manager or owner of the trees who may have suggestions for projects they would like to see carried out. Trees *are* felled in the course of habitat work or for safety reasons so you may find that you can study these and examine research avenues, such as looking at the ages of trees and the amount a tree has grown in particular years. Do not forget that simple estimations are often good enough to give a guide to responses.

Tree basal area

The basal area of a stand can be calculated by summing the basal areas of all of the trees in a stand. The basal area (ba) of a tree at breast height can be estimated as:

$$\mathrm{ba} = \pi \times \mathrm{dbh}^2/40000$$

where the dbh is the diameter at breast height measured in centimetres and the basal area is in square metres.

Height of trees

It may be necessary to measure tree heights. Although a surrogate for height can sometimes be obtained from the diameter at breast height (dbh – see p. 90), this is only accurate if it can be validated using fallen or cut trees of the same species to check the relationship between girth and height. It is better to measure tree height directly by a trigonometric method (Figure 3.10). This involves using a clinometer to sight up to the top of the tree to measure the angle between your eye and the top of the tree, and then trigonometry can be used to calculate the tree height. Note that by convention, the lengths of felled trees are measured in straight lines from end to end (i.e. ignoring any curvature of the log).

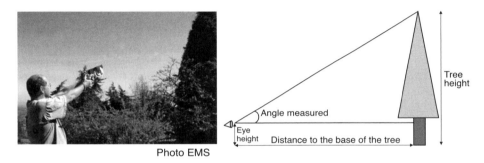

Photo EMS

Figure 3.10 Using a clinometer
The angle from the horizontal to the top of a tree is measured. Using this angle and the distance to the bottom of the tree, you can calculate the height of the tree:

tree height = (distance to base of tree × tangent of angle measured) + your eye height

Note: you must add on your eye height since you have taken the angle from here and not the ground. It is worth remembering that the tangent of an angle of 45° is 1 and therefore if you can move away from the tree so that the angle you make to the top is 45° then the height of the tree will be the distance that you are from the base plus your eye height. This can be particularly useful for trees growing in open parkland, but is less likely to be feasible in dense woodland.

Timber volume

Timber volume is a measure of yield that is used by agro-foresters in coniferous plantations and that can also be applied to broadleaf trees. Timber volume can be measured for single trees, stands and fallen timber by using the dbh and the height of the tree (for conifers) or height of the timber (for broadleaf trees). The height of timber is taken as the distance from the base of the tree to the point where the diameter is 7 cm (in broadleaf trees this is where the stem tends to become indistinguishable). For an individual tree, a tree tariff number is obtained from tariff charts (which can be found in Matthews and Mackie 2006) for the species within specified height ranges. This is then used in association with the dbh to get an estimate of the timber volume from tariff number alignment charts.

Growth

To establish patterns of growth, the dbh needs to be measured over a long period of time (over 5 years and preferably 10–20 years depending on the species). Permanent quadrats can be established and stands of trees mapped to scale, noting their state of health (e.g. bark damage, presence of dead limbs, etc). Each year during the non-growth phase (i.e. winter), each tree can be relocated and the dbh of all trees in the stands remeasured. Coring can be used to look at past growth patterns, but permission is needed from landowners for such sampling.

During the winter, deciduous trees stop growing, and a dormant terminal bud forms at the end of each branch. The following spring, when the twigs elongate from this bud, the scales covering the bud drop off and leave scars (known as terminal bud scars or girdle scars) in a ring around the twig. By working backwards from the tip of a twig, it is possible to distinguish the previous years' terminal bud scars and to measure the length of twig between them (giving the amount of twig growth in a particular year).

Age

The age of a tree is very difficult to establish unless the complete series of tree rings can be counted. You can use one of three simple methods to calculate age. The first, using local maps to establish the age of the woodland, is likely to be the least accurate. Second, age can be interpolated from recent growth patterns (e.g. using tree ring measures from fallen or felled trees to age living trees on the basis of the diameter at breast height). The third and most accurate method involves coring the tree and counting the rings to give the age of the tree. Note that you need to seek permission from the landowner before going ahead and be aware that some trees suffer from coring if it is not properly executed. White (1998) gives further details for large and veteran trees.

4
Sampling Mobile Organisms

The obvious problem that ecologists face when sampling animals is that most are mobile. Movement in animals also means that it is sometimes hard to decide what the population is that is being studied. Mobile populations may not be limited to a defined area such as a field or woodland, but may be present because they are using a resource (e.g. a spring for water). Depending on the monitoring technique used, the population sampled may be those animals that are particularly active and hence trapped or seen more easily (e.g. males looking for mates, individuals seeking food). Smaller animals, including many invertebrates, may be readily captured and are often killed to facilitate counting. For larger animals, it is usually not ethical to kill individuals for such purposes and instead researchers rely upon sightings, trapping and subsequent release, and indirect evidence (e.g. tracks and signs). This is skilful work: one only has to think of the problems of identifying groups of very similar species (e.g. warblers), or distinguishing signs that are very similar between species (e.g. deer droppings) to realise the difficulties that can be faced. You need to be aware of your limitations and choose a field in which you have some experience or can rely on an expert for help. Capturing, handling and even observing animals should be done with as little impact as possible to avoid distress to the animal. It is important that any captures are returned to the field as soon as is practicable. See MELP (1998a), CCAC (2003) and ASAB (2006) for further advice on the treatment of animals in research.

Sampling mobile organisms can be quite challenging even when your project is based in your home country where you are familiar with legal and logistical arrangements. The complications of using often expensive and sometimes sophisticated equipment and techniques may be offset by linking your work to established projects. This can be especially helpful when working overseas, and when you are relatively unfamiliar with the customs and laws of the country in question. Such links can be particularly useful when working on vertebrates and those invertebrates that are either protected in law (as with bird-wing butterflies or tarantula spiders) or are hard to sample (e.g. live within forest canopies). In this chapter, we first discuss the general issues associated with sampling mobile organisms and then deal with sampling issues by animal group.

General issues

It is usually important to standardise the counts or captures made, especially if comparisons will be made between sites or times of day (or year). This can be achieved by ensuring that the methods stay constant and that any researchers involved are fully trained and know exactly how to implement the technique. However, there are occasions where differences in the situation experienced (e.g. in habitat structure) can mean that one technique is not as easily or appropriately employed in areas that are to be compared. Here, care is needed to use strategies that will enable comparisons. One method is the catch per unit effort (CPUE), which is a relative counting system used for trapping animals (especially in fish stocks). CPUE requires that the catching mechanisms and the time over which they are set are equal. For example, if you catch fish using pot nets in a river to investigate large-scale habitat differences between fish species, to use CPUE the pot traps must be of the same design, set and collected at the same time and there must be no barrier to trapping any fish species. You must ensure that if you are working in different river habitats, there is the same access to the traps between highly vegetated and open waters. Modifications may be required; for example, you may be advised to clear an area around any traps where the density of vegetation may prevent effective catching. The choice of how to locate your sampling points also requires consideration. Common methods include sampling within quadrats or blocks of land, or using transects (see Chapter 3 for a discussion of these methods). Point counts can also be used to locate, count and observe animals, especially for those species that are relatively easy to see or hear. The positioning of blocks, transects or sampling points can be determined systematically, by using random coordinates or by a systematic random method (see Chapter 1 for further details).

When observing animals, either from a distance, or in the hand or a trap following capture, it is worth making some general assessments of their condition. These might include the general condition of the animal. Indicators of condition include activity level, presence of abnormalities such as the loss of a limb, signs of disease or fungal infection, presence of tumours, cuts or other damage, presence of any ectoparasites, and, if stomach contents are taken or the animal dies or is killed during capture, the presence of endoparasites. Any dead animals or parts of the animals removed during the capture process (e.g. toes during toe clipping) could be kept for DNA analysis. Specimens for DNA (or RNA) analysis should be kept in separate containers to prevent cross-contamination, transported frozen (in liquid nitrogen or dry ice in a suitable container), dried (using moisture-indicating silica gel to prevent rehydration), or in a preservative such as ethanol or propylene glycerol. For example, many scientists dry specimens of insects and plants, whereas tissue and blood samples tend to be transported in non-denatured 70% ethanol or physiological buffer. There is also a range of specialist products (e.g. RNAlater) and an increasing dialogue about other alternatives (e.g. other alcohols, acetone, salt solutions and detergents). Samples should be frozen at an appropriate temperature as soon as possible (e.g. at $-80\,°C$ for long-term storage). Freezing in the field is difficult, although there are small fridges that can be used with dry ice and can be posted or transported relatively easily. See Vink et al. (2005) and Williams (2007) for some discussion on this topic.

Animals can be photographed as a permanent record to enable identification to be checked later, and any queries about the specimen resolved at a later date. Where possible, the age, sex, length and mass of the animal can also be taken. For many small animals the mass can be easily measured by containing the animal in a soft cloth bag and using a spring balance to take the total mass (the mass of the bag can be subtracted later). Other measurements depend on the group involved: for tailless amphibians (e.g. adult frogs and toads) the length is usually snout to vent (anus/cloaca); for tailed amphibians such as newts and salamanders, reptiles and small mammals the snout to vent and total snout to tail lengths can be taken separately; for birds, the wing length, tail length, tarsus length, breeding condition and stage of moult can be noted; for mammals, tooth condition, breeding condition and species-specific aspects (e.g. antler length) may also be useful.

Many animals (e.g. spiders, reptiles, birds and small mammals) hide in burrows and holes. These can be searched using endoscopes to not only enable you to ascertain that there is an animal worth searching for, but also help you to avoid surprising a potentially poisonous or aggressive animal such as a snake or scorpion.

Distribution of mobile organisms

In a similar way to static organisms, mobile species can also be distributed evenly in space (homogeneous), or be randomly distributed or clumped (aggregated or heterogeneous). Plotless sampling is a reasonably effective method of exploring this (see Chapter 3, Box 3.3). It is also possible to utilise this technique to look at the distribution temporally rather than spatially. Here organisms are counted in blocks of time (time periods) rather than blocks of space (quadrats or other spatially allocated blocks).

Direct observation

This is the most straightforward method of monitoring animals and is widely used to estimate the size of flocks, herds and shoals. However, this frequently requires considerable skill and experience, especially when animals are moving away from you or in a random manner. Counting live vertebrates should only be attempted after a prolonged practice period as initial estimations are confounded by inaccuracies. Direct counts are also widely used by entomologists on dead invertebrates and some conspicuous live invertebrates (e.g. snails). Counting dead invertebrates can give an absolute measure of the captured individuals but may not provide population sizes or densities (see the discussion on pitfall traps on p. 139). Counts can be made along transects, at specified points and within quadrats. It is important to choose the correct format before the research project begins as each of these approaches has their own idiosyncratic problems. The best methods depend on the behaviour and conspicuousness of the species concerned and the nature of the habitat being surveyed.

Behaviour

One extra dimension of mobile animals is that they exhibit a range of behaviours that may differ between habitats, or because of individual factors (e.g. age, sex, degree of motivation), or biotic factors (e.g. the threat of predation or the number of competitors) or abiotic factors (e.g. season, weather or pollution level). There are several different approaches to the study of behaviour (e.g. using Tinbergen's classification – see Martin and Bateson 2007):

- causation (what factors influence the onset or characteristics of the behaviour?);

- development (how does an individual animal develop the behaviour?);

- function (what is the behaviour for?);

- evolution (how did the behaviour evolve?).

The following section summarises some of the important aspects of studying behaviour in the field. For more detailed coverage, consult Lehner (1996) and Martin and Bateson (2007), and see Case Study 4.1 for a discussion of researching behaviour on semi-wild populations. Before proceeding with behavioural studies, it is important to make preliminary observations of the system that you will study. This will enable you to be able to decide how you will describe the behaviour(s) involved, choose and define categories of behaviour, as well as choose appropriate species and groups. This section considers behavioural work in the field and for animals that are in captivity temporarily for behavioural assessment. In planning such studies, you need to determine whether you are recording discrete events (short-term actions such as lying down, standing up) and/or states (longer times of activities such as periods of sleep). You may also wish to identify whether there are any bouts of events (i.e. clusters of the behaviours). Although there are many different behavioural questions that can be asked, common ones include:

- Using choice tests to examine preferences or differences in response to stimuli. This may include visual, olfactory, or auditory stimuli, which may be natural or artificial in origin.

- Identifying groups (i.e. clusters or spatial patterns of animals). This can involve measuring the distances that individuals are from each other with those found close together being identified as group members, whilst those further apart are from separate clusters.

- Examining dominance hierarchies (e.g. whether animals out-compete others for resources, e.g. food or space).

- Investigating associations between animals (using indices of association) and the proximity of individuals to one another (e.g. mother and offspring relationships).

- Describing individual distinctiveness in terms of behaviours, e.g. aggressiveness, curiosity, etc. (as well as in order to be able to recognise individuals).

Case Study 4.1 Cracking the chemical code in mandrills

Dr Jo Setchell is a lecturer in Evolutionary Anthropology at Durham University and Editor-in-Chief of the *International Journal of Primatology*. Her research takes a strongly integrative approach, combining behaviour, morphology and demographic studies with genetics, endocrinology and chemistry to address questions relating to reproductive strategies, life history, sexual selection and signalling in primates. Her case study examines the problems of integrating field-based sampling with chemical analysis in the laboratory.

Photo JS
A male mandrill

Model organism and sampling problems faced

Jo works on mandrills, a large monkey that lives in deep rainforest in Central Africa. Mandrills are extremely sexually dimorphic: adult males are over three times the mass of females, and are brightly coloured, with red, blue, pink and purple colouration on their faces, rumps and genitalia. Mandrills are unusual in possessing a sternal scent gland and males invest much time and energy in scent-marking. Working on mandrills poses a variety of problems: they have yet to be habituated in the wild, although there are fascinating studies of their behavioural ecology, using radio-collaring technology. Work on zoo populations also poses the problem that such animals are in artificial situations and their behaviours may be different to those of wild populations. Jo also faced the additional problems of needing fairly sophisticated chemical analysis of biological samples (odour samples), and therefore required access to analytical equipment and techniques. Jo collaborates widely, and in this case study she describes some of her work on odour profiles and their relationship to age, dominance status, health and genotype.

How the problem was overcome

Jo decided to work on a semi-free-ranging colony, which provides a useful approximation to natural conditions, with benefits of long-term study of individuals and captures, enabling collection of morphological data, blood and other biological samples. She began by analysing the odours produced by males, comparing these with the signaller's features: sex, age, dominance status, endocrine status (via faecal testosterone and glucocorticoids), health (via faecal parasite analyses) and genotype. Jo used routine captures to build up a bank of odour samples using dry, sterile cotton swabs to collect secretions from sternal glands, putting the swabs in sterile glass vials, freezing them in liquid nitrogen, and storing them at $-80\,°C$. She collected control swabs (exposed to the same environment, but not rubbed against the animal), and hair samples from the sternal gland (soaked with secretion) and elsewhere on the body (no secretion). Shipping the frozen samples back to the UK for analysis involved the completion of various paperwork.

Jo collaborated with Stefano Vaglio, who was also interested in mandrill odour. Stefano was collaborating with a mass spectrometry laboratory but had trouble getting samples from zoo-housed animals (mandrills marked filter paper, but destroyed it before collection). The team exploited recent methodological developments (dynamic headspace extraction to concentrate volatile chemicals and solid-phase microextraction for hair samples) to detect compounds in sternal-gland secretions, and tentatively identified them by comparing the spectra with those of a mass spectral database. They used relative rather than total abundance of the compounds that comprised odour profiles controls to control for differences in the amount of secretion produced. They analysed the samples over a short period to minimise interassay variability, using controls to identify compounds that did not derive from mandrills. Mandrill odour profiles, like those of other mammals, are composed of many different compounds, so they used principle components analysis to reduce these to a smaller number of uncorrelated principal components, followed by discriminant function analyses to test whether groups could be reliably discriminated between (e.g. males and females) and matrix analyses to compare the genetic distance between individuals with the chemical distance.

Advice for students wanting to study chemical communication in primates

The important message from Jo's work is recognising that collaboration can be key to solving problems scientifically. You should collaborate with experts in chemical analysis from the beginning to design sample collection methods, run pilot studies and explore analysis methods. If you can, collect multiple samples from each animal (simultaneously and over time). Anyone integrating different disciplines should make sure that they are thoroughly grounded in the new field, so they can communicate with collaborators productively. Respect people you work with, at all levels, in the field and laboratory, making sure that you agree in advance who contributes what, and what each person expects from the collaboration. More generally, if you get the chance to study any taxa in the field, particularly those we know little about, collect all the data and samples you can. Sadly, yours might be the only study of your population (or species). Stay alert to new possibilities (e.g. an Old World monkey that scent-marks). Finally, enjoy the opportunity to conduct fieldwork – it can be a life-changing experience.

Jo collaborates widely to understand whether individual genotypes might be communicated via odour. None of her work would be possible without the primate centre, and the team of enthusiastic collaborators with whom she has had the pleasure of working including Kristen M. Abbott, Marie J. E. Charpentier, Leslie A. Knapp, Tessa E. Smith and John Waterhouse. Particularly, Jo would like to thank her PhD supervisor, Alan Dixson, for introducing her to mandrills, and E. Jean Wickings, who was responsible for the colony of mandrills at the Centre International de Recherches Médicales de Franceville, Gabon.

There are a number of different ways of recording behaviour. In practical terms this can involve scoring or writing descriptions of the behaviours, or using checklists or timings. This can be accomplished directly by the observer writing on record sheets or entering data directly into a portable computer or data logger. Alternatively, animals can be videoed and the behaviours monitored later using software to interpret distances, activities, etc., or by the researcher transcribing from the video. Event recorders can log behaviours using a variety of mechanical or electronic methods of identifying behaviours and recording them. Event recorders and video may prevent problems of observers interfering with behaviour; however, these systems can be quite expensive and may not be as flexible or sensitive as human observers. You should also consider what measurement you wish to record. This may include the latent period (the time until a specific behaviour occurs), the frequency (the number of occurrences of a behaviour per unit time), the duration (the length of time a behaviour lasts) and the intensity or strength of a behaviour (the amount of food eaten, distance travelled, number of steps taken, the number of grooming movements).

You also need to determine when to record the behaviours. You could record behaviours as they occur, or use focal sampling (examining animals in turn for a specified period or until the behaviour ceases), or scan sampling (where the whole group is scanned at set intervals and all animals engaged in particular behaviours are recorded) or behaviour sampling (where the whole group is observed and all instances of specified – relatively rare – behaviours are recorded).

The period of time being surveyed is also important. This may include using continuous sampling (which covers the whole sampling period but is time-consuming and may increase observer fatigue) or periodic sampling (where the recording period is divided into sessions or periods). The time period chosen needs to be short enough to be accurate in terms of the behaviour being monitored and long enough to cover all the behaviour(s) being measured. Periodic sampling can include monitoring periods separated by periods of non-recording.

Choice tests can be based on the simultaneous exposure of individuals to the stimuli or successive tests where animals are exposed to different stimuli in rotation. Decisions must also be made about whether animals will be tested on a single occasion or if repeated testing will take place. In addition, how the degree of response is recorded should be determined. This may involve noting whether a response occurs or not (all or nothing), or by measuring the intensity of the response (e.g. the number of responses, or the speed of any response).

When examining the development of behaviours with age, either cross-sectional or longitudinal studies can be used. The former examines a number of different age classes at one period in time, while the latter follows a cohort as it ages and hence reduces the impacts of some confounding variables but, by definition, requires longer time-scales for the research.

Monitoring behaviour can be particularly problematic for those species that are secretive, at low densities, or found in difficult to observe habitats (e.g. burrows or tree canopies) or times of the day (e.g. those that are nocturnal) or year (e.g. those that hibernate in winter). Studying behaviour in the field is inherently problematic since it is usually difficult to assign cause and effect (unlike in more focused laboratory-based experiments). No matter what behaviours you are sampling, you will need to be aware of particular problems that may need to be addressed (Box 4.1).

Box 4.1 Avoiding problems in behavioural studies

There are a number of different problems that can be faced in behavioural studies, including avoiding bias, and issues with measurement and analysis.

Bias

- Prevent bias that may occur if animals are retested (e.g. to gain average or maximum responses to stimuli) to such an extent that they ignore the stimulus;

- ensure that the time of day and season is appropriate for the behaviour studied;

- avoid order effects (where stimuli have different responses depending on the order in which they were presented) by altering the order in which stimuli are presented during the study;

- reduce observer effects (where the presence of the researcher interferes with normal behaviours) by screening the observer from the animals being studied;

- reduce experimenter bias (e.g. unconscious influences on the scoring of the behaviour being measured by having well defined definitions of scores).

Measurement

- Design further studies to support your findings by repeating the study (literal replication) or using another design that complements the first study (constructive replication);

- avoid floor and ceiling effects (where scores cluster at either end of the scale being measured) by more precise scale design;

- take care to avoid anthropomorphism of animals' reasons for exhibiting behaviours;

- determine which behaviours may be combined later (e.g. 'browsing' and 'grazing' can be combined as 'feeding', but 'walking' cannot later be separated into 'slow' and 'fast walking');

- train observers so they are consistent in their approach (i.e. in how they score particular behaviours) and do not change their approach as they become more skilled or fatigued;

- use short recording periods to avoid observer drift where observers change their interpretation of a behaviour (this is especially problematic if the behaviour is not well defined);

- check the reliability of measurements (i.e. that measurements are repeatable, consistent, and sensitive enough to capture appropriate differences in behaviour;

- ensure the validity of measurements so they are unbiased, specifically reflect behaviours being measured, and are appropriate surrogates of any underlying scientific aspects being monitored;

- use controls where possible;

- work within any ethical and legislative frameworks (e.g. the choice of sites and species, and the avoidance of pain or discomfort).

Analysis
- Use replicates to ensure that enough data are gathered, avoiding too large a sample size (use pilot studies to assess the variability of behaviours and how different comparisons are);

- analyse independent data points (i.e. do not repeat measurements on a small number of animals in order to increase the size of the dataset – pseudoreplication – see Chapter 1);

- examine interactions between conditions that may produce different responses in combination than they do separately to avoid too simplistic a view of behaviours;

- take account of group effects (e.g. where several animals from each of different families are used, their relatedness should be accounted for in the statistical analysis);

- phylogenetic relatedness should be included in cross-species studies.

Indirect methods

In the absence of appropriate methods of direct observation, indirect signs can be used to identify whether a particular animal is or has been present. Such methods can also sometimes be used to estimate population sizes or densities. The signs used depend on the species being surveyed and include the presence of runways and tracks (especially where substrates are soft, e.g. in mud, sand and snow). The presence of dug areas of soil, holes, burrows and nests may also be species specific, as can any eggs or eggshells found. **Note:** that disturbing animals' homes, young and/or eggs is frequently restricted by law. Cast skins of invertebrates and reptiles, hair of mammals and birds' feathers may also be used to identify many species. Dung and bird pellets can not only indicate the occurrence of the animals that produced them, but can also be used to identify what individuals have been eating. **Note:** that the use of dung in indicating abundance may be problematic in some species since rates of defaecation can depend on the food type consumed as well as the abundance of animals. The corpses of animals on roads can indicate the presence of a species in an area, while evidence of attacks by predatory birds or mammals can demonstrate the presence of both the predator and the prey.

Capture techniques

It is often necessary to catch animals in order to count, measure and/or mark them to enable their movements to be monitored, or for populations and densities to be estimated. Whatever the reason, care must be taken to use appropriate techniques that are efficient for the species in question, and that reduce or remove any chance of catching non-target species or unnecessarily harming individuals. Where a technique leads to the death of individuals, it is important to consider not only the legal framework in which you are operating, but also the ethics of killing the animals. For many vertebrate species, capture (even of live specimens that will subsequently be released) may be governed by the law and you must make sure that this is observed, obtaining licences where appropriate.

Much equipment used in capturing animals is expensive and should be well maintained and protected from damage in the field by either wild and domesticated animals or from vandals. It is possible to make your own equipment, especially where this is not available commercially. If you do go down this route, you need to make sure that it is fully tested and fit for purpose. Case Study 4.2 describes the development of one such piece of equipment. Whether the apparatus that you use is mainstream or home-made, all equipment used in a survey should be of the same type and be capable of functioning in a similar way, in order to ensure that collections can be compared. Traps should be set for the same length of time, ideally for the same time periods. Similarly, the same sampling effort should be used over similar time periods for any more active capture techniques (e.g. hand-searching).

You may find that you need to store specimens or parts of specimens for identification and verification purposes. This can include DNA analysis (see p. 96 for further details). Soft bodied invertebrates (e.g. insect larvae, worms and molluscs) should be stored in preservative (usually methanol) which is regularly topped up or replaced. Hard bodied animals can be pinned through the body and mounted in cases which contain a fumigant to prevent attack by insects and other pests. Small specimens may need to be mounted on card or acetate, which is then pinned into the case. Species with large and/or fragile wings (e.g. moths and butterflies) may be kept in envelopes or set so that the wings are spread as in life. These are then pinned through the body and mounted in cases. See Mahoney (1966), Oldroyd (1970) and Lincoln and Sheals (1979) for further details of preserving invertebrates.

When preserving vertebrates (or even parts of vertebrates) it is important to ensure that it is legal to own such items. For example, it is illegal in some countries to own some vertebrate skins and bones, and birds eggs. Skeletal material can be cleaned (usually by boiling and removing any flesh) and kept in labelled zip lock bags. Be aware that vigorous boiling of delicate specimens, including many amphibians, reptiles such as snakes and small birds, can result in a pile of undistinguishable bones. Study skins of mammals and birds are cleaned of as much flesh as possible and dried and cured before

Case Study 4.2 Barnacle larva trap

Christopher D. Todd is Professor of Marine Biology in the School of Biology at St Andrews where he has been for the past 31 years. He has broad interests in intertidal community ecology and the ecology of wild salmonid fish. His case study details the process of developing novel equipment to sample small animals under difficult environmental conditions.

Photo CDT

Larva traps *in situ* on a mounting sheet

Model organism and sampling problems faced

Most benthic marine invertebrates produce free-swimming, planktonic larvae that can confer dispersal and colonization advantages to species that are sessile or sedentary as adults. Barnacles are major space-dominating invertebrates on wave-swept rocky shores. There can be tens of thousands of adults per square metre, and settlement densities of post-larval juveniles also attain extraordinary levels (tens per cm^2 per tide). Early post-settlement mortality of newly settled cyprid larvae (the last larval stage before adulthood) is typically high. Ecologists have long discussed the relative importance of 'pre-settlement' versus 'post-settlement' factors in driving population and community structure on a local scale. In addressing this controversy, Chris and his team have sought to quantify larval input (and its variation) to the intertidal substratum using novel trapping methods.

Barnacle cyprid larvae are typically under a millimetre long and capturing them with nets requires fine mesh that clogs easily with suspended particles. Sampling devices must be small, robust (to withstand heavy wave crash) and able to sample larvae over a tide cycle without clogging. For years, such difficulties were implicitly acknowledged but not actually addressed, with daily counts at low tide of newly settled larvae being used as a proxy for larval supply. However, settled larvae may not attach and metamorphose, but detach, become planktonic again and settle elsewhere.

How the problem was overcome

Chris developed a passive trap using the principle that a cylinder with a height–aperture ratio of 12 or more has 'dead space' at its bottom where fluid remains static (irrespective of how fast water flows past the aperture, or how heavy wave crash may be). On irregular rocky shores there are complex local flow hydrodynamics affecting barnacle larvae. He needed a trap that could be replicated within 10 cm of one another and without leaking during low tide. He searched for a proprietary item that could be modified and found the Greiner Cellstar skirted sample tube was ideal: being polypropylene it could be cut/machined, it has a leak-proof screw-cap, and some versions have a conical bottom with no surrounding skirt. Traps are set in slots on Perspex mounting sheets cable-tied to ropes around the rock, where the substratum is approximately vertical. The mounts enable settlement tiles to be attached.

Early results were encouraging but Chris recognised that he needed an effective killing solution. He used 4 M urea solution to retain larvae because it was not toxic to people, but caused larvae to rapidly cease swimming on contact. During tidal immersion, the urea solution is washed from the trap. If the solution is stained using bromophenol blue dye, then larval capture and urea retention efficiency at varying levels of wave crash can be measured using the absorbance of retrieved urea solution in a spectrophotometer. Chris can calculate urea dilution with seawater to assess trap efficiency in retaining larvae during different levels of wave crash.

A 'spiral staircase' of internal baffles and a conical aperture opening were two improvements that were made to enhance larval retention without hindering influx. With these modifications, the trap appears equally effective over the full range of wave action. Now Chris and his team routinely capture several hundred larvae per trap day. These high sample densities provide reliable data for assessing the relationship between planktonic larval supply and successful settlement on adjacent settlement panels.

Advice for students wanting to develop novel techniques
Different species respond differently to given trap apertures: *Balanus crenatus* was captured most efficiently with conical entrances of 1 cm^2 area cf. 0.5 cm^2 for *Semibalanus balanoides*. Preliminary tests are necessary to check this prior to final experiments in your locality. Picking and sorting samples is tedious, especially when particles are suspended in the water column, but there is no substitute for careful sorting by hand, using a good dissecting microscope. Undertaking experiments on rocky shores has attendant health and safety requirements and fieldwork can be physically demanding and is best undertaken in groups or pairs.

When devising practical solutions to conceptual challenges, look around your laboratory or DIY store to see how you can modify proprietary items for your purposes. You will be surprised how often seemingly intractable problems can be solved with imagination and lateral thought. The advantages offered by the tubes that Chris uses are that they are cheap, made of plastic (easily cut/machined), can be set in protective resin, have a leak-proof seal and conical elements already cast in them; they are also produced to a high standard of repeatability of quality and dimensions. Finally, although you may not seek to sell the devices you develop, it is courteous to inform the manufacturer of the (unusual!) use to which you are putting their product.

This work was supported by the NERC (grant: NER/A/S/2002/00957). Numerous colleagues helped develop the larval trap, including Chris Andrews, Adrian Gude, Joe Phelan and Birgit Weinmann.

being stuffed with cotton wool. These need to be kept in boxes which contain a fumigant to prevent insect or other attack. Dung samples and bird pellets can be dried and kept in airtight containers. Cast skins of reptiles, feathers and samples of mammal fur should be placed in envelopes or plastic zip lock bags, again with regular checks for signs of attack from mould. See Mahoney (1966) for further details of preserving vertebrate specimens.

Marking individuals

Mark–release–recapture methods can be used to estimate population size. This is based on the assumption that if a sample of animals is marked and released, these will naturally mix with the rest of the population. When a second collection is made the proportion of recaptured (marked) animals to the total capture size reflects the size of the total population (for details see Chapter 5, p. 252). There are a number of assumptions of mark–release–recapture that should be met:

- the behaviour of marked animals is not affected;

- the likelihood of survival of marked animals is no different to that of unmarked individuals;

- marked animals become completely mixed in the population;

- the probability of capturing marked animals is the same as that of capturing any member of the population;

- sampling time is small in relation to the total time of the study;

- sampling is at discrete time intervals (although some methods do not require this assumption).

Marking techniques are discussed in each of the separate animal sections as this is often taxon specific. However, as a general point, mutilation (e.g. removing toes on small mammals or reptiles or amphibians to provide individual marks) is not advocated: non-invasive methods (e.g. spot painting) are favoured as (in addition to ethical considerations) these are less likely to affect the behaviour of the animals. There may be legal constraints to marking vertebrates and some invertebrates (e.g. octopus). Depending upon the country in which you are working, a licence may be required if the animal will be under any pain, distress or lasting harm. However, ringing, marking, tagging (e.g. ear notching, freeze branding) are often not regulated procedures (unless for regulated species as in many birds) and researchers may not require a licence for this. For animals of sufficient size, an individual number can be written on using indelible ink or paint (e.g. on the wings of butterflies, backs of lizards, etc.). For smaller animals, if individual recognition is important, then a code (either involving the position or, where the animals are too small for positions to be easily separated, different colours of spots of paint or ink) may be useful (Figure 4.1). Similar patterns can be used when fur clipping small mammals, where using three clips in six possible positions can result in 41 unique combinations of individual marks: six individual clips, 15 combinations of two clips and 20 combinations of three clips.

Where animals have obvious natural markings or other features that can enable individuals to be recognised (e.g. horn shapes, ear notches, etc.), then these can be used either through direct observation or by using photographs to monitor the animals. The ears and horns of rhinos, patterns on spotted cats such as jaguars (noting that each side of

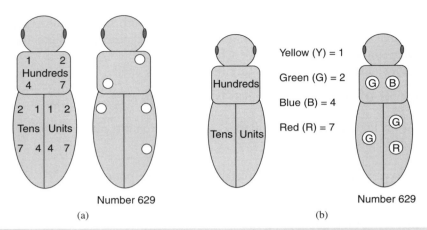

Figure 4.1 Use of ink or paint spots to identify individual invertebrates
Using the position of the spots (a) or using colour of the spots (b), 999 individuals can be identified using a maximum of six spots per animal.

an animal will be different to the other) and the belly patterns on newts, are all examples of markings that have been successfully employed in identifying individual animals.

More detailed discussion of mark release recapture can be found in Pollock et al. (1990), Sutherland (1996), Hayek and Buzas (1998) and Southwood and Henderson (2000). Marking techniques are discussed by Woodbury et al. (1956), Stonehouse (1978) and Nietfeld, Barrett and Silvy (1996), whereas Murray and Fuller (2000) provide a critique of marking methods and their impacts on the animals being marked.

> **Note:** this handbook does not conclusively outline the legal position or interpretation of any act or regulation. In all cases (especially of vertebrate marking) it is the responsibility of the researcher to check to see if the procedure being considered requires training and/or a licence.

Radio-Tracking

A relatively sophisticated technique that can be applied to most animals over a few centimetres in length is radio-tracking. This subject is complex and the technology is continually changing, hence the following overview should be supplemented by more technical reading (see Table 4.1 for a summary of some of the factors that should be considered). Many animals, including mammals, birds, reptiles, fish and even some invertebrates, lend themselves to radio-tracking to establish home-ranges, habitat use or migration routes. This specialist methodology requires that a radio transmitter is fitted to animals on a collar, clip or suction device, for example, or glued onto the animal using biodegradable adhesive or implanted within the animal. Animals are tracked either manually in the field or remotely. The size, type and power of the transmitter is an important issue and much depends on the size and behaviour of the host animal as well as the size of the budget allocated to a project. Transmitters may be encased in resin to prevent damage from moisture and from chewing by the animal. Detailed planning and information is required if radio-tracking is to be used (to assess the maximum transmitter weight, long-term signal duration, etc.). For slow moving and/or relatively static animals living in relatively small habitats (e.g. snails and burrow-living spiders), light-weight tags (e.g. passive integrated transponders – PITs) that use radio-frequency identification (RFID) technology can be attached to animals. Such tags respond only when a signal is sent to them, meaning that batteries are only needed on the combined transmitter/receiver held by the researcher and not on the attachment on the animal (in a similar way to security tags on books, etc., in shops). This will substantially reduce the weight, but also means that the transmitter/receiver must be placed close to the animal in order to obtain a response.

Table 4.1 Some considerations in the choice of radio-tracking equipment

Device	Comment
Implanted transmitters	If animals are streamlined (e.g. snakes or fish) or frequently push through tight burrows or dense vegetation, then implanted transmitters may be most suitable. These also enable animals to shed skins without losing the transmitter. This method has been implicated in relatively few cases of damage to some animals.
Attachment of external transmitters	Collars are used for many animals. These become a problem if the animal can remove it or grows too quickly for collars to be suitable. Alternatives include attaching transmitters to harnesses, leg bands or gluing them directly onto the fur, feathers, scales or carapaces of the animal.
Weight of the transmitter	The transmitter and its aerial, battery, etc., should weigh less than 5% of the animal's body weight (and even less than this for animals where extra weight would cause significant problems – e.g. birds, bats, fish and other aquatic organisms).
Life span of the transmitter	Most of the lifespan of the transmitter is linked to the battery and hence the weight restriction. Battery life can be extended in some transmitters by setting shorter and more intermittent pulse rates, begin pulsing after a delay (a week or so to enable the animal to thoroughly mix with the population), or pulse only when required (at night rather than during the day, or to coincide with field work – e.g. for one day every two weeks). The pulse strength will also influence battery life – weaker pulses use less power but are more difficult to detect.
Type of antenna used	Longer antennae increase the range of the signal. If long trailing whip antennae are not possible, loops or helical antennae may be considered although care should be taken to keep them away from the animal's body and the electronic components of the transmitters since this can cause interference.
Influence of habitat	The power of the signal may be reduced by the density of the vegetation (e.g. trees) and rocky outcrops in rural habitats or buildings in urban habitats.

Most radio-tracking currently uses VHF (very high frequency) transmitters. However, modern GPS-based systems can also be employed. These usually have a receiver attached to the animal that either downloads information about the animal's position (obtained from GPS satellites) at set intervals (e.g. hourly) onto a chip on the collar that can be retrieved later, or sends the data via another satellite to the researcher. Such systems are expensive, have short battery lives and are currently only suitable for larger mammals. If solar cells can be used to recharge the batteries, larger birds may also be studied using this technique.

Radio-tracking technology has also been used to find den sites by tagging prey items used as bait that are then taken back to the animal's den. Transmitters can also be used to monitor traps to inform researchers when they have been sprung and hence need emptying, reducing the amount of time that animals are in traps and avoiding needless visits to empty traps. Further discussion of radio-tracking can be found in White and Garrott (1990), MELP (1998b) and Mech (2002).

Invertebrates

Photo EMS

Invertebrates are one of the most rewarding groups to work with, owing to their relative abundance (thus providing large enough sample sizes for statistical analysis) and relative ease of capture (sometimes even in relatively poor conditions, as in the winter in higher latitudes areas or summer in hotter climes). Although some equipment for working with invertebrates can be specialised and expensive, simple collecting nets, sample vessels and a hand lens are sufficient for many surveys. The most suitable hand lenses have a combination of lens strengths, usually $10\times$ and $20\times$ (more expensive hand lenses are larger in diameter and much easier to use in the field). Hand-searching is one of the most widely used techniques and can also be effective to collect invertebrates that live on animals, or in nests and burrows (e.g. fleas, ticks and lice). Fibre-optic viewers and cameras can be useful in monitoring animals that live in holes or burrows (e.g. tarantulas, ants in their nests, termites in their mounds, etc.). The techniques employed to survey invertebrates depend on both the habitat in which they live, as well as the ecology and behaviour of the species involved. Hence, this section examines separately those invertebrates that live in aquatic systems, soil, on the ground and in vegetation (including dead and decaying matter), and those that are airborne. Additional information can be found in McGavin (1997), Southwood and Henderson (2000) and Gibb and Oseto (2006).

Invertebrates are an exceptionally large group requiring special attention. A distinction between insects and other invertebrates is often made: insects are a subgroup of invertebrate comprising those that almost always have six legs as adults, whereas most other mature non-insect invertebrates do not have six legs. A large portion of non-insect orders are marine, but there are some large terrestrial orders, including spiders (Araneae), ticks and mites (Acari), and land molluscs (Gastropoda). Identification of invertebrates can be a rather specialist occupation, even with some more obvious species (e.g. butterflies). Field guides, identification keys and monographs for specialists can all help with this (e.g. see Chapter 2, Box 2.4).

In Britain there is a code for collecting insects and other invertebrates that was produced in consultation with a large number of interested organisations.[1] This code can be a useful ethical guide to invertebrate studies. With the exception of groups such as cephalopods in some countries, there are usually few, if any, restrictions on handling and marking invertebrates unless they are particularly endangered. In Britain for example there are currently under 200 invertebrates afforded protection under European legislation, a tiny proportion (<0.001%) of the total number of invertebrate species.[2] If you wish to work with protected species, you may need to apply for an appropriate licence.

Direct observation

Where invertebrates are large enough to be seen from a reasonable distance, or are relatively static and unlikely to move far if disturbed, direct observation may be useful. Butterflies in open countryside, snails colonising plants or rocks, and dragonflies and damselflies hunting over the surface of ponds are all examples of groups that can be observed with relative ease. The density of the habitat and behaviour of the species being surveyed are crucial elements in whether direct observation is appropriate. Large flying insects that are reasonably restricted to open habitats tend to be the easiest to survey directly.

Butterfly census method

Long-term monitoring of flying insect numbers has been standardised through methods such as the butterfly census method that is employed in the UK butterfly monitoring scheme (UKBMS). The protocol was developed in the 1960s by Pollard who wished to investigate annual and seasonal differences in butterfly numbers (see Pollard and Yates 1993 for further details). The method is very simple (see Box 4.2) and not dissimilar to transect-based bird counts (see p. 207) and can also be adapted for use in surveying other flying insects (e.g. hoverflies, dragonflies and damselflies). This method can be adapted for smaller scale surveys (either in terms of the period of sampling and/or the area to be sampled). In essence, choosing appropriate times of day and weather conditions as well as standardising the transect length, the walking speed, the distance from the observer that is monitored and avoiding double counting are the important elements for this method. To cover large areas, 'W'- or 'M'-shaped transects can be used as a systematic sampling technique to try to cover heterogeneity in the sampling area (Figure 4.2). Whatever the path taken, butterflies (or other flying insects, e.g. dragonflies, damselflies, bumble bees, hoverflies, etc.) that are flying or resting within a set distance of the path (e.g. 5 m either side) can be counted and the density calculated (Box 4.3).

---

[1]http://www.amentsoc.org/publications/online/collecting-code.html
[2]http://www.naturalengland.org.uk/Images/specieslistofannex_tcm6-3714.pdf

Box 4.2 Butterfly census method

For standardised butterfly censuses that follow the UKBMS protocol, first establish a fixed transect approximately 2–4 km in length that includes habitats which are in partial shade and/or full sunshine. Densities of flying insects are low in areas where it is dark or heavily shaded so avoid including such areas. The weather should meet specified criteria – e.g. in the UK it is suggested that for butterfly census transects the shade temperature is at least 17 °C (or between 13 °C and 17 °C if most of the walk is in sunshine). In northern upland sites lower temperatures (e.g. those over 11 °C) may be acceptable. Wind is also critical and butterflies are not usually counted during periods where the wind speed is greater than force 5 ('small trees in leaf begin to sway'), except in woodlands where butterflies are frequently found flying in sheltered spots. Once the transect is established, the standard approach is to walk this each week from April to September inclusive, between 10:45 and 15:45 in the UK. Butterflies should be counted up to 5 m to either side and ahead of you, avoiding high flyers or wayward animals. You must make every attempt to avoid double counting of individuals. The method assumes that you can identify at least the commonest species, but in areas of doubt particularly with some skippers and whites, then individuals are assumed to be more common species.

Note: you can divide the transect into a maximum of 15 sections based on natural boundaries and features and count butterfly numbers separately for each section. By doing this, you can compile information relating to local habitat differences.

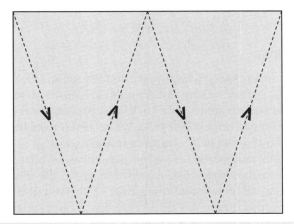

Figure 4.2 'W'-shaped transect walk
Similar designs using 'M'-shaped or triangular walks can be used, the principle being to cover an area of land in an objective way while sampling as much heterogeneity as possible.

> **Box 4.3 Calculating the density of flying insects from census walks**
>
> The density can be estimated as:
>
> $$\text{Density} = \frac{\text{the number of animals}}{\text{(the length of the path} \times \text{the width surveyed)}}$$
>
> So if 25 butterflies are seen along a 1500 m transect, observing all those within 5 m each side of the path, then:
>
> $$\text{Density} = \frac{25}{1500 \times (5+5)} = \frac{25}{15000} = 0.00167 \text{ m}^{-2} \quad (\text{i.e. } 1666.7 \text{ km}^{-2})$$

Indirect methods

Invertebrates do not tend to leave tracks as they pass through a habitat; however, there are a number of signs that can be used to identify the presence of some groups. Many arthropods (e.g. spiders, woodlice, dragonflies, cicadas) leave exuvia (remains of the exoskeleton) following moulting, and insects that undergo complete metamorphosis (egg, to larva, to pupa, to adult) may leave behind empty pupal cases. Sometimes such remains can be in obvious places: on the sides of ponds or emergent vegetation in the case of dragonflies; on tree trunks for cicadas; and pupal cases hanging by threads from vegetation. The eggs of some invertebrates can also be found (e.g. those of snails and slugs may be seen in clumps under logs and stones). Earthworms may leave casts on the surface of the ground. Some insects (e.g. crickets and cicadas) produce sounds that attract mates, and that can also be used to identify their presence and may enable identification to species level.

Using insect sounds

Orthoptera (grasshoppers and crickets) make audible sounds that are species specific and include clicking, rasping, ticking and drumming. Identification is relatively easy as there is usually a relatively limited number of species present at any one time and in any one place (up to five on average in the UK). There are an increasing number of field guides that include audio-files of cricket and grasshopper songs to assist in identification (e.g. Ragge and Reynolds 1998) as well as sound libraries such as that at the British Library.[3] Singing animals can either be recorded *in situ* or the sound used to identify their location so they can be collected using a net or by hand. It is helpful to carry an audio player during the survey to facilitate access to appropriate guides and/or a recorder to enable you to bring audio samples back for confirmation. The use of a microphone fitted to a parabolic reflector can help to concentrate the sound when recording (Figure 4.3). Playing back recordings may also encourage males to sing. This

---

[3] https://sounds.bl.uk/Browse.aspx?category=Environment&collection=British-wildlife-recordings

method can be used to count animals based on point counts or transect counts in a similar way to that used for bird recording (see p. 210). Some researchers have developed sound traps that comprise loudspeakers (playing a particular species' song) covered by a sticky trap to catch the attracted females (and in some species other males). It should be noted that surveying can only take place in the appropriate season (e.g. between June to September in the northern hemisphere) on hot sunny days. You should aim to map the calls of the insects, noting whether they were isolated or not and whether they were producing a calling or a courtship song. Similar techniques can work with other insects that produce sounds (e.g. cicadas).

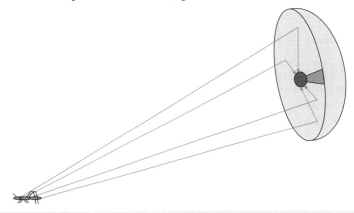

Figure 4.3 Parabolic reflector concentrating sound onto the central microphone

Capture techniques

Most research on invertebrates necessitates the capture of individuals, with the best technique depending as much on the habitat in which they live as the ecology and behaviour of the species concerned. A wide range of methods can be employed, the most commonly used are associated with five major habitat types: aquatic, soil-living, ground-active, plant associated and airborne. As with all techniques it is important to understand any limitations, especially when comparing different sites. There are two groups of techniques: active and passive. The former involves the researcher in hunting and actively trapping the animal by hand, using nets or some other physical intervention. The latter includes the setting of traps that may or not be baited. One immediate problem is that relatively immobile animals will be caught less frequently in passive traps than will more mobile ones. This does not necessarily just differ between species, since males and females of some species may differ in the amount to which they move around (males may be hunting for females that may be more sedentary, or females may be searching for egg laying sites, while males may be less active). Similarly, animals with different feeding behaviours may move around more or less frequently. Animals of different sizes may be caught less effectively than others using some techniques or may find it easier to escape once captured. Therefore, it is necessary to understand aspects of the target animal's ecology and behaviour at different times (of day or the year) as well as understanding the limitations of any trapping methods used. Likewise the effectiveness of active searching or netting may vary between different sites if the species is easier to find in one type of area (say an open habitat such as a meadow) than it is in another (e.g. a shaded area such as a woodland). In addition, smaller animals and those that are more cryptic will be harder to see than larger, more conspicuous species. So long as these considerations are taken into account when planning, analysing and interpreting the study, you should still be able to gather useful data.

Killing and preserving invertebrates

It may be necessary to kill those animals captured, either because it is part of the capture technique (when insecticides or killing agents are used in traps) or in order to facilitate identification. From an ethical point of view, as few animals as possible should be killed and then using the fastest and most efficient technique. Although there are some methods that work for a large number of taxa, it is important to recognise differences, especially between some of the major groups of invertebrates (Table 4.2). Preservation of invertebrates is important in order to ensure that any collections do not degenerate. Many insects can be kept dry in the presence of a preservative that prevents other animals or fungi from attacking the specimen. Other species need to be kept in a preserving solution (which may be toxic and must be handled with care). More careful attention is needed where specimens are needed for DNA analysis (see p. 96 for further details). For further information, see Lincoln and Sheals (1979) for invertebrates in general, Oldroyd (1970) for insects, Millar, Uys and Urban (2000) for insects and arachnids, and Gibb and Oseto (2006) for arthropods.

Table 4.2 Summary of killing and preservation techniques for commonly studied invertebrates

Taxon	Killing method	Preservation method
Annelids	In general, worms should be anesthetised before killing to prevent distortion (immersed in water to which alcohol is slowly added to 5–10%). They are transferred to 5% formalin for at least 24 hours to kill them	Worms can usually be kept in either 5% formalin or 70–90% ethyl or methyl alcohol (leeches should not be kept in alcohol since it tends to remove any colouration)
Molluscs	Most species of terrestrial, freshwater and shallow sea molluscs are killed in 5% formalin or 70–90% ethyl or methyl alcohol after relaxing for 24 hours in water (boiled water for terrestrial species) to which has been added a small amount of menthol (nudibranchs can be frozen and the ice containing them thawed in 5% formalin)	Most mollusc species are kept in 5% formalin (particularly suitable for larger species) or 70–90% alcohol. If snail shells are kept dried, the soft parts may need removing from larger species (with a pin following boiling in water)

(continued)

Table 4.2 (*Continued*)

Taxon	Killing method	Preservation method
Arthropods	Butterflies and large moths can be killed by pinching the thorax between the finger and thumb Most insects and many other arthropods can be killed in killing bottles containing ethyl acetate (this dissolves plastic containers and can discolour some green insects) or potassium cyanide (which is more toxic to the researcher than ethyl acetate and can discolour yellow bees and wasps) on a pad of blotting or filter paper or a bed of plaster of Paris (large animals can take some time to die) Freezing is suitable for many arthropods Near boiling water can be used to kill scorpions and many insect larvae	Large winged species are stored dry in (waxed) paper envelopes or absorbent paper (they must be thoroughly dry before storing and preferably stored in boxes containing a fungicide) Insects may be set (laid out to display the wings) and pinned onto cork (usually through the body or for small animals glued with water soluble glue to small pieces of card that are then pinned) and stored in boxes (containing an insecticide) Insect larvae and many arthropods, are stored in formalin or (more usually) ethyl or methyl alcohol (70–90%). This is not suitable where wing veination is important for identification, or for hairy animals, or those with wing scales, or waxy blooms on the body Very small animals are pinned in gelatine capsules or mounted on microscope slides (after soft internal tissues are dissolved by immersion in potassium or sodium hydroxide)

Marking individuals

The recognition of individuals of many adult insects can be achieved by using spots of paint or ink from permanent markers or even numbers drawn onto the wings. Care needs to be taken to ensure that marking does not damage the wing and that marks do not otherwise influence the behaviour and survivorship of the individual. Hard-bodied invertebrates (e.g. some insects, crustaceans, arachnids and snails) can be marked using dots of paint. For beetles with waxy wing-cases, flexible paints (e.g. cellulose dope) can help to prevent marks from chipping off over time. Longer term marks have been made by shaving a small mark onto the surface of the wing-case, which is then covered by cyanoacrylate glue to prevent fungal infections in the damaged area. Where long-term monitoring is intended, trials may be needed to assure yourself that marks will last the study period. Small transponders can be glued to the backs of large snails and beetles, whereas larger animals such as tarantulas may be chipped with passive integrated transponders (PITs) that can be read from a short distance. See Case Study 4.3 for a discussion of using PITs to mark tarantulas. Soft-bodied animals (e.g. insect larvae,

Case Study 4.3 Tarantula distribution and behaviour

Dr Emma Shaw is an Ecology lecturer at Manchester Metropolitan University who has over 10 years' experience of research on spiders. Emma has two main areas of arachnological interest: the role of spiders in integrated pest management programmes and the conservation of CITES protected tarantulas in Central America. This case study describes her work on the movements of tarantulas over a number of years and the challenges found in extracting them from their burrows, marking them individually and subsequently recapturing them.

Photo SM
Emma holding a female tarantula

Photo EMS
Field anaesthetisation chamber

Model organisms and sampling problems faced

Emma works on the CITES protected *Brachypelma vagans* (Ausserer, 1875) or Mexican red-rump tarantula, which is found in parts of the Yucatan peninsula of Mexico, Belize and Guatemala. The spiders have bright red abdomens, but are velvety-black elsewhere. This colouration and their relatively docile nature has led to their over-collection for the pet trade, resulting in them being placed on the CITES list that protects all species in the *Brachypelma* genus.

These spiders live in burrows that they, for the most part, dig themselves. This can make finding individuals difficult, particularly as they are nocturnal and spend the day underground. A major issue in the conservation of this species is the paucity of detailed ecological knowledge. We do know that they tend to dig burrows in open areas because of the lack of roots, but nothing is known about their movements from their natal burrow. *B. vagans* females live in burrows throughout their life cycle (up to 15–20 years) and seem to aggregate in large groups, whereas the males leave the burrow once maturity is reached and become much more dispersive.

Emma works on a large population in the Chiquibul forest reserve, at Las Cuevas Research station, Belize. There are approximately 150 burrows within a clearing at the research station and Emma has been mapping their distribution for 4 years. This annual mapping has revealed that a great deal of movement occurs, but to estimate the level of dispersal has posed great problems. Usual ways of marking animals, including using paint or external tracking devices, are unsuitable for tarantulas as they moult annually as adults, so Emma used internal radio-frequency identification (RFID) tags to provide each spider with a unique ID number. This raised practical problems, not least that insertion of the tag takes place in the field. Also *B. vagans* have very large and sharp fangs and urticating hairs that are flicked from the spiders' abdomen when agitated. Although not deadly, the venom from a bite causes significant discomfort and the irritation from the hairs can be quite long lasting.

How the problem was overcome

Finding burrows is relatively straightforward because they can be located during daylight hours with ease, some even having silk over the entrance making them even easier to spot. Once located, Emma marks the burrows with flags so they can be found again at night. She uses a red headlamp to reduce the number of insects flying at her face, and also because spiders cannot detect red light which minimises disturbance. To determine the sex and life stages, Emma extracts the spiders from their burrows using a 'fishing technique' with a blade of grass to mimic the movements of prey in the burrow. When the spider is fully emerged out of the burrow, the entrance is then blocked. Some individuals are easily tempted out; others cause great frustration by not responding.

Steve Reichling from the USA has developed the technique of inserting RFID tags into the abdomen of tarantulas that necessitates anaesthetising spiders in the field. Emma adapted this method and designed a field anaesthetisation chamber that contained vinegar and baking powder, which when mixed produce CO_2. This technique allowed time to safely make the incision required to insert the tag. Following this procedure, spiders were kept with plenty of water available, since dehydration following the loss of haemolymph can be fatal. Emma successfully tagged 22 tarantulas in 2008 and 18 in 2009 but found only two tagged individuals in 2009 and only five in 2010. She also found significant redistribution of burrows within the clearing from one year to the next. The next stage will be to establish levels of emigration and recruitment.

Advice for students wanting to study tarantulas

Many students want to work on tarantulas, and rightly so, since they are magnificent creatures, but rarely do students understand the time, money and problems involved. There are the obvious costs of actually getting to field sites overseas as well as costs associated with acquiring research permits. Since *B. vagans* has CITES protection, extra permits are required if samples need to be taken. Such permits not only cost money but can also take quite some time to be issued so advance planning is essential.

The weather affects the success of catching tarantulas as they do not like the rain, often resisting attempts to extract them under wet conditions. Care must be taken when sampling from burrows as other animals may be present: Emma has come face to face with a rather unhappy snake, and a burrow containing a tarantula being eaten by a scorpion. Tarantula work can be long, laborious (and sometimes unrewarding) and should not be attempted alone.

Despite the bad press that tarantulas can get, they are very misunderstood. Emma is hoping to address some of the shortfalls in knowledge of this amazing group of spiders and encourage people around the world to look at them in a more favourable light.

earthworms and slugs) can sometimes be stained using a vital dye (e.g. Nile blue) or a UV fluorescent dye. Hairy species (e.g. some caterpillars and many flying insects) can be dusted with fluorescent powders that show up strongly under UV light. Freeze branding and injected transponders have also been used with slugs. Visible implant elastomer (VIE) tags have been successfully used to mark earthworms. This is a fluid that when injected into the animal sets into a soft polymer that can be seen through the cuticle. See Southwood and Henderson (2000) and Hagler and Jackson (2001) for further details of marking invertebrates.

Capturing aquatic invertebrates

Aquatic invertebrates can be found at times of the year when other invertebrates are not active and hence are popular subjects for study. In addition, some species have been used as indicators for pollution (see Chapter 2, Box 2.3). One of the major concerns in sampling aquatic invertebrates is the efficiency of sampling in water, which can be subject to sampling errors and is generally thought of as inefficient. You need to be aware of the limitations of any study, choosing more than one method if the whole community is under investigation. The techniques that follow include many of the standard methods for investigating aquatic invertebrates. New (1998), Southwood and Henderson (2000) and Hauer and Lamberti (2006) give further details about such techniques. Care should be taken to choose techniques appropriate for particular species (e.g. depending on whether the animals are active in the water column or buried in the substratum) and for particular habitats (e.g. depending on whether the water is flowing strongly or is static). Both biotic (living) and abiotic (non-living) factors in flowing waters can change over quite short timescales, which can be problematic if you wish to examine freshwater invertebrates alongside associated environmental data (see Chapter 2). Case Study 4.4 discusses one research project that resolved the timescales over which biotic and abiotic factors in stream environments could be monitored.

Case Study 4.4 Stream invertebrates

Dr Amanda Arnold is an ecologist working at the Institute of Evolutionary Biology, University of Edinburgh, who has studied freshwater invertebrates both as a student and researcher for the last 10 years. She has a particular interest in the contributions of small-bodied organisms to running water invertebrate assemblages. This case study describes the selection of a sampling protocol for small aquatic invertebrates and associated environmental variables.

Photo MKD
Amanda taking a benthic sample using a Surber Sampler

Model organism and sampling problem faced

Freshwater invertebrates are ubiquitous within rivers and streams and it is possible to observe individuals in shallow waters and collect them with the most basic of implements (e.g. a tea strainer or plastic cup). However, the most abundant and diverse components of freshwater invertebrate assemblages are largely invisible to the naked eye. These consist of invertebrates (meiofauna) less than 0.5 mm in size that become visible when viewed with a hand lens and/or under a microscope. Some such organisms, the permanent meiofauna, remain small bodied throughout their life cycle. Examples include copepods, cladocerans, nematodes and rotifers. These organisms are interesting because they are a little understood fauna of running waters.

A common problem in ecology is identifying patterns in the abundance and distribution of organisms and ascertaining the environmental and biological factors that drive these patterns, a situation that is also true for the permanent meiofauna of streams. In most freshwater invertebrate populations, predictable changes in abundance and distribution occur throughout the year. However, small-bodied organisms such as the permanent meiofauna often live for short periods of time and pass through multiple generations, so peaks in abundance of these taxa may be linked to when favourable conditions occur. Amanda decided to carry out a temporal survey to quantify the distribution, abundances and composition of permanent meiofauna relative to temperature, water level and organic matter. These variables were chosen because temperature is important for growth and reproduction; very low or high water levels can cause mortality; and organic matter provides a source of energy aiding growth and reproduction.

While designing the survey, Amanda faced a number of challenges. Changes in environmental variables occur over a variety of time-scales. For example, temperature changes every second but differences also occur at seasonal intervals throughout the year (e.g. between summer and winter). In addition, environmental variables do not always follow obvious patterns, and sudden changes in physicochemical parameters can occur. For instance, water levels in streams often rise and fall in dramatically short periods of time in response to precipitation. The life cycles of stream invertebrates are widely thought to be adapted to some aspect of this temporal framework. Amanda's main problem was in determining how often to sample meiofauna and environmental variables relative to each other. Furthermore, to account for the non-uniform distribution of invertebrates and organic matter within the stream bed, it was important to establish what number of sampling units to collect and how to collect them in an unbiased manner.

How the problem was overcome

Amanda decided that given the generally short lifespans of many of the permanent meiofauna that she should take samples every 2 weeks within a stratified random design that would allow her to pick up changes in population size. She also decided that the study should last for at least a year to take into account responses of the meiofauna to the full range of seasonal variation in environmental variables. As environmental variables were hypothesised to constrain abundances they needed to be measured regularly in a timeframe that was relevant or synchronised to the life cycle of the organisms. She decided organic matter could be sampled when the invertebrates were collected as this might relate to their distribution. To identify between- and within-season variation in temperature and water level and any unpredictable changes in these parameters, Amanda decided to record these on an hourly basis throughout the sampling programme. On each sampling occasion she collected 10 benthic sample replicates to provide an acceptable sample mean allowing for 20% error. The stratified random design was employed to partition the collection of samples in order to control for sampling bias.

Advice to students wanting to study aquatic systems

Running waters can be highly unpredictable systems to work in; both high and low flows can severely alter the distribution and abundances of organisms and can delay or destroy field surveys and experiments that may have taken months to plan and set up. Following collection of samples, processing and identification of organisms can also be a lengthy process. However, samples are generally easy to collect, equipment is inexpensive and there are an excellent range of taxonomic keys covering most UK freshwater taxa. In addition, unpredictable disturbances in these systems can provide interesting avenues for research. In short, they are highly rewarding environments to work in because they provide opportunities to investigate both fundamental and applied research on a large range of fauna and flora in a highly dynamic environment.

Netting

Netting is the simplest of all sampling methods used to catch aquatic invertebrates (Figure 4.4). Pond nets can be used to catch pelagic (those found within the water column) and surface-active species; however, their use needs to be standardised (i.e. moving the net a set number of sweeps through the water body at the same depth using the same net each time). Sampling error may occur when trying to control the area sampled. Experiment with netting between canes to standardise the lateral movement and attach a thin, banded, fibreglass rod to the rim of the net or handle to estimate the depth of netting. If benthic invertebrates (bottom-dwellers) are required then nets designed for use on the bottom of the water body should be used (e.g. kick nets – see p. 124).

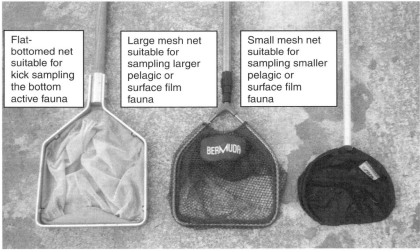

Photo CPW

Figure 4.4 Flat-bottomed pond nets suitable for catching surface, pelagic and bottom-active invertebrates

Netting is appropriate for many types of surface dwelling and nektonic (swimming) invertebrates. However, mobile heteropteran bugs (e.g. water skaters) and beetles (e.g. whirligig beetles) may avoid nets and move away as the water body is disturbed. Mosquito larvae are known to swim downwards to avoid capture and only slowly return to the surface. A specially designed Belleville mosquito sampler is effective against this behaviour (Figure 4.5).

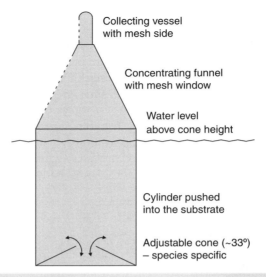

Figure 4.5 Belleville mosquito larvae sampler
The cylinder (without the funnel and collecting chamber attached) is pushed firmly through the water column and into the top of the substrate. This is then left *in situ* for 5–20 minutes before connecting the concentrating funnel and collecting vessel. The whole apparatus is then inverted, concentrating the larvae in the funnel and collecting vessel.

D-shaped nets can be used to collect animals on the bottom of the waterbody. The flat edge of the D allows the net to rest more snugly against the bottom than does a round dip-net. Although such nets can be used to scrape animals off the bottom, a better technique is kick sampling (Figure 4.6). This is one of the simplest methods for sampling the benthic fauna from the stream bed, involving standing upstream of the net and kicking the substratum so as to dislodge invertebrates that then float downstream to be caught in a net (usually of about 0.9-mm mesh size). The best nets for kick sampling are those with a rigid enough handle and frame so that it can act as a support when kicking with one foot in flowing water. Specialised kick screens can also be used (Figure 4.7). These are simply pieces of netting (1 m × 1 m of 0.5 or 0.6 mm netting) with a pole fitted at each side and a weighted line along the bottom to prevent the net from floating in the water column. These are used in a sweeping motion through the water or as a drift net to catch floating invertebrates. Where the water is shallow, they may be used downstream of kick sampling to collect any animals that have been disturbed, lifting the net with a forward sweeping motion to retain the catch.

Photos AML

Figure 4.6 Using a kick net and sorting the sample
The researcher stands upstream of the net and kicks the substrate to dislodge invertebrates so that they float downstream and are caught in the net. The foot must be kept close to the aperture of the net to ensure that dislodged animals are not simply swept around the outside of the net. To standardise the catch effort, either time the process (the normal time for each sample is 3 minutes) or standardise the number of kicks (at 20 for example). The net can then be upturned into a white tray so that the dark animals can be seen more easily when being sorted.

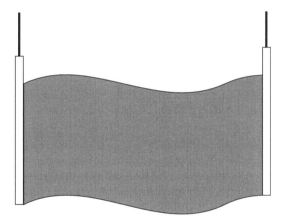

Figure 4.7 Kick screen
Ensure that the bottom of the net is as close to the river bed as possible. When removing the net, lift it out against the current to prevent loss of the catch. Variations on this include designs with a collecting vessel in the middle of the net to collect plankton.

If it is especially important to standardise the catch by area, then it is worth considering using a Surber sampler (Figure 4.8). This involves sampling within a standardised area (using a quadrat linked to a net and lifting stones to remove any animals beneath or attached to them, which are then swept into the collecting vessel). The efficiency of this technique may be low in some situations depending on the balance between catching free-floating animals and losing animals from the disturbed area in the quadrat. The type of animals, substratum and water flow can all influence the efficiency. Before sampling, it is important to decide whether stones that straddle the quadrat are to be sampled or not. Neither kick sampling nor Surber sampling can be used to estimate population size as they are only relative methods that tend to under-record those species that are usually firmly attached to stones. If sampling gravel or cobble river beds, a Hess sampler (Figure 4.9) can be useful. This is an open cylinder that is pushed firmly into the river bed and the substratum within it disturbed and scrubbed *in situ* so that any animals are washed downstream into the integral net. A mesh covered window on the upstream side enables water to flow through the apparatus but prevents animals from escaping or entering the cylinder.

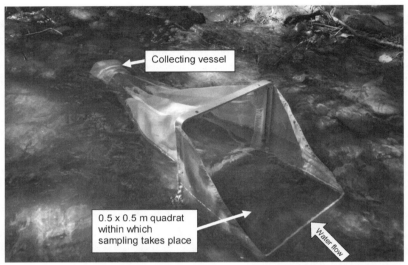

Figure 4.8 Surber sampler

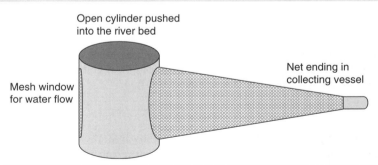

Figure 4.9 Hess sampler

To sample organisms at the surface and within the water column, drift nets can be suspended in the water stream. They can either be fixed to the bottom of a shallow stream using weights or pegs (Figure 4.10), or fixed at a specified depth using floats and anchors, and the mesh size is chosen to be appropriate to the organisms being studied.

They can also be used in groups to sample from across the profile of a river. Similarly plankton nets (cone-shaped nets of fine mesh cloth with a removable tube at the end in which the plankton collect) can be suspended or towed in the water column to collect microscopic organisms (Figure 4.11).

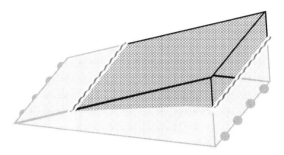

Figure 4.10 Drift net
Weights (dark spheres) hold the net at the correct position in the water column.

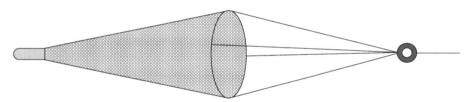

Figure 4.11 Plankton net
When the net is towed or suspended in the current, plankton are collected in the vessel at the end of the net.

Suction sampling

Where burrow entrances can be seen in wet sand or mud, it is possible to either extract animals by digging them out, or by using a manual suction sampler (Figure 4.12). This is simply placed over the hole and the handle drawn up creating a vacuum inside that pulls the animal out of its burrow.

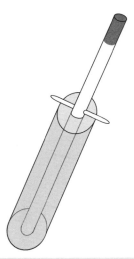

Figure 4.12 Suction sampler for animals in burrows

Benthic coring

Coring is appropriate for muddy or sandy ecosystems particularly in estuarine and coastal habitats. Coring can take place from a boat or while wading. Corers are metal or plastic pipes (with a handle at the top of the pipe) that are pushed quickly into the substratum. Depths taken are usually not more than 50 cm for sandy environments and around 10–20 cm for muddy areas. The speed of coring is quite important as animals will flee as soon as they are disturbed. It is good practice to go a little deeper into the substratum than that required for your investigation. An extra depth of about 5 cm acts as a bung holding in the sample and if any of this unwanted sample falls out then it will not affect the required sample. Adding a rubber or cork bung in the top of the corer creates a vacuum that holds the sample within the corer until required for sorting. If the samples are particularly wet then a metal plate should be inserted at the base of the corer – this can be difficult and some of the sample may be lost in the process. Benthic coring can be standardised by controlling the depth (marking the outside edge to indicate the sampling depth – plus 5 cm extra). A more sophisticated technique (freeze-core sampling) involves injecting liquid nitrogen into the substratum via a probe, pulling out the resulting frozen block of sediment and then extracting the animals that have been captured within it (Gordon et al. 2004).

Drags, dredges and grabs

In deep waters, where kick sampling is not practical, drags, dredges and grabs may be used to take samples. A drag is a net that is pulled along the bottom disturbing the river bed with tines or blades (e.g. the naturalist's dredge – Figure 4.13). Grabs are effectively double buckets that close together on reaching the river bed either via a manual (Ekman dredge) or counter weight mechanism (Peterson grab) – see Figure 4.14. Although useful, these are specialised pieces of equipment that are expensive to buy and operate because they normally require the use of a boat and some assistance. Furthermore they are unsuitable for studying the zonational sequence and stratification of invertebrates as the samples are heavily disturbed on lifting.

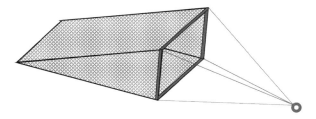

Figure 4.13 Naturalist's dredge
The mouth of this net is made of metal and acts as a scraper to scoop up material when towed along the river bed.

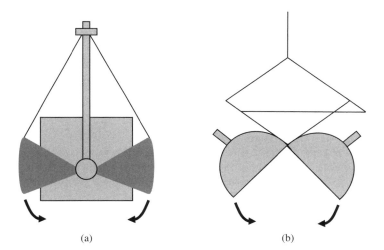

Figure 4.14 Grabs for collecting benthic animals
(a) Ekman grab with a remote-operated closure mechanism; (b) Peterson grab closes automatically when the device hits the bottom and releases a trigger.

Wet extraction

Sieving mud and sand is an option when samples have been taken from cores, grabs or drags. However, wet extraction is sometimes favoured because unlike sieving these do not require constant attendance. One such method is the Baermann funnel that employs heat to extract the animals and is especially suited for extracting nematode and enchytraeid worms and insect larvae (Figure 4.15). Other wet extraction techniques have been used to increase the efficiency of the collecting nematode worms. One such method, the Whitehead and Hemming tray, involves increasing the surface area for extraction by placing the soil sample on tissue paper that is supported on a mesh tray. The tray stands in water at room temperature just up to the level of the tissue. Nematodes migrate through the tissue and, after 24 hours, can be concentrated using a funnel system similar to that in the Baermann apparatus. The Bidlingmayer sand extractor extracts a wider range of aquatic and semi-aquatic animals and can be made very simply, requiring only water, sand, a deep tray and some metal gauze (Figure 4.16). Sand extractors are cost and time effective although they may not be appropriate for all groups of invertebrates.

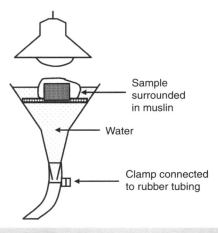

Figure 4.15 The Baermann funnel
A small sample is wrapped in muslin and placed on gauze. This is set into a funnel that ends in a piece of tubing that can be closed by a clamp. With the clamp closed, the funnel is filled with warm water up to the level of the sample. The apparatus is left (usually overnight) while the animals migrate out of the sample and into the water, collecting in the neck of the funnel, from where they can be drawn off by opening the clamp. A light above the whole apparatus will both illuminate and heat the sample.

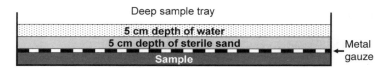

Figure 4.16 Bidlingmayer sand extractor
The sample is spread over the base of a tray and completely covered with gauze that is then covered with 5 cm of sterile sand, over which is poured water to about 5 cm above the sand. The extractor is left for 24–48 hours. Animals rise from the sample and colonise the sand that is scraped or lifted off using an inserted plate. To remove animals, the sand is placed in a black tray and shaken, picking out each animal as it appears. To make sorting easier in white trays add a dye (e.g. green) to colour the animals but leave the sand unaffected.

Artificial substrate samplers

There are a number of traps that exploit the way in which many aquatic organisms colonise bare surfaces. In all cases, appropriate substrates for the species under investigation are suspended in the water column and removed after a set time period (e.g. for 2 weeks). Any organisms that have colonised in the meantime can be removed (by scraping them off the substrate) and identified. Samplers based on glass plates can also be used for the collection of algae. Basket samplers are wire baskets containing rocks (e.g. made of limestone, concrete or porcelain) of an even size. Similarly, rock bags made of nylon mesh can also be used. Hester–Dendy multiplate samplers are made in a variety of sizes and enable multiple plates of metal or wood to be colonised (Figure 4.17). These can be used with different spacing between the plates to enable different species to be able to colonise. Mesh bags containing leaf litter can also be efficient in collecting aquatic invertebrates (Figure 4.18).

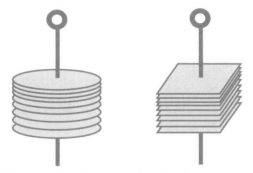

Figure 4.17 Hester–Dendy multiplate samplers
These are left *in situ* for about 2 weeks, during which time they will be colonised by a wide variety of invertebrates. Different sized spacers are used to create different sized gaps between the plates in order to collect a range of animals.

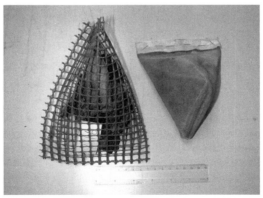

Photo MR

Figure 4.18 Mesh bags containing leaf litter used to collect aquatic invertebrates
Bags made of large mesh netting (left) will collect larger animals than those made of fine mesh netting (right).

Baited traps

Freshwater crustaceans (e.g. crayfish and mitten crabs) can be captured using mesh traps that have funnel entrances (Figure 4.19) and are baited with fish or chicken necks. **Note**: that a licence is required to catch crayfish in the UK and that introduced species of crayfish and the introduced Chinese mitten crab must not be re-released into any waters following capture.

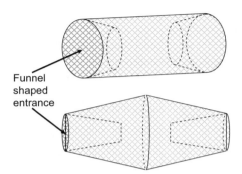

Figure 4.19 Crayfish traps
The funnel entrances allow the animals to enter easily but are more difficult to escape from, thus helping to retain animals once they are caught.

Capturing soil-living invertebrates

Sampling invertebrates from soil is a relatively simple affair, although the extraction from the associated debris can take a considerable length of time. Soil invertebrates are frequently sampled by a collection of leaf litter from quadrats or from soil cores. Ideally these methods should be standardised, but in reality this is very hard to achieve. Although it is easy to define a sampling area in the horizontal plane, controlling the depth of sampling is more difficult. Although variation in soil sampling depth may be unavoidable, invertebrate density can be corrected if the dry weight of the sample is taken, although it may take many hours to process even small numbers of samples. Other works that give more detail regarding sampling invertebrates from soil, litter and plant roots include New (1998), Southwood and Henderson (2000) and Gange (2005).

Dry sieving

Dry sieving is rarely a suitable option unless you are only interested in large animals and their immature stages (>20 mm). However, even then, ensuring that mobile invertebrates do not escape before the litter sample is in the sieve is something of a skill, and this method can be ineffective for most groups except for larger insect larvae and eggs. Gardener's sieves are commonly used for dry sieving, although stacks of sieves of decreasing size can be used to separate out smaller components (Figure 4.20).

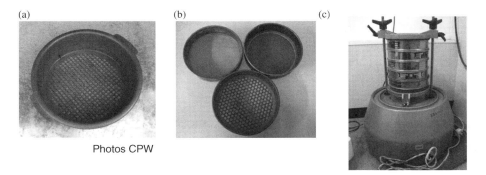

Photos CPW

Figure 4.20 Soil sieves
(a) Gardener's soil sieve for separating coarse fractions; (b) sieve set for separating increasingly fine fractions; (c) sieve set stacked on a sieve shaker with the largest mesh size at the top, used to separate fractions from soils.

Floatation and phase-separation

Floatation involves placing the sample in an aqueous solution of magnesium sulphate. Alternative solutions of table salt (sodium chloride), sugar (sucrose), potassium bromide or zinc chloride have been also been used, although less commonly. Although beetles will float in pure water, most other invertebrates need the specific gravity of the solution to be raised before they float towards the surface. For standardisation, it is essential that the specific gravity of the solution is the same for each sample being extracted (a simple hydrometer can help here). A specific gravity of 1.2 is recommended for magnesium sulphate solutions, although extraction from clay soils may require stronger specific gravities.

Once the sample is in the floatation solution it is then agitated either with air bubbles or with a stirring device for 2 minutes to free animals from the matrix. Animals can be lifted out with a small tea strainer into a tray with clear water for the final separation and cleaning process before identification. For an alternative to this labour intensive method employ the rather elaborate Ladell can method (see Southwood and Henderson 2000). Floatation may vary in its success at retrieving animals in a standardised way if soil types vary dramatically over the study area. It is also very messy and labour intensive.

Phase-separation is a technique for collecting very small animals from plant material (e.g. mosses) and involves the use of two solutions that do not readily mix (e.g. kerosene (paraffin) and ethyl alcohol). Here the sample is placed in a jar that is over half filled with 95% ethyl alcohol. A few centimetres depth of kerosene is then added and the jar shaken to mix the solutions. The jar is left to settle, gently tapping it to allow any bubbles to surface. Invertebrates tend to be concentrated at the boundary between the two solutions. Most of the kerosene can then be pipetted off and discarded. The boundary layer can then be separated off and the invertebrates collected. Repeating this process on any debris left in the jar can increase the efficiency of the technique.

Tullgren funnels as a method of dry extraction

These are also called Berlese–Tullgren funnels to reflect the fact that Tullgren modified the original Berlese funnels by adding a light as a heat source. This is perhaps the most widely used of all soil extraction methods, probably because it is clean, does not require constant attendance and is effective. The principle of the Tullgren funnel is simple: invertebrates move away from heat to avoid desiccation. The Tullgren funnel design exploits this behaviour by placing a hot lamp source over a sample of soil or leaf litter that rests on gauze in a funnel, at the end of which is a pot containing a preservative and killing solution. Animals move away from the heat of the lamp until they fall into the pot below (Figure 4.21).

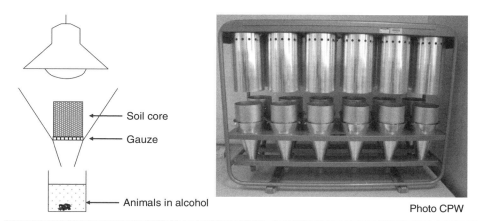

Figure 4.21 Tullgren funnels
The soil core or leaf litter should be dried slowly to prevent animals from dying *in situ*. More sophisticated designs include a temperature control to enable the heat to be increased slowly over the extraction period. The process can take several days and even weeks and is most effective for small, mobile animals that can resist desiccation such as mites and adult beetles. The funnels come in different sizes from those that will extract small arthropods such as mites from small soil cores, to larger designs that can extract adult beetles from large leaf litter samples. Maintaining a higher humidity level for longer will aid the extraction of animals that are more vulnerable to desiccation such as Collembola and insect larvae.

Some researchers recommend that when using soil cores, it is best to invert the core so that the upper part (the A horizon) of the soil rests on the gauze. This allows natural movement out of the core towards the preservative. **Note:** avoid the soil or litter touching the sides of the Tullgren funnel as animals can get caught in the condensation that settles there and the sides will get hot and may kill invertebrates on contact. On refined designs, a double funnel has been used to partially alleviate this problem. Cooled surroundings can also create a more effective temperature gradient within the apparatus. A similar method, the Kempson bowl extractor, also exploits this by immersing the sides of bowls (within which the samples are housed) in a cold water bath (Figure 4.22).

There are a few drawbacks to using such methods, notably that immobile stages (e.g. eggs) and large animals that cannot fit through the gauze are not extracted. Furthermore, extraction is dependent on the type of soil, its moisture level, and the behaviour and condition of the animals. Phototactic animals may not be extracted but will die from desiccation nearest the light. In such cases it is worth considering using a modified Tullgren funnel with a light source below it.

A version of a Tullgren funnel that is useful in the field is the Winkler sampler (also called the Moczarsky–Winkler selector) – see Figure 4.23. Here the sample is placed in a mesh bag within a cloth bag that tapers from the top to the bottom,

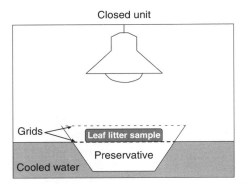

Figure 4.22 Kempson bowl extractor
The sample is placed between two grids, the upper of which prevents animals from escaping, whilst the lower one reduces the amount of debris that drops into the preservative (usually trisodium orthophosphate or ethyl alcohol). Some designs include extra layers of net between the sample and the grid to further retain the leaf litter.

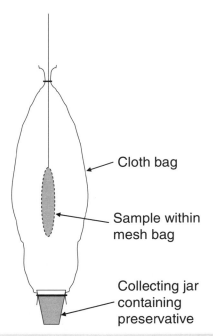

Figure 4.23 Winkler sampler
Samples may be dry-sieved first using a fairly coarse sieve (e.g. Figure 4.20a) before placing in the mesh bag. The apparatus should be suspended away from direct sunlight. Migration of the majority of the animals from the sample may take several days.

ending in a jar of preservative (usually ethyl alcohol). The bag is closed and left in a shaded location while the sample dries out naturally and animals migrate down into the collection jar. This is a useful method if you are sampling in the field for sustained periods.

The use of lights to dry out samples and repel invertebrates from leaf litter or soil can also be used in quite crude designs to reasonable effect. This can be achieved by placing the sample at the top of an inclined tray over which a light has been positioned (Figure 4.24). Animals will slowly migrate away from the light and collect in the bottom of the tray where there is a killing solution (e.g. methyl alcohol). Such designs can be useful in extracting animals from samples, but care should be taken in using them to estimate abundance since the method is not as efficient as either Tullgren funnels or Winkler samplers.

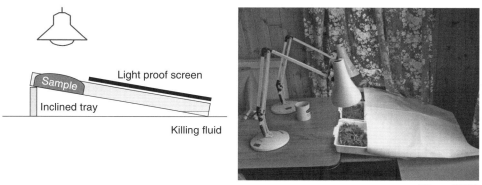

Photo CPW

Figure 4.24 Simple inclined tray light-based separator
As the sample dries out, animals migrate away from the light source and down to a preservative. The steeper the angle of inclination of the tray, the faster animals will migrate, but the more debris will be found in the preservative. Efficiency is increased by covering the lower part of the tray with light-proof card or plastic.

Chemical extraction

Chemical extraction is useful when extracting cranefly larvae (leather jackets) and earthworms directly from leaf litter or soil. The chemicals used are an irritant to the animals concerned that causes many of them to rise to the surface to avoid contact. Potassium permanganate in aqueous solution (1.5 g per litre) or more commonly a 0.2% aqueous formalin solution (three applications of about 9 litres per $0.5\,\text{m}^2$) is applied to the soil surface using a garden sprayer or watering can. Formalin is less toxic to the worms than potassium permanganate and hence preferred. For a standardised approach, both the sampling area needs to be defined (e.g. 1 m^2), and the quantity of liquid (e.g. 5 litres) sprayed onto the area controlled. Furthermore, a time limit can be set (15 minutes is normally sufficient) for the collection of animals on the surface to avoid any sampling biases. By controlling the area, solution concentration and volume, and time limit, the density of animals per square metre can compared between samples. The only variables that cannot be controlled are the porosity of the soil and the soil type, which are limitations to any study of this kind.

Many researchers use an aqueous solution of mustard powder as an irritant with which to extract earthworms, as well as slugs and possibly some insect larvae, living in the upper regions of the soil. Suggestions vary as to the best application rate. Most studies recommend between 5 litres and 10 litres of 0.5–1% aqueous solution per 0.25 m^2 applied either as a single application or as two applications applied 10 minutes apart. For example, Lawrence and Bowers (2002) recommend dissolving 50 g of mustard powder in 100 ml of water and stirring to release the active ingredient (allyl isothiocyanate). After being allowed to stand for at least 4 hours, the paste is made up to 7 litres with water that is then evenly sprayed over an area of 42 cm × 42 cm. Some studies have reported difficulty in mixing mustard powder into water. The different experiences found in dissolving mustard powder in water may depend on the brand and age of powder used. Possible solutions include first dissolving the powder in 5% acetic acid overnight and then diluting it with water or slowly mixing the powder with a small amount of water to get a thick paste that is left for at least several hours to mature before making the final solution up with water before use. Pouring the solution into frames to stop it flowing laterally across the ground will help to standardise the approach. Edwards and Bohlen (1996) discuss these methods in more detail.

Electrical extraction

It is possible to extract earthworms using electricity, although, because of the nature of this method, it should be very carefully considered and experienced researchers should always be involved. Many descriptions of this technique involve the use of a single electrode with an appropriately insulated handle passing 220–240 V at 3–5 A directly into the ground. However, some researchers have successfully employed low amperage (3 mA), DC transformers and used two electrodes. The success of the technique depends on the dampness of soil; if the surface is dry the current may drive worms down rather than bring them up. Generally the method is thought to be better for deep-living worms. It has been reported that those exiting near to the electrode(s) may die. **Note: some commercially available electrical worm probes designed for fishermen to extract worms have been recalled following shocks and even deaths during their use.**

Capturing ground-active invertebrates

A wide range of invertebrates live on the surface of the ground, associated with the upper horizons of the soil, within leaf litter, under logs and stones that are resting on the ground, in dead wood or dung, or are found living or resting on the ground flora. Hand-searching can be an efficient (if time-consuming) way of sampling such animals. To standardise captures, the effort used (e.g. the number of researchers and time employed per sampler) should be kept constant at different sites. Where logs and stones are turned, it is important to return them to their previous position following sampling. Lifting bark from rotting wood to sample the animals underneath should be approached with care and bark replaced where possible. The benefits of using direct searching is that you can be selective about the animals that are taken. The method is less biased towards more active animals, although mobile animals may be disturbed by the collector and escape whereas secretive animals, including those found under leaves, may be missed. Laying down stones, logs and other pieces of wood can encourage a number of groups of surface-active invertebrates to colonise, which can be subsequently collected by hand-searching. Sampling invertebrates from dung should be approached in a similar way to extraction from soil and other substrates using desiccation methods (e.g. Tullgren funnels – p. 134), hand-searching and floatation (p. 134). Pitfall traps baited with dung may also be effective.

Bear in mind that it may take some time to develop the appropriate technique to efficiently find and catch the wide range of animals found in soils and other sediments. It is worthwhile ensuring that all participants are properly trained before data are recorded to avoid an increase in invertebrate numbers found as the researchers' skills improve. Further details of appropriate techniques can be found in New (1998), Southwood and Henderson (2000) and Leather (2005).

Pitfall traps

One of the most commonly employed techniques for collecting surface-active invertebrates is pitfall trapping. Pitfall traps are containers made of plastic (e.g. vending machine cups), glass (jam jars) or metal (tins) sunk into the ground, usually containing a preserving and/or killing solution. Animals moving across the ground surface fall into the trap and are prevented from escape by the solution. Several solutions may be used:

- aqueous solution of 5% formalin;

- 70–100% aqueous solution of alcohol, normally either ethyl alcohol or methyl alcohol, sometimes with the addition of 5–10% glycerol, which prevents the animals from drying out completely even if the alcohol evaporates;

- 25–100% aqueous solution of anti-freeze (active ingredient ethylene glycol or preferably propylene glycol, which is less toxic for those handling it) – use 70% if the traps are to be left for more than a week, or are in danger of being diluted by rainwater.

Whatever the solution you use, add a few drops of detergent to break the surface tension so that the animals sink. Water should not be used on its own since captured animals tend not to drown quickly. **Note:** All chemicals should be handled with care (see Chapter 1 on health and safety). Formalin and ethylene glycol are especially problematic since they are skin irritants. You should always wear waterproof gloves when filling or emptying traps and consider using protective glasses and having a portable eye wash available. Consult immediate medical advice if any such solution is ingested. Some animals may be attracted or repulsed by particular solutions (or even different concentrations of the same solution) and so this should be kept as constant as possible between sites and across sampling periods. Records should be made if traps have dried out or been flooded during the sampling period since this will impact on the efficiency of captures. Where traps are to be emptied after very short periods of trapping (e.g. twice per day in comparisons between nocturnal and diurnal activity), or where live animals are required (e.g. for mark–release–recapture methods of population estimation), then traps can be left dry. In such cases, it is important to place some gardener's hydroleca (or some vegetation) at the bottom of the trap to provide a little protection from desiccation and (limited) refuges for animals so as to reduce in-trap predation. Traps may be baited with dung or carrion to collect animals that feed on these items. In either case, the bait can be suspended in the centre of the trap so that the animals fall past it and accumulate at the bottom (usually within a killing and/or preserving solution). Case Study 4.5 describes one research study that used pitfall traps to collect dung beetles in tropical forests.

To set pitfall traps so that they can be standardised, a standard number (e.g. five pitfall traps per site) should be placed in a set pattern (e.g. in a line) at a set distance from each other (e.g. 2 m apart) and collected and reset at regular intervals (e.g. every 2 weeks) – see Figure 4.25. Where ground vegetation differs substantially between sites, it can be cleared around the trap so that the sward is short ($\sim$1 cm high) checking that there is no barrier to movement. Make sure that each trap is flush with the soil surface around the perimeter of the trap. Set and collect the traps on the same day across all sites: determine the collection days before you start the project and stick to them as analysis becomes complex if the catch time is allowed to vary. A fortnight is a commonly used period for leaving traps between collections, although in very wet or very dry habitats this time should be reduced.

Glass is one of the most efficient materials for pitfall traps since they are very smooth sided, but are expensive to buy and replace and a hazard if broken in the field. Many researchers use plastic cups: the same size and type must be used for each site. Trap efficiency can be enhanced by choosing a steeply sided and tall cup without horizontal

Case Study 4.5 Collecting insects in Costa Rica

Dr Erica McAlister is one of the Curators of Diptera at the Natural History Museum, London. As part of her job, Erica collects new material from different parts of the world. Her case study explores the problems of using a wide range of equipment in a collecting site overseas and in taking large collections of animals back to the UK for processing.

Photo EM

Erica sampling invertebrates in a tropical forest

Model organism and sampling problems faced

As curator, Erica is required to add to the collections at the Natural History Museum in London but she regularly faces problems when exporting material back to the UK. Her work in Costa Rica is typical of the experiences many face when working abroad (and in the UK) where there is often a requirement to have permits from local government officials to carry out fieldwork of any kind. Erica also had to surmount numerous practical problems that were exacerbated by the lack of infrastructure and freely available data. Because of the lack of reliable maps, Erica faced problems in allocating grid references to samples and had to think of ways of maintaining a reliable electricity supply for her light traps while stopping tropical rainstorms washing away her experiments. In this case study Erica describes two experiments and the general problems of working in a remote, tropical environment.

How the problem was overcome

Erica planned two types of experiment: a faunistic study of tropical forest diversity and a specific study on the relationship between dung beetle species richness and habitat disturbance. For the dung beetle study, Erica used pitfall traps, which consisted of a plastic cup containing a killing solution (usually alcohol) purchased from a local hardware store. However, she had to modify the traps to prevent violent downpours from flooding them and larger amphibians, etc., from falling into them. Erica made a roof for each trap out of cardboard, using nails for support. Although this successfully reduced flooding and prevented unwanted amphibian deaths, the roofs were attractive to the local capuchin monkey population, which ran away with them and ate them! Having catalogued her samples, she faced another hurdle: since the pitfall trap samples were stored in alcohol, they could not be taken on a plane due to restrictions on carrying liquids in the hold. To

> circumvent this, Erica drained the alcohol from the sample tubes just before she flew back and replaced the liquid with tissue soaked in alcohol to maintain an acceptable temporary level of preservation. Back in the UK, Erica refilled the tubes for long-term storage without incurring any noticeable damage to the samples.
>
> Erica also planned to use light traps for the faunistic study but knew that she would not have a mains electrical supply. Using two car batteries she was able to ensure that the light trap was operational for much of her stay. She overcame the problem of storage of moths in alcohol (where they would tend to lose their colour) by pinning her light trap samples in the field.
>
> In both of these studies, Erica had to allocate reliable grid references to each sample taken and transfer this information to her tubes and field notebook. Because of the lack of reliable mapped data, she used a global positioning system (GPS) and overlaid these coordinates on a basic local map. However, she found that taking GPS positions could be time-consuming due to the dense canopy cover that restricted the satellite signal.
>
> **Advice for students wanting to study insects overseas**
> Erica recommends that you prepare well, knowing which equipment you can obtain in the host country and which you have to import (and whether that would be allowed under airline and customs rules). Be careful to take care of important items, including your specimens, data and notes. It is essential to keep recording sheets in a clean, waterproof environment, so consider using plastic bags and silica gel to prevent your valuable notebooks from becoming damaged. Be aware that laptops and microscopes may become damaged by the high level of humidity in tropical forests, so investigate whether it is possible to at least purchase a waterproof case or, ideally, rugged equipment designed for field use. If the worst happens, use a dry room back home to dry out equipment, samples and notebooks, and scan your documents if they are threatening to fall apart. When collecting material for DNA samples, the best storage method will depend on the type of insect; for flies Erica recommends beem capsules or Eppendorf tubes (with pierced lids) stored in a zip-locked bag with silica in to desiccate the specimens to prevent the DNA from degrading or becoming contaminated. Use tissues soaked in alcohol to maintain preservation while samples are transported rather than cotton wool as the fibres become tangled around insects. Lastly, make sure that you obtain all necessary permits. Try to make contact with a national or local society before you leave for fieldwork because they may know the process of getting permits quickly. Erica's golden rule is to keep field notebooks safe: separate these from the main bulk of the luggage by carrying them with you as hand luggage.

ridges, especially near the lip. Damaged and old cups should be replaced. Some researchers use two cups (one inside the other) to make emptying easier: the cup in contact with the soil surface is permanently fixed and never removed during the study and holds the second cup containing the solution. This double-cup method prevents soil falling into the hole and having to re-dig each time when resetting the trap. The outer cup should have a hole in its base to allow drainage and preventing the inner cup from floating out if water gets between the two cups. If dry traps are used, then both inner and outer cups need drainage holes. **Note:** at the end of the study remember to collect all traps as they could carry on collecting animals for some time after they have been forgotten.

CAPTURING GROUND-ACTIVE INVERTEBRATES

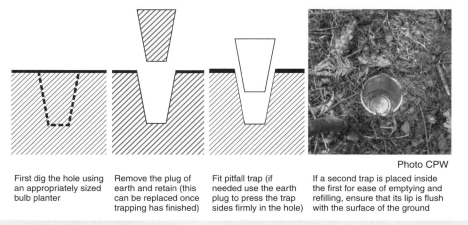

First dig the hole using an appropriately sized bulb planter

Remove the plug of earth and retain (this can be replaced once trapping has finished)

Fit pitfall trap (if needed use the earth plug to press the trap sides firmly in the hole)

If a second trap is placed inside the first for ease of emptying and refilling, ensure that its lip is flush with the surface of the ground

Photo CPW

Figure 4.25 Setting pitfall traps

Pitfall trapping has been used to sample a wide variety of invertebrate groups, particularly spiders, mites, beetles (especially ground beetles – Carabidae), springtails, centipedes, millipedes, ants and woodlice. Pitfall traps also attract flying insects but these animals should be disregarded as their capture depends on micro-scale differences in vegetation structure, trap prominence and location that cannot be controlled for effectively in such studies. Unfortunately pitfall traps are not selective in what they catch, and small mammals, reptiles, amphibians and even small birds can be killed alongside target organisms. Sometimes this is due to the accidental capture of surface-active vertebrates, although animals such as shrews deliberately target the traps to prey on struggling invertebrates. Accidental capture of vertebrates can conflict with the conservation aims of a site such as those relating to protected species. Although different techniques have been tried to prevent such captures (including placing covers a couple of centimetres above the trap, or placing chicken mesh just above the level of the fluid), these are rarely successful in reducing vertebrate capture and may prevent some large invertebrates from being collected. If inadvertent capture of vertebrates is a problem, then trapping at that site should be abandoned.

Placing covers over traps helps to prevent rain from diluting the solution or flooding the traps (especially in areas or times of the year where rain is expected). If using rain covers, which can be as simple as a piece of hardboard supported on 6 inch nails, sufficient space must be left above the trap so as not to impede larger species. Human interference is usually a minor problem: do not make the traps too obvious (they could be painted a brownish colour if they contrast too much with the surrounding ground and vegetation) and only mark one trap with a cane for relocating the study site. Animal interference can also be problematic with inquisitive species (e.g. crows and geese) picking traps out of the holes.

Such disturbances should be recorded to help to interpret your results. In the case of missing traps, catches can be expressed as the number of individuals per trap collected. However, such an adjustment cannot be made for the number of species caught, and interpolation may be required to help correct for this (Box 4.4).

Box 4.4 Taking account of missing traps

Suppose we have a study where 4 out of 10 traps are missing and we have captured 240 individual animals from 10 species. The number of individuals expected in all 10 traps could be estimated as:

$$\frac{\text{Total number of invertebrates found} \times \text{number of traps set}}{\text{Number of traps collected}} = \frac{240 \times 10}{6} = 400 \text{ expected}$$

The above formula would not necessarily work for the number of species since the number of species found in two traps is highly unlikely to be double that found in one (many will be the same species in the two traps). One way of dealing with this is to look at the accumulation of species trap by trap to estimate how many extra would have been likely if all the traps had been recovered. Here we simply look at the number of species found in the first trap collected and then how many extra were found in the next one, then the one after that, etc. If we plot this, we could extrapolate the curve formed to see what we would expect in all of the traps set. In the example shown, there would be expected to be 11 species had all 10 traps been recovered (this compares to about 17 species if the formula above had been used). Even with such methods, care should be taken if the number of traps missing is a high proportion of those originally set.

Trap no.	No. new species	Cumulative no. species
1	4	4
2	1	5
3	2	7
4	0	7
5	2	9
6	1	10
7	?	?
8	?	?
9	?	?
10	?	?

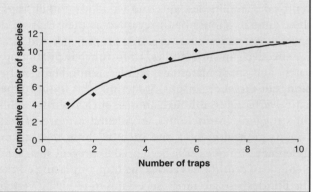

Most ecologists agree that pitfall traps are a measure of 'active trappability density' and are not absolute in their sampling efficiency. This active trappability density is a function of an animal's search for food, mates and resting places as well as its pre- and post-copulatory behaviour and differences in its habitat preference. For many insects, pitfall traps are more effective at collecting active adults, frequently misrepresenting the real population structure of a community due to the lack of less active immature animals. Perhaps the most significant of all are the differences between species – although one species may be sampled effectively by pitfall traps, another closely related species can be under-represented or even absent from the samples. For example, the hunting spider *Pardosa pullata* is a common ground-active spider frequently caught in pitfall traps, whereas *Pardosa nigriceps* is a rarely trapped hunting spider more commonly found climbing in the substrata of vegetation.

Despite a few flaws and uncertainties, researchers remain cautiously in favour of pitfall traps because they produce results that are statistically viable while remaining cheap to operate. They have also been shown to be a reliable method of collecting animals such as spiders over a long period of time as their use is not strongly weather dependent, unlike G-Vacs (see p. 148), for example.

Pitfall traps may be linked using barriers or fences. These comprise long thin pieces of wood, metal or plastic, partially buried into the ground with either end touching the edge of a pitfall trap to extend the trapping edge of the traps. Double arrays of such barriers can be used to surround a resource (log, stone, patch of dung) to investigate the surface-active animals that move into and out of the area enclosed (Figure 4.26). To get a more accurate estimate of the movements into and out of a resource such as dung, a comparative control plot should be used that does not have dung in the centre and that therefore measures the background movements of animals in either direction.

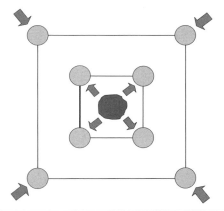

Figure 4.26 Barriers used with pitfall traps
Barriers can be set up to surround a resource such as a dung patch or log or stone in a double layer so that animals captured in the inner traps represent (mainly) those leaving the resource. Arrows indicate the direction of animal movement. Animals captured in the outer traps are (mainly) those moving towards the resource.

Some authors refer to 'H' traps in pitfall trap designs. This variation can be used to assess the possible direction of movement between areas, as illustrated in Figure 4.27. Collecting data in this way can help to study the movement of arthropods onto a site. However, this type of design assumes that movement is linear when in fact many animals may take a more random, less direct route. As some of this discussion has illustrated, although pitfall trapping is a relatively simple technique, there are many factors that have to be taken into account in its employment (Table 4.3).

Table 4.3 Factors to consider when using pitfall traps

Factor	Considerations
Size of traps	Depth (deeper retain the catch better but are more difficult to set)
	Circumference (larger catch more but are more difficult to set)
Colour of traps	Contrast (may invoke avoidance or attraction behaviours)
	Surface texture (if painted may provide footholds for escape)
Type of traps	Plain or barrier (barriers increase the catch and can enclose areas but are more difficult to set)
	'H' trap to assess the direction of movements of animals
	Type and size of barrier (higher retain better but are more obvious to vandals)
Type of material	Health and safety (e.g. glass breaks more easily)
	Ease of carrying (e.g. weight of glass traps)
	Surface texture (grooved or pitted sides may increase escapes)
Shape of traps	Ease of setting (those similar in size to bulb planters are easier to set)
	Ease of carrying (e.g. those fitting together are easier to carry)
	Ease of escape (shallower traps are easier for animals to escape from)
Number of traps	Coverage (more traps cover the area better but are more time consuming to process)
	Individual traps or multiples (e.g. trapping in groups provides more animals per trapping unit)
Arrangement of traps	Coverage (appropriate spatial distribution enables heterogeneity to be accounted for)
	Interference (traps too close together may interfere in terms of overlap of trapping extent)
	Pseudoreplication may result if traps at one site are used as replicates
How set	Level with soil surface (to prevent animals crawling round the lip)
	Catches in different vegetation densities may be influenced by relative difficulty of movement
Dealing with flooding/rain	Drainage holes (may be useful especially on waterlogged soils)
	Rain shields (may help prevent flooding and dilution of killing solutions)
	Weighting down traps in waterlogged ground can prevent traps from floating out of the holes

(continued)

Table 4.3 (*Continued*)

Factor	Considerations
Preventing collapse of hole when collecting	Rigid containers are better than highly flexible ones if soils are crumbly
	Use of exterior sheath (e.g. outer trap) can prevent holes collapsing when emptying traps)
Wet or dry	Live or dead collections (the former are necessary for mark–release–recapture techniques)
	Water or killing solution/preservative (depends on the time between collections)
	Avoidance or attraction (some solutions attract/repel species differentially)
	Health and safety (many solutions are toxic and/or corrosive)
Baited or non-baited	Deliberate to target particular animals (choice of bait is important here)
	Accidental (e.g. mammals dying in the traps can act as baits)
When to sample	Time between collections (depends on weather conditions and logistic constraints)
	Collect on same or different days (ideally former but latter may be needed if sites are distant)
Preventing capture of non-target animals	Placement (avoiding some habitats at some times of year)
	Interception (using covers or mesh above the fluid layer)
What is being sampled	Catches depend upon activity AND abundance (not just abundance)
	Differential activities (e.g. males of some species of spiders are more mobile than females)
	Distance animals may move into traps (not always sure of limits of trappable population)
Dealing with trap loss	Individuals can be counted per trap and scaled up to account for missing traps
	Statistical interpolation can be used to estimate the numbers of species lost
Ethical considerations	Are too many trapped animals being killed?
Over-trapping	Is there evidence of population depletion over time?
Comparison of different habitats	Different vegetation types inhibit animal movement to different extents
	Other environmental variables such as shade can influence the species caught

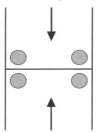

Figure 4.27 Bird's-eye view of an H trap
Barriers are made in the shape of a letter 'H' from metal or plastic lengths (50 cm long with 15 cm high sides buried 3 cm into the ground). At each of the four corners in the H is a single pitfall trap placed to collect animals coming from one direction only. Animals captured in the upper set of traps will tend to have originated from the top of the site, whilst those caught in the lower pair of traps will tend to have come from the bottom of the site. Animals caught in the two traps at the top of the H should be kept separate from those caught in the traps at the bottom when the collection is taken away for identification.

Suction samplers

Suction samplers have long been used in entomological studies. Typically these have either been old style Dietrick (D-Vac) or Burkhard machines (lawnmower engines on a backpack). These types of sampler are expensive, heavy and cumbersome and there is some suggestion that they have low suction speeds (see Stewart and Wright 1995). New developments have led to designs such as the Vortis and the Univac[4] samplers that have higher suction speeds but are still heavy, awkward and very expensive. Many researchers now use modified garden vacuum cleaners (G-Vacs) with a net inside to catch the invertebrates as they are sucked from grassland and heathland swards (Figure 4.28). These are cheaper to buy and maintain, are much more portable, have high suction speeds ($\sim 16\,\mathrm{m\,s^{-1}}$), and are quick and easy to use. However, suction samplers cannot be used during wet weather and so should only be used on dry vegetation. Even on dry summer days, it is rare that sampling can take place in the early morning when dew is clinging to the vegetation. Suction sampling is particularly useful in homogenous vegetation in which there are a number of small inactive species that are less likely to be sampled by pitfall trapping (see p. 139). They may also give a good reflection of the immature invertebrate population, which is poorly sampled by other methods. Additionally, suction samplers have the advantage over sweep-nets (see p. 154) as they can sample species that occur very low down in the vegetation.

---

[4]http://www.burkard.co.uk

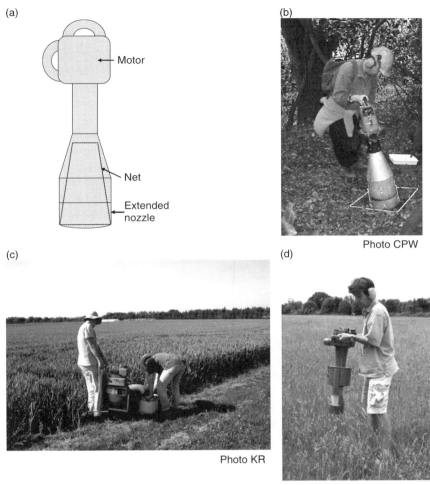

Figure 4.28 Suction samplers
(a) and (b) G-vac based on a modified garden vacuum cleaner; (c) D-vac sampler; (d) Vortis sampler. Once a G-vac sample is finished, the machine is then turned upside-down and contents of the net can be emptied into a labelled plastic bag. Catches from suction samplers can be standardised by surveying from within a quadrat, by placing the nozzle on the ground at 1 m intervals for a specified time (e.g. 10 seconds) and/or frequency (for one commonly used G-vac machine, the Ryobi RSV 3100 with a nozzle diameter of 13 cm, 75 sucks equals 1 m^2).

However, although G-Vacs collect a large number of specimens, they should not be considered to be an absolute method. One study found that suction samplers incur an edge effect caused by suction at the sides of the nozzle. This edge effect could also be caused by differences in the diameter of the nozzle: the greater the diameter, the smaller the edge effect. Further sampling errors may also occur when using some machines in dense vegetation that interrupts airflow and causes a filter to develop under which animals can hide and avoid capture. However, machines with high nozzle airspeeds may be relatively unaffected by dense vegetation.

Emergence traps

Many insects emerge from larval and pupal states within a range of substrates, including soil, dead wood and leaf litter. Some insects (including some beetles) are saproxylic, spending their larval stage within rotten or dead wood. It is usually not practical to hand-capture such insects when emerging from deadwood. Since many emerging insects are positively phototactic (i.e. move towards light), if deadwood is placed in darkness except for one or more small holes letting in the light, then insects will tend to move towards these points of light. The holes can lead to collecting tubes that may contain a killing solution. Emergence traps collect insects, mostly beetles and flies, that can occur at low densities: this can make analysis difficult unless a large number of traps are used. Furthermore, like barrier traps (see p. 145), emergence traps are usually fairly conspicuous and may be disturbed by vandals.

Where deadwood can be removed (with permission) from a habitat, then the darkness can be maintained by placing the wood in a sealed box. Where deadwood must remain *in situ* then blackout cloth can be used, supported with chicken wire so that it does not touch the wood but allows free movement of the insects. In both cases, a few holes are made in the material leading to a collecting chamber. To prevent insects moving back towards the deadwood, baffles can be used within the tubing (Figure 4.29). This technique can also be used to collect animals emerging from soil, leaf litter and dung. A different approach is to use an Owen trap that looks rather like a triangular tent with a built-in mesh floor and a collecting bottle at the top (Figure 4.30).

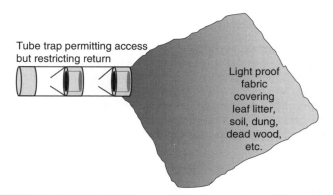

Figure 4.29 Emergence trap
The fabric part of the trap should be light proof and would normally entirely cover the deadwood, etc. from which animals are emerging. The tube traps can be connected to collecting containers containing a killing and preserving fluid.

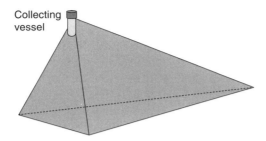

Figure 4.30   Owen emergence trap
Owen traps feature a built in floor and can be used in the field or laboratory to collect animals from debris such as dung, rotting wood or leaf litter that are placed within the trap. Similar types of emergence traps that do not have a floor can be used *in situ* to collect animals emerging from the ground surface on which the trap stands. In both versions, emerging animals move upwards and are caught in the collecting vessel.

Capturing invertebrates from plants

Many invertebrates are found on vegetation either because plants provide food directly (for herbivores) or indirectly (for predators feeding on herbivores). Other animals may simply be using the foliage as a refuge from predation or adverse climatic conditions. Some species (e.g. aphids and the caterpillars of some butterflies and moths) can be found in quite large numbers on individual plants. Case Study 4.6 describes a research project that monitored the caterpillars of one species of butterfly that live and feed communally. Whatever the reason, the links between plants and invertebrates can be very strong and, in some cases (e.g. for pollinators), symbiotic. Capturing animals from low-level vegetation is usually relatively straightforward as long as the plant concerned does not have thorns and is not too dense to sample within. High in the canopy can be more problematic and most techniques used for this are only accessible if you are working with an established research group. Examples of these techniques include the use of platforms, walkways, towers and cranes (see Ozanne 2005 for further details).

Case Study 4.6 Butterfly life cycles

Robin Curtis is a PhD student at the Zoological Society of London who has studied British butterflies for over 15 years. His main research interest is in conserving species and habitats in relation to climate change and habitat fragmentation. His case study involves catching, marking and measuring small animals in the field in order to monitor the population levels of a rare butterfly species.

Photo RC

Weighing an adult butterfly in the field

Model organisms and sampling problems faced

Butterflies are ideal study organisms. They are active during the most sociable hours (10:00–16:00) and only in reasonable weather. They are relatively easy to study and rarer species inhabit some of the most ecologically important habitats. Butterflies also show rapid responses to changes in the environment, making them ideal indicators of ecosystem health.

There were two main sampling problems; calculating accurate population levels and determining the movement of individuals between colonies. Most butterflies are monitored by a transect method where observers count the number and species of butterflies within an imaginary 5 m box on a fixed route on a weekly basis. This methodology underpins the Butterfly Monitoring Scheme (BMS) dataset that has provided important information on changes in population trends and distributions. However, Robin wanted to study one of the rarest butterflies in the UK, the Glanville fritillary, which only occurs on the south coast of the Isle of Wight and is not recorded on any transect routes. He also needed an accurate way of marking individual butterflies and calculating how far they had travelled on a daily basis.

How the problem was overcome
The larvae of the Glanville fritillary are quite unusual in that they live communally within a silken 'web'. These webs are extremely conspicuous, but only for 3 weeks in late March. During this period Robin visited all known colonies and counted the number of webs and the average number of larvae in each web to estimate population levels. The methodology he used was quite straightforward, but he found counting larvae extremely monotonous. The main problem was to survey all the sites accurately within the scheduled time period. As all the main colonies occur on landslips that are subject to regular slippages, he also needed to ensure that someone knew where he was working and to be careful to contact them at the end of each day. Over several seasons he found the dramatic fluctuations in numbers so intriguing that he continued the survey for 15 years – it is now the longest running, most accurate survey of any butterfly in the world.

To evaluate movement of adults between colonies, Robin undertook a mark-release-recapture study during each spring. Butterflies were caught, given a unique number by using a combination of three coloured felt-tip dots on different parts of the wing, and their wing length was measured using callipers. He also recorded each individual's location with a GPS (as a waypoint), and used carbon dioxide to anaesthetise the butterfly, so he could weigh it prior to release (to calculate fitness). Robin said, 'I really had no idea whether weighing them would work, but I purchased some cheap (£35) jewellery scales that were accurate to one hundredth of a gram. The first time I put a butterfly on the scales it didn't register at all and I thought I had wasted my money. However, I then realised I hadn't calibrated them (!), and after this the system worked well. I was amazed to find that males weighed five hundredths of a gram and females twice as much.' The use of a hand-held GPS in conjunction with geographic information system (GIS) meant that it was relatively straightforward to download the waypoints at the end of each day and calculate distances that individuals had moved.

Advice for students wanting to study butterflies
Many people would assume that studying butterflies is an idyllic way to spend the day. However, the realities of mark-release-recapture are that it needs to be conducted every day in suitable weather, and after 6 hours of sitting in the sun marking butterflies without a break the novelty quickly wears off. You should also be aware that studies of larval preferences and oviposition localities may provide more important insights than those on adult butterflies. Because of the short lifespan of most butterflies, the most important thing is to be organised and plan your fieldwork schedule well in advance.

Despite this, butterflies offer huge potential for research. You can collect a lot of high-quality data within hours without lots of expensive equipment, and there are many people available for you to contact for advice. Also, the vast amount of data accessible from the BMS means that even if your season is a wash out, the limited amount of data you have collected can contribute to the understanding and conservation of some of Britain's brightest gems.

Pootering

A pooter or aspirator is a device that is used to collect small arthropods (less than 3 mm wide and up to 6 mm long). A pooter consists of a collecting tube from which two tubes emerge (Figure 4.31); one tube is used to collect the arthropod and the other is for the mouth.

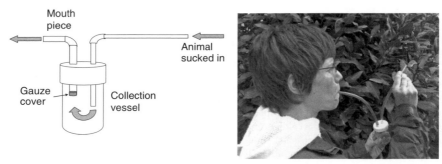

Photo CPW

Figure 4.31 Pooter used to suck up small invertebrates
The user sucks through the mouth tube, creating suction while placing the collecting tube over the arthropod. The animal is then sucked into the collection vessel ready for identification. If you have a number of empty collection chambers the same size as the pooter chamber, the top can simply be removed and placed on an empty chamber ready for the next collection. When making (or choosing) a pooter, you should ensure that the tube you suck through is wider than the tube placed over the animal (this increases the suction force) and that there is some gauze over the end of the non-mouth end of the sucking tube to prevent small animals from being ingested. **Note:** this does not prevent spores or bacteria from being ingested and some entomologists prefer to use a version that uses a bulb to create the suction, or that creates the suction by blowing down the mouth piece, or by using a moistened paintbrush to pick up small animals. This is important if collecting from noxious environments such as dung or cadavers.

There are several ways in which pootering can be standardised: by time, biomass and/or area. For example, 5 minutes could be set as the time limit to collect animals from a series of bushes. Alternatively, insects from one entire stand of vegetation may be collected and expressed as a function of dry weight (see the discussion on biomass in Chapter 3). This is a rather labour intensive and not recommended if quick results are needed. Two quick and widely used methods are sampling the vegetation using quadrats or line transect counts (see Chapter 3) to collect all the animals in a given area. Belt transects can be used, but this is less common and perhaps only suitable for flying insects (see p. 112).

Sweep netting

Sweep nets are only really useful on long vegetation (> 30 cm) where they can be used to collect insects from the canopy and, to a lesser extent from the subcanopy of grasses and tall(ish) herbs and ferns. Sweep nets have strengthened rims and sturdy handles to enable them to be swept through vegetation, disturbing and collecting animals *en route* (Figure 4.32). It is important that the vegetation is dry when sampling. Sampling using sweep nets can be standardised in one of four ways making sure that the same type of net (of the same size) is used:

- large quadrats or blocks of vegetation can be entirely swept of insects;

- the number of sweeps can be used to standardise the regime (e.g. 30 sweeps in each habitat);

- line transects (see Chapter 3) can be walked making a sweep at each step at a constant speed;

- timed sweep netting of an area can be made providing that the walking speed and style of sweeping (i.e. the arm length employed and swing length taken) is kept constant.

(a)   (b)

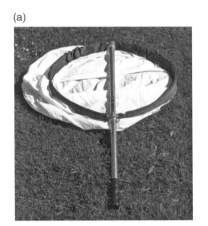

Photo CPW   Photo EMS

Figure 4.32 Sweep net
(a) Sweep net and (b) sweep netting invertebrates from a bush. Sweeping is usually done by grasping the net firmly in both hands and using a figure of eight motion to force the net through the vegetation while walking at a steady pace. To have the best chance of retaining the catch, the end of the net should be draped over the rim after sweeping to prevent animals from escaping (Figure 4.35c). Once the sweep netting sample has been taken, animals can be collected using a pooter (see p. 154).

Sweep netting is a quick and easy method to use but the catch may be low and lacking diversity if the vegetation is short or the area being investigated is small. This will have statistical repercussions that may be minimised by trialling the method in the habitat you wish to study. It may also be useful to complement this method with others (e.g. suction sampling – see p. 148) since many groups lower down in the vegetation may not be collected in large numbers by sweep netting. In some circumstances, sweep netting may knock canopy invertebrates off the vegetation to below the level at which they can be collected. This unwanted effect is dependent on the structure and type of vegetation: when comparing samples collected from different habitats be aware that comparisons may be problematic if such differences are too great.

Beating

Beating is used to collect invertebrates from branches of trees and shrubs by hitting the woody part of the plant with a stick while placing a sheet, collecting tray or a beating tray underneath the part being beaten (Figure 4.33). Beating trays are usually about $0.5\,m^2$ and catches can be standardised per area of collection, provided the number of times each branch is tapped is also kept constant. Beating has limited use, being restricted to low-level woody vegetation and may not be suitable for highly mobile flying insects that escape quite easily, but it can be a cheap and easy method of sampling shrubs and the subcanopies of trees.

(a)  (b)

Photo CPW   Photo EMS

Figure 4.33 Beating trays
(a) Black and white versions and (b) in use beating invertebrates from a tree. Animals are obtained by striking branches firmly with a stick and catching whatever falls onto the beating tray. Although these are slightly curved, which helps to retain the catch, many animals caught will be mobile and you should not wait too long before collecting the animals with soft forceps or a pooter (see p. 154). Beating trays are usually white so that dark animals show up against them. As they get soiled with use, this can lead to problems in ease of sorting animals from dirt. Black beating trays can be useful if much of the catch is expected to be light in colour.

Fogging

Otherwise known as chemical knockdown, fogging gets its name from the fog created by the kerosene or diesel (known as the carrier), which is provided by a fogging unit, and that carries an insecticide into the canopy of a tree where it kills or knocks down insects. The insecticides are often pyrethroid based (e.g. Pyranone, Reslin E or K-Othrin). Reslin E and similar chemicals tend to be favoured because they are harmless to vertebrates and break down in minutes in direct sunlight leaving no toxic residues. Chemical knockdown is a very specialised technique requiring dedicated equipment operated by an experienced team of researchers. Further details are given in Stork, Adis and Didham (1997) and Ozanne (2005).

For efficient fogging, the wind should be at a minimum and it is common to sample insects with foggers at daybreak, on dry mornings after a rainless night (leaves should not be wet before beginning sampling). To catch the insects that fall from the trees, several $1\,m^2$ trays, sheets or conical traps are suspended underneath the branches being fogged. The traps are normally left for 2 hours (known as 'drop-time') before they are collected. If using trays or sheets, the insects collected have to be pootered immediately or brushed into a bottle. Conical traps look like inverted witches' hats with a collecting tube at the bottom filled with a killing and preserving fluid such as methyl alcohol: these can be left unattended if theft or interference is not an issue. Fogging from the ground works best on trees under 30 m, but for trees taller than this a rope and pulley system is needed to winch the fogging machine into the tree subcanopy (Figure 4.34).

Photo NES

Figure 4.34 Fogging in rainforest
The fogging illustrated was undertaken in Borneo using a Swingfog machine hauled up on rope and located about 45 m above the ground. The tree (probably *Shorea* sp., Dipterocarpaceae) was about 75 m high. The insecticide, which was specially formulated by the Welcome Research Laboratories so that it had a knockdown agent but was not designed to kill the animals, was released using radio control.

The benefits of using fogging to sample canopies probably outweigh the disadvantages. Fogging is not dependent on 'trap behaviour' and is relatively unselective in what it catches, active or not, thus increasing the representation of the catch. The only groups that seem poorly represented in the samples are leaf miners and wood borers. Furthermore, it is not as destructive as some other canopy sampling methods, such as branch clipping. One consideration for most entomologists is the time taken to sort samples: fogging provides clean samples with very little debris and thus the sort time into major taxonomic groups is relatively quick. However, it is expensive, labour intensive and limited by weather conditions. If handled incorrectly, the fog can drift killing non-target invertebrates on adjacent trees. Similar techniques can be used to sample invertebrates from lower vegetation than tree canopies. Ultra-low volume (ULV) insecticide sprayers have been used to fog patches of vegetation, including nettles, bracken, etc.

Capturing airborne invertebrates

Most adult insects fly, and thus we may need methods for sampling individuals in active flight. Where species are relatively obvious, they can be monitored using transect or point counting (see Case Study 4.7 for example). Where capture is necessary, one of the simplest techniques is using a net (Figure 4.35a) either while the animal is airborne, or by placing a net over the animal while it is resting on the ground or on vegetation (Figure 4.35b). Once caught, animals can be retained in the net by draping the end of the net over the edge of the frame (Figure 4.35c). Although most airborne invertebrates are insects, there are other groups that are passive flyers. Small spiders, for example, may actively seek movement via wind currents by climbing above the ground and extending a line of silk that catches the wind and drags the animal behind it (so called ballooning). In fact, surveys from planes have found spiders and other aerial plankton at considerable heights above the ground. Although some techniques such as interception traps may collect these relatively free-floating species, most techniques are targeted at species that are active flyers. Most studies concentrate on trapping or surveying species that are close to the ground; an obvious problem if the targets are those species that fly high in or above the forest canopy.

Case Study 4.7 The birds and the bees

Dr Jenny Jacobs has worked on various research projects investigating agricultural ecosystems at Rothamsted Research for the last 9 years. The main focus of her work has been pollination and she recently completed a PhD at Rothamsted Research in association with the University of Stirling and the Game and Wildlife Conservation Trust. For her PhD she researched the role of flower-visiting insects in the pollination of fruit-bearing hedgerow plants and explored the consequences for frugivorous farmland birds.

Photo JRB

Bumble bee on blackthorn

Model system and sampling problems faced

Hedgerows provide shelter and food for a large number of animal species ranging from insects to small mammals and birds. The challenge Jenny faced was how to manage her fieldwork to provide a comprehensive picture of hedgerow ecosystems, considering that they contain multiple species of fruit-bearing plants, pollinating insects and birds. Only by running an intensive, year-round programme of field sampling was it possible to understand the links between insects visiting the flowers of fruit-bearing hedgerow plants, the availability of fruits for farmland birds and the abundance of frugivorous birds in relation to fruit supply. Practically, this was further complicated by having two field sites, 75 miles apart.

Jenny needed to monitor the activity of pollinators visiting hedgerow flowers. There are several texts available on the subject of sampling for pollination ecology studies from which Jenny was able to select her methods. The difficulty was determining which methods to use to obtain sufficient data for a system she was studying for the first time. In this example, there were many flowers on a hedge, with the possibility that pollinator densities could be high. Conversely, pollinator density could have been low since the flowering period of some of the plants studied were either early in the year such as *Prunus spinosa* L. (blackthorn), or late in the year such as *Hedera helix* L. (ivy) when cold weather could reduce insect activity. Furthermore, pollinating insects are mobile, flighty and sometimes difficult to identify on the wing. She experienced similar challenges with monitoring the activity of birds in hedges over the winter.

How the problems were overcome
To determine the best method for surveying pollinator densities, Jenny spent a little time trialling methods such as counting the number of insects seen visiting flowers while walking along a transect, and observing insects visiting quadrats of flowers for a fixed amount of time. In terms of taxonomic identification of pollinators in the field, Jenny found it was easiest to group insects according to their morphology initially. Depending on her knowledge she could group some further according to species, family or order. She also collected examples to identify to a higher level using keys, and gained help with this by consulting experts and attending identification courses. During the winter she trialled methods of surveying birds such as point counts and transect walks, and practised her identification of birds by sight and call by joining a local bird group.

Advice for students wanting to study a multicomponent ecosystem
Jenny advises that the researcher should be prepared to do a lot of field sampling. In order to obtain sufficient data, this requires good planning to optimise the use of available time. During the initial stages of a project, finding appropriate field sites can also be time-consuming, and the researcher should always remember to contact the relevant landowners before starting fieldwork. Although it may take up some time at the start of a project, Jenny strongly advises doing small-scale pilot studies. Even just a couple of hours of observation of pollinator activity would help the researcher choose a suitable method of sampling. Pilot studies may also help establish the best time of day to sample (e.g. pollinators are generally most active during warm weather). However, Jenny found that the weather was unpredictable, plus the large number of hedges and distance between her two sites meant she needed to sample in all weathers to collect enough data. Projects with intensive field sampling may require the researcher to work in all weather conditions, for long hours, with early-morning starts and late finishes, but gaining insight into the processes operating in an ecosystem is immensely rewarding. Jenny also advises not to be shy of contacting experts and hobbyists for their help with the taxonomic identification of organisms, since they are often delighted to share their knowledge and enthusiasm.

The project was funded by the Biotechnology and Biological Sciences Research Council, UK, and The Game and Wildlife Conservation Trust, UK. Jenny would like to thank her supervisors for their guidance and support: Dr Juliet Osborne and Dr Ian Denholm (Rothamsted Research), Professor Dave Goulson (University of Stirling) and Dr Chris Stoate (Game and Wildlife Conservation Trust). She is also grateful to the farm teams at the Allerton Research and Educational Trust, and Rothamsted Research, and to local Hertfordshire farmers, all of whom allowed her to study their hedges and dictate their hedge-cutting regimes.

Photos CPW

Figure 4.35 Nets for catching airborne insects
(a) Types of nets: (from left to right) kite net, butterfly net, folding net on extension handle and folding net - open (above) and folded (below); (b) catching flying insects that are found resting (**Note:** the end of the net is held in the air since many insects fly upwards when disturbed and can then be isolated in the end of the net); (c) retaining insects when captured by draping the net over the rim. When catching butterflies and other highly mobile insects in flight, the net needs to be flicked over the rim while it is still in motion, using a skilled flick of the wrist.

Another type of trap designed to sample animals from above vegetation are suction traps that are targeted at insect pests such as aphids. One example, the Rothamsted suction trap (Figure 4.36), is 12.2 m high and samples 0.75 m^3 air per second at 16 sites across the UK. A network of such traps has been employed at over 70 sites across Europe with some having been established for over 40 years (data and further information are available from the Rothamsted website[5]).

New (1998), Southwood and Henderson (2000) and Young (2005) give further details about the monitoring of flying insects and other airborne invertebrates.

---

[5]http://www.rothamsted.ac.uk/insect-survey/STAphidBulletin.php

Photo JRB

Figure 4.36 Rothamsted suction traps

Sticky traps

One of the most basic types of sticky trap involves old-fashioned fly papers that can be unfurled and hung from trees or poles to catch any insects that land on them. Some of these have sugary coatings to act as attractants and can be efficient where the likelihood of wind wrapping them around each other or the vegetation can be avoided. Other types of sticky traps are sold for agricultural use in monitoring crop pests and usually consist of bright yellow pieces of hardboard or plastic coated with a non-setting glue. Clear sticky traps are best for sampling if an attractant effect is not desired. In crop studies, these can be positioned above the height of the vegetation on poles in sets with one placed horizontally while the others are placed vertically but pointing at right angles to each other (Figure 4.37). Using single-sided traps placed horizontally with the sticky side down may reduce the number of predators and parasitoids caught if it is pest species that are the main target. Some suppliers sell blue sticky traps for collecting thrips that are suggested to catch relatively few beneficial insects (e.g. bees). It is difficult to extract captured animals from such sticky traps for identification purpose and this can become especially difficult if they become covered by debris in windy conditions. Some glues can be softened using alcohol or other solvents (e.g. ethyl acetate). **Note:** some sticky traps are sticky enough to even catch vertebrates and have been employed for lizards and small mammals (e.g. mice and other rodents).

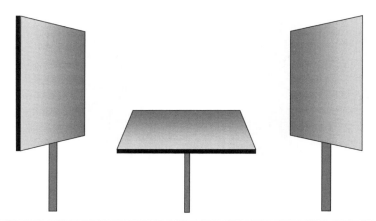

Figure 4.37 Positioning of sticky traps

Using attractants

A number of different baits are attractive to different groups of insects, some examples of which are given in Table 4.4. Having selected an appropriate bait for the target group, the baits are placed at appropriate times of the day (i.e. sugaring at dusk, rotting fruit at sunrise, etc.) to catch the insects at their most active phase. These baits should be visited regularly (every hour or so) because wasps and ants can interfere with the solutions and birds may predate on the attracted insects.

Table 4.4 Examples of baits and target insect groups

Bait	Example target species
Rotting fruit	Moths, butterflies, ants, flies, beetles
Urine	Some butterflies
Dung	Beetles, flies, some butterflies
Carrion	Beetles, flies
Dry ice (frozen carbon dioxide)	Termites, fleas, mosquitoes, some ticks
Fish (including tinned tuna)	Ants, some butterflies
Rotting eggs	Flies
Peanut butter	Ants
Turpentine	Some wood–boring beetles
Sugar solutions, e.g.	
Beer and molasses	Some beetles (especially some cerembycid [longhorn] beetles)
Red wine, sugar and sweet rotting fruit (e.g. strawberries or bananas)	Moths (especially noctuiids), butterflies, ants, some beetles, flies, orthopterans, etc.
Pheromones	Flies, moths, cockroaches

Very simple bottle traps can be designed to hold insects to avoid the unwanted effects of predation, etc. The principle of the trap should be that the bait is placed at the bottom of the trap with a large void between this and the top of the trap (i.e. make the trap tall). Insects, particularly flies, move towards light and will collect in the upper part of this void when they have finished consuming some of the bait (Figure 4.38). Alternatively, you can experiment with bait mounted onto a sticky trap or a combination trap using water traps (see p. 172) with an added attractant.

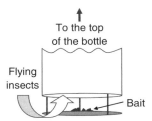

Figure 4.38 Bottle trap for flies and other flying insects
These can be made quite simply, for example by cutting the base off a 2 litre plastic drinks bottle. A circular disk, which acts as a stage on which the bait can be placed, is attached by three uprights to the bottle, leaving a 2 cm gap to allow access to the bait by passing insects.

Pheromones can also be used as baits. These are available commercially for a wide range of pest species (mainly Lepidoptera, Diptera and Dictyoptera). For some Lepidoptera, adult males can be collected using pheromones from virgin females. Pheromones are species specific and are used (sometimes in combination with pesticides to kill the animals caught) to monitor pests (e.g. in the home, in grain stores or in glasshouses). Other groups such as cockroaches and some beetles (e.g. weevils) can also be caught in such traps.

A wide variety of traps are available that are targeted at different species and animals of different sizes using pheromones and a range of other attractants (Figure 4.39). Common trap types used are funnel traps for larger species (over 12 mm) and Delta traps for smaller species or when relatively small numbers are expected. Using a combination of chemical baits, some adult Diptera of both sexes (e.g. fruit flies) can be caught in funnel type traps called McPhail traps. For some Lepidoptera, if live animals are required, then assembly traps should be used (Figure 4.40). These employ virgin females that are placed in a mesh container in the middle of the trap. The trap is placed across the flight path of the males and they assemble within the trap, failing to escape because of baffles built into the body of the trap.

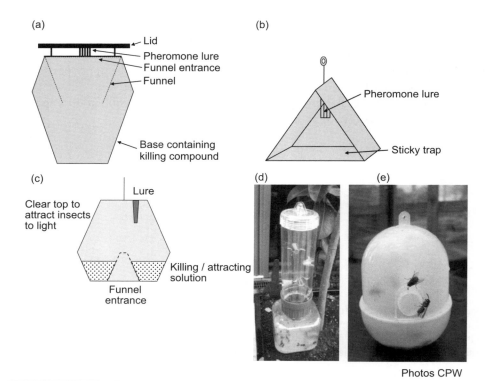

Figure 4.39 Attractant-based traps
(a) Funnel trap; (b) Delta trap; (c) McPhail trap; (d) cylinder wasp trap; (e) bell-shaped wasp trap. The traps can be placed on the ground, or hung from vegetation. Baits, which should not reach the opening of the trap, can include the types of substances listed in Table 4.4. If liquids are used and dead animals are required, try adding a few drops of detergent to lower the surface tension and help to drown the animals. Collections, especially of living animals, should be made regularly and you should be careful to avoid being stung by any wasps that are alive within the trap.

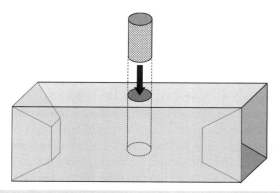

Figure 4.40 Assembly trap
Virgin females are placed within the mesh container that is inserted into the body of the trap (see arrow); males 'assemble' in the body of the trap.

Refuges

Artificial refuges can be used to collect animals that normally hide in crevices. Small insects and spiders that live in crevices in bark can be collected by tying 15 cm wide pieces of corrugated cardboard around tree trunks. Those bees and wasps that nest in small holes can be collected from trap-nests (pieces of wood in which holes have been drilled, or bundles of pieces of garden canes tied together – Figure 4.41). However, colonisation can take a whole season and is thus not an ideal method for short-term studies.

Photo CPW

Figure 4.41 Trap-nests for bees and wasps

Flight interception (window and malaise) traps

Interception using window traps can be a useful method of collecting flying insects. Window traps are made of either fine black netting or, for smaller traps, clear plastic sheet. The window part of the trap is suspended at 90° to the ground using supporting poles and guy ropes (Figure 4.42). Insects fly into the 'window' and fall down, being collected in trays or guttering filled with water with a little detergent to reduce the surface tension.

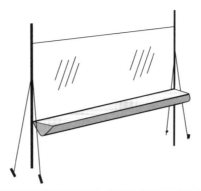

Figure 4.42 Window trap
Animals hit the window – made of Perspex or netting – and fall into the gutter that contains preservative.

If you are not careful, window traps can suffer from many problems, not least because they are obtrusive and attract vandalism. Furthermore, if left in open habitats they are susceptible to wind damage. The collection trays may collect too much rainwater and overflow if not checked regularly. Insects in flight can be patchy and their movements are weather dependent and thus window traps may not collect sufficient numbers for statistical analysis on all sampling occasions in all locations. This has some implications for the long-term monitoring of insects in flight. Window traps are also not cheap (even if made by user) making replication an issue. Window traps are only used for larger insects, including some beetles (carabids, scarabs, chafers, etc.), since flies and other small insects will not be effectively sampled by this method. However, larger beetles may be inefficiently caught in these traps, sometimes having a tendency to bounce off the mesh or plastic away from the collecting trays.

An alternative method is to use malaise traps, which look a little like tents with a central partition but without the two side panels (Figure 4.43). Malaise traps suffer some of the same problems as window traps (i.e. they may attract vandals, cannot be used in wet or windy weather, have similar problems with respect to replication, and are expensive). Malaise traps are particularly effective when placed in the flight paths of insects, for example along a woodland ride. However, the efficiency of malaise traps is dramatically reduced in windy conditions and they are rather ineffective at sampling beetles and true bugs (Hemiptera). It is strongly recommended that if you use either window traps or malaise traps, you supplement them with an additional trapping method effective for the group under investigation (e.g. sweep netting for flies and wasps in long vegetation).

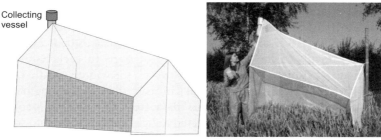

Photo JRB

Figure 4.43 Malaise trap
Flies, wasps and other insects hit the centre partition and fly upwards and are caught in the roof of the tent. The roof slopes upwards and animals naturally move in that direction. At the highest point, a hole leading to a sealed container outside the tent collects the insects.

Light traps

Light traps are widely used to collect night-flying moths and other insects that are attracted by bright lights. The choice, design, use and suppliers of moth traps are described by Waring (1994) and Fry and Waring (2001). At its simplest, illuminating a white sheet or light-coloured wall can result in the aggregation of large numbers of animals (Figure 4.44). More sophisticated designs comprise UV traps similar to those used to control pest species in food outlets, only instead of zapping them, a sticky card behind the light collects the animals. Light traps may also be used in combination with attractants. CDC (Centers for Disease Control) traps utilise low-wattage light traps, sometimes with dry ice as a source of carbon dioxide to act as an attractant for mosquitoes. The most commonly employed light traps are those designed for trapping night-flying moths, although they can also be effective at catching a number of other insect groups, including chafers, dung beetles and caddisflies.

(a)  (b)

Photos CPW

Figure 4.44 Simple light traps for insects
(a) Moths accumulating around a light on a wall; (b) a white sheet prepared under a light to collect insects at night.

The use of light traps is weather dependent and is not always suitable for long-term monitoring for this reason. Light traps attract night-active moths and hence reflect their activity and not necessarily their abundance. Because of their expense and the fact that it is difficult to have a high density of traps in each habitat sampled, replication may be difficult to achieve. However, providing weather conditions remain stable between nights (be sure to take measurements), the pragmatic solution may be to use a series of consecutive nights' data, treating each night's collection as a replicate. The reason why moths are attracted to light traps is subject to some debate, with ideas ranging from moths confusing the light with that of celestial bodies such as the moon, which they use as a migration cue, to the idea that the light is seen differently by different parts of a moth's compound eye creating areas of dark and light, and the moth is actually trying to fly towards the darker patches it perceives. Whatever the reason, large catches can be obtained under certain circumstances, especially when weather conditions are favourable (warm temperatures and low wind speeds) and on nights when moonlight is low. Lights close together can be less successful, especially if in relatively well-lit areas, compared to single traps in otherwise quite dark sites. Even quite poor lights can attract moths from reasonable distances (over 100 m).

There is an increasing number of types of moth trap available, providing both portable and more fixed options (Figure 4.45). Portable traps tend to be less efficient, not least because they are usually small and rely on battery-powered lights. Larger, mains-operated light traps can collect catches of orders of magnitude more than portable traps. There are several common types of light used: mercury vapour (MV) bulbs, black light bulbs, actinic tubes, and black light or UV tubes (Figure 4.46). The first two of these require a mains electricity supply (although a generator can be used in the field). In general, light sources such as MV bulbs and (to a much lesser extent) actinic tubes that emit both UV and visible light tend to attract more moths. Black lights that emit mainly or only UV light can be useful if being used in areas such as gardens where visible light would be annoying for neighbours. The Goodden light comprises several clustered lamps (in one tube) that shine green and is claimed to provide similar spectral characteristics to MV lights, but that can be run from a battery. Where mains supplies are used, care must be taken to ensure that water does not affect the electrics or that rain does not get onto the surface of the bulb, which can become very hot and can shatter if wetted. Use of external cabling and a rain cover for the bulb can be important in these situations. All traps should be set around dusk (some can be fitted with a photocell that switches it on at an appropriate light level) and then collected at dawn. Moths may become active once they warm up and birds (e.g. magpies) have been observed collecting them as they try to escape in the early morning. Moths can escape from most types of traps and collections at hourly intervals overnight have found many times the number that are caught in traps that are left until morning.

Figure 4.45 Examples of moth traps
(a) Rothamsted trap with mains-run mercury vapour bulb; (b) Robinson trap with mains-run mercury vapour bulb; (c) modified Skinner trap with mains-run black light; (d) Gladiator box trap with battery-run actinic ring tube; (e) portable Heath trap with battery-run actinic tube; (f) portable, mesh Johnson Safari trap with battery-run actinic looped tube; (g) portable, mesh Johnson Ranger trap with battery-run actinic looped tube; (h) portable, mesh Moonlander trap with battery-run Goodden light.

Photo CPW

Figure 4.46 Different types of light used for moth traps
From left to right: mercury vapour bulb; black light bulb; Gooden light; actinic ring tube; actinic looped tube; actinic tube.

There are various factors that should be taken into account when choosing moth traps (Table 4.5). Rothamsted traps have become standard traps in the UK for long-term fixed site monitoring and have been used across a wide network of sites for around 50 years (currently around 90 sites across the UK[6]). These traps use MV lights powered by mains electricity and trap moths in a relatively small trap that contains a killing agent. The trap is designed for small captures that will not adversely impact on localised populations when used over long time periods. The Robinson trap is a very successful trap commonly collecting 500–1000 moths per night in the peak season, but is expensive, bulky and needs an electricity supply or generator. The MV bulbs are expensive and become very hot while running and are easy to break. Egg boxes are usually placed within the trap within which moths can roost once trapped.

Skinner traps are portable traps based on a box-like design that can utilise bulbs or tubes run from either mains supplies or batteries. They usually have no base and are used in the field by placing them on a sheet of cloth or plastic on which egg boxes can be placed and within which moths can roost. Box traps made from plastic storage boxes can also be portable, with the added advantage that all of the light fittings, electrics and egg boxes for roosting can be carried within the box between sites. Heath traps are cheap, portable and can run from a battery or mains supply. However, they are not good at attracting many

---

[6]http://www.rothamsted.ac.uk/insect-survey/LTTrapSites.php

moths or retaining them and pooled collections from several traps may need to be made to cover even the common macro-species. They are cold-operating, usually being fitted with actinic tubes, although they may be supplied with different designs of light that may not be comparable. Some have clear baffles round the light whereas others have white baffles. Fry and Waring (2001) report that some changes to the electronics of the trap may halve captures when on battery compared with previous models (although interestingly not when used on mains). Safari traps and Ranger traps are portable traps made out of mesh cloth held in place using metal or plastic hoops and rods. They are usually fitted with tubes run on batteries, although mains driven bulbs can also be used. Some designs are built to sit on the ground, whereas others can be hung from trees. The Moonlander trap is a new design that is portable and mesh-based, which has the entrance hole underneath and the light hanging down inside the trap from the top. It also comes with its own foam inner liner that can be used instead of the egg boxes used in most other portable designs. Although some lepidopterists claim that this design is highly successful, others have captured very few moths with it in comparison to other portable traps.

Table 4.5 Factors to consider when choosing light traps to collect moths

Factor	Considerations
Types of light	Mercury vapour versus black light versus fluorescent tubes versus actinic tube versus Gooden tube (see Figure 4.46)
	Mercury vapour health and safety issues of burning and skin/eye damage
	Output, e.g. 12 V versus 6 V
Type of power source	Mains, generator, battery
Style of trap	Robinson versus Heath, Skinner, etc.
	Availability and cost
Size	Portability
	Shadow casting
	Retention
Number used	Logistics versus replication
	Pseudoreplication
Position	Line of flight
	Possible shading
What population is being sampled?	What is the immediate extent of the light's influence?
	From how far away do moths come?
Ethical considerations	Are too many trapped animals killed?
Capture comparability	What differences occur in different vegetation types?
	What differences occur in different climatic conditions?

Rotary traps

Rotary traps (sometimes referred to as whirligig traps) comprise nets mounted on arms that rotate on a motorised spindle (Figure 4.47) to catch ballooning spiders and flying insects (e.g. beetles, mosquitoes and aphids). The device is mounted on a pole set between 1–10 m in height, and rotates in a circular motion at a speed of about 16 km per hour. Some designs have only one net with a counterbalance on the opposite arm. Where there is a danger of animals escaping once caught, baffles can be fitted inside the nets. Rotary traps operate independently of wind conditions less than or equal to the operating speed. This is particularly useful in woodlands where the wind speed is variable, uncertain and usually less than 0.5 km per hour. However, despite their usefulness, rotary traps are usually custom built and thus expensive.

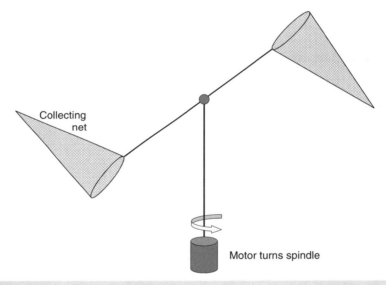

Figure 4.47 Rotary trap

Water traps

Water traps are cheap and convenient methods for collecting flying insects, although they are really most effective at catching flies and wasps. At their simplest, water traps are shallow bowls partially filled with water or some form of killing agent. The colour of the trap is particularly important: light and bright colours that contrast with the background tend to be most effective; yellow is normally favoured over white because it attracts Hymenoptera as well as Diptera. However, if trapping biting flies and wood borers, black or red traps should be used. It can be useful to employ a range of colours (including white and yellow) if the purpose is to document the whole community.

Water traps can be set at various heights within a habitat to sample different insect communities although they need to be properly secured particularly if they are left in exposed situations. One problem associated with water traps is interference from grazing stock and birds: both may disturb the catch either by rubbing against pole mounts, drinking the water or even (with birds) eating caught insects from the trap. To overcome some of these problems a lid can be fitted as shown in Figure 4.48 (an interception device fitted inside the trap can help to ensure that insects fall into the water).

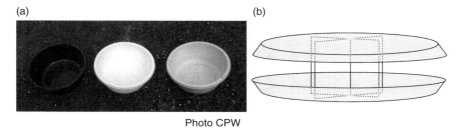

Photo CPW

Figure 4.48 Water traps
(a) Coloured water traps; (b) trap with lid and integral baffles to intercept insects that then fall into the lower bowl.

Fish

Photo CPW

This section concentrates on the study of freshwater fish, although the techniques employed may be applicable to fish sampled in many shallow waters, rock pools, lagoons and around coral reefs. The capture and/or study of fish from deep ocean waters is beyond the scope of this book (for techniques used to sample marine species, see for example Sparre and Venema 1999 and Cadima et al. 2005). Working in more controlled environments (e.g. fish farms and hatcheries) can be productive, as can monitoring the fish that are brought to market (especially in countries where there are few restrictions on the sizes and species being caught).

Many of the techniques described in this section are quite sophisticated, are limited by legal issues, involve expensive equipment, and/or have significant health and safety implications. As such, for many researchers, smaller species or observational studies are likely to be the mainstay of their research, unless they are linked to an established university team or government agency. If you are lucky enough to be given access to data from such organisations, they are likely to have been collected using the types of standard methods described here and in texts including Backiel and Welcomme (1980), English, Wilkinson and Baker (1997), Coad (1998), Southwood and Henderson (2000), Gabriel et al. (2005), and Bonar, Hubert and Willis (2009).

Where rivers, streams or lakes are shallow and if the waters are relatively clear, some observations may be made from the bank, or if deeper waters are involved, from boats or by snorkelling or scuba diving. Fish can be captured by hand, using hand nets or spear guns while wading, snorkelling or scuba diving. Alternatively, a slurp gun (Figure 4.49) can be used to suck up small fish (e.g. from crevices and holes). These techniques are often more suitable for shallow marine systems such as coral reefs rather than freshwater systems where the water tends to be less clear. A range of other techniques (e.g. electrofishing or poisoning) can also be employed by skilled workers to stun or kill fish (this usually requires a licence). The use of explosives is not recommended and is often illegal. Fish can be attracted by the use of appropriate baits or, at night, lights for more efficient netting or trapping. Baits include many different types of foodstuffs either attached to lines, within traps or simply scattered into the water. The use of lights at night (including torches shone on the water, illuminated lures on lines, lights strung along nets or even lights left inside traps) can be particularly effective but, because of this, is controlled by licensing in some countries. Other attractions include the use of decoy fish that are sometimes used in ice fishing or spear-fishing. Notwithstanding this range of methods, fish are most often collected

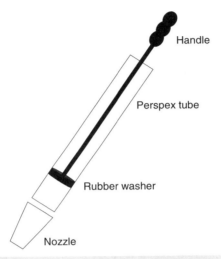

Figure 4.49 Slurp gun
The nozzle can be added or removed depending on the size of the animal.

using lines, or nets or traps.

Depending on the purpose of the study, it may be necessary to kill captured fish. As in all cases of collection and killing of animals, ethical considerations should be at the forefront to ensure that the collection is absolutely necessary and that the minimum number of specimens is taken (there are statistical methods to help calculate minimum necessary sample sizes – see Krebs 1999). Once these considerations have been satisfied and killing is still deemed necessary, the researcher must ensure that the animals suffer as little as possible. Fish may be killed by immersion in solutions such as formalin (aqueous solutions of formaldehyde); however, this can cause distress and animals should be anaesthetised first. Remember that collecting specimens in many countries requires a licence and that permits are frequently required to export specimens from one country and import them into another. Identification of fish can present problems when trying to identify fry (juvenile fish), although many adult fish can be named in the field.

One of the problems with studying fish is that unless the animals are brought into the laboratory, it is difficult to observe their behaviour, and bringing them back to the laboratory is likely to impact on their behaviour. Observations by researchers using scuba or snorkel gear (Figure 4.50) can help in making behavioural observations of fish (although if you plan to do this, you will need to ensure that your presence does not influence their behaviour).

Photo CPW

Figure 4.50 Using snorkel gear to observe fish
Snorkelling, like much field work, should not be carried out alone. Care should be taken, especially in shallow water, that your presence does not influence the behaviour of animals under observation, and that you avoid bumping against coral, which can harm both it and you. Underwater note pads or slates are necessary for collecting data, and waterproof guides may be available to assist with identification.

Direct observation

Fish can be observed from the bankside and under water, and some fish such as migrating salmonids can be observed jumping up rivers through weir systems. However, this may only be possible in clear shallow waters, often those with a calcareous river bed, where the depth of the water does not exceed 1 m. The weather should be clear and bright but without too much reflection on the water surface, and the wind speed should be slight to avoid rippling on the water surface. Fish scare easily so approach the site slowly creeping towards the riverbank on all fours. Allow the fish to settle and use a similar point counting method as described for birds to estimate numbers (p. 207). **Note:** be vigilant to avoid double counting as fish can change position with other individuals in a shoal with ease. Photography can help here: if a shoal is photographed (or videoed), the numbers of fish can be estimated later.

Indirect methods

It may be useful to survey not only adult fish, but also the eggs, for assessments of productivity. Different species have different strategies for producing eggs: from producing large quantities that are scattered widely, to painstakingly gluing fewer eggs onto vegetation, and from immediate abandonment of the eggs to guarding spawning sites. Fish eggs are either buried (as in salmonids, e.g. trout), attached to vegetation, rocks and other substrata (as in esocids, e.g. pike), or are free-floating (often near the bottom, i.e. semi-buoyant) in freshwater fish (as in some cyprinids, e.g. minnows). Eggs can be counted, or visually estimated using a scoring system, or measured as a volumetric estimate if the species lays eggs in masses (as in minnows). Depending on where eggs are laid, nets of mesh between 0.3 and 0.8 mm can be used to collect them, although they will collect other material as well, thus samples will need sorting. Areas where fish appear to be guarding their spawning sites should be targeted with nets. Alternatively, emergence traps can be used to cover the bed of the water body where fish eggs are found so that when the eggs hatch, the fry can be counted. Such traps need be no more than containers or fine nets secured to the bottom: use a sheet of Perspex or similar material slid underneath the trap to allow you to remove the trap from the bottom without losing the fry. Substrata and vegetation can be collected in spawning grounds and examined for eggs, or artificial materials can be introduced in the expectation that fish may lay on these.

Capture techniques

There are a wide range of techniques employed for different types of fish, ranging from active netting and hand-capture to more passive nets and traps. Catching fish using spear-fishing is another traditional method where fish can be speared from the bank, from boats, through holes in ice, or by the hunter swimming, diving, snorkelling or scuba diving. Modern spear guns (Figure 4.51a) can be very powerful and are restricted or banned in some areas of some countries: check with the appropriate authorities before using one. Angling (Figure 4.51b) can be employed for larger specimens but is a time-consuming and a rather hit or miss technique. Baited lines can be set either vertically using weights, or attached to floats to suspend the hooks at particular depths. Such lines can be very short (comprising only a few hooks) or stretch for several kilometres at sea. All such lines should be checked regularly. The types of hook and baits used are particular for the targeted catch. Neither line catching nor spear-fishing are useful for estimating abundance, but can be used to provide specimens in order to collect gut contents, parasite loads, etc. The most commonly employed, and arguably the most effective for ecological studies, capture technique is netting or trapping. Once captured, fish can be measured (length and weight) and examined for condition (including parasite load) and gut contents may be collected. In temperate climates, as fish grow, the scales of many species grow in size, producing a series of growth rings that can be used to age the fish.

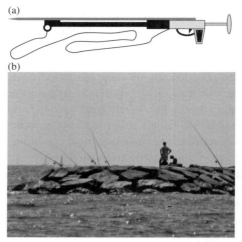

Figure 4.51 Sport fishing techniques
(a) Spear gun; (b) coastal fishing using rods and lines.

Nets and traps

There are many different types of nets that are used in trapping fish (see Figure 4.52 and Table 4.6 for examples). Nets can be used from the banks of water bodies, by researchers wading through the water, swimming, snorkelling or scuba diving, or by being towed behind boats. All nets are fairly expensive to buy, time-consuming to use, and costly to repair. They can easily become snagged and damaged or lost. It is recommended that a hydrographical survey of the study site is carried out prior to use, to identify any potential snag areas. Since fish have the potential to be damaged or killed by any kind of nets, sometimes electrofishing or trapping may be considered as more humane alternatives. Many netting methods catch large numbers of fish and such nets should be checked at appropriate intervals to prevent damage to the animals. Since most modern nets are made of materials that do not rot, care must be taken to ensure that they are properly secured to prevent them from coming adrift and continuing to catch fish for some considerable time.

FISH 179

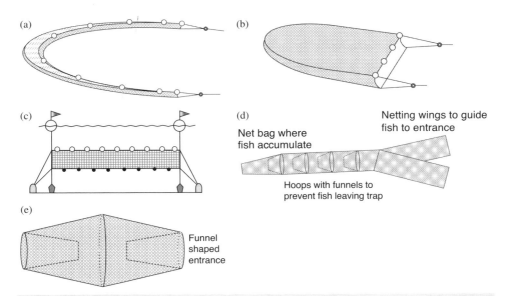

Figure 4.52 Examples of nets and traps
(a) Seine net; (b) trawl net; (c) gill net; (d) fyke net; (e) minnow pot.
Open circles indicate floats, whereas dark circles are weights.

Table 4.6 Summary of different types of net

Net type	Comments
Seine nets (Figure 4.52a)	Designed to be operated by two people in open waters normally for fish that are bottom feeders. Generally about 50 m long, this net is hard to use, especially when emptying the catch, which can be a slow business. Fish may also escape the net if the operators do not encircle the area to prevent escape. Seine nets can also be placed downstream and collect fish that are encouraged to swim away from researchers splashing upstream.
Trawling nets (Figure 4.52b)	Used to sample the river bed (known as a beam trawl) or, if floats are mounted to the upper edge of the net, then they may also collect midwater (pelagic) species. Trawling nets are normally towed by boats and are best confined to large freshwater lakes and rivers.
Dredges	These are heavier than trawl nets and are designed to be used on rough bottoms of water bodies.

(continued)

Table 4.6 (*Continued*)

Net type	Comments
Gill netting (Figure 4.52c)	This works more like a trap in that fish swim into the net and become stuck, not being able to more forwards or backwards. They have the advantage over some other nets in that they can be used to sample different depths by setting up the net with weights and floats. However, bottom-feeders are often missed using this method. Gill nets can also be used to selectively trap different age classes of fish depending upon the mesh size used. These nets cannot be left unattended: fish should be collected at regular intervals and rubbish removed regularly. It is unfortunate that gill netting may kill some fish, which can be especially problematic in mark–release–recapture studies, as well as from an ethical point of view.
Trammel nets	Can be used for fish that are otherwise hard to gill net. Trammel nets have three layers of netting with a fine inner net surrounded by coarser layers: fish strike the net taking the fine inner net through the coarse outer netting creating a pocket that surrounds and traps them.
Fyke nets (Figure 4.52d)	Fyke nets are very similar in appearance to pot traps, except that a leading net is placed at the entrance to the trap guiding fish into the funnel.
Pound nets	These are larger versions of fyke nets and can be used as floating nets to catch surface and pelagic fish, or can be anchored to the bottom to catch bottom-dwellers.
Lift nets	These are square nets that are lifted from underneath the fish that may be attracted to bait placed on the net.
Cast nets	These are widely used by fishermen in shallow waters around the world. When the cast net is thrown it opens up, but as soon as contact is made with the water, weights draw the net together. The user then pulls the net and this creates a pocket in which the fish are trapped. Cast nets are used mostly in shallow saline waters where they capture shrimp, small fish and mullet.
Dip nets	These are net bags (of various shapes and sizes) attached to the end of a long pole. They are especially useful for catching small fish in ponds and rock pools.
Push nets	These are larger than dip nets and are pushed ahead of the user.
Enclosure nets	Such nets work by suddenly enclosing a volume of water and hence trapping any fish present. These nets include those thrown or dropped into the water as well as those designed to pop-up within the water body.

Trapping is a fairly versatile method of catching most species of fish, especially for nocturnal species, or those occurring at low densities. Many trap designs are based on the same principle as fyke traps (Figure 4.52d): entry is easy through a funnel opening but escape is unlikely and disorientating for most species. A number of different types of traps are used for different species, under different conditions and at different depths. They include pot traps (often used for bottom-dwellers), fyke nets, hoop nets, eel traps, pot traps, flume nets, weir nets, cod traps and bag nets. Most of these operate on the same principle of having one or more entrances shaped like a funnel that the fish can swim into but find difficult to escape from (similar to minnow traps – see Figure 4.52e). Most types of trap can be baited, weighted and marked with a buoy : pot traps are simply cast overboard but fyke nets may need pegging in. Traps are usually more efficient when left overnight. Traps are usually set at the minimum between dawn until dusk for a one-off survey, and are checked daily to avoid larger predatory species from preying on others. If using a series of traps make sure that they are of standard dimensions to allow comparison between them and that there is no obstruction to fish movement around the traps: clear vegetation (under permission if necessary).

Collecting fish larvae

Many fish larvae are free swimming and can be captured (albeit not easily or necessarily efficiently) using towed nets, push nets, modified lift nets or hand nets. In some species the larvae attach to vegetation using a special organ on the head (as in esocids, e.g. pike). However, they are easily disturbed by sampling attempts and are possibly more readily collected using nets than by taking samples of the vegetation. Juvenile fish can usually be captured using similar methods to those used for adults, remembering of course that mesh sizes may need to be adjusted to account for their smaller size.

Electrofishing

Electrofishing can be particularly useful in cloudy waters where observation is not possible or in vegetated river habitats where it is difficult to operate a net. However, it is a very specialist and potentially dangerous technique involving passing an electric current through water via electrodes (anode and cathode), stunning the fish, disorientating them and allowing easy capture. Stunned fish are collected with nets and placed in aerated water containers to recover for between a few seconds and a minute or two for large fish. Full recovery can take several hours (up to 12 hours have been suggested). Electrofishing can be used in both point sampling or transects, but much depends on the size of the river and whether or not a boat is used. The capture rate of fish depends on the conductivity, clarity, temperature and depth of the water. In deep waters over 2 m or in turbid currents, electrofishing is not an efficient method.

Marking individuals

Fish can be used in mark–release–recapture studies and may be marked using tags, fin-clipping, tattooing or freeze-branding. However, an intensive regime may be needed to catch and recapture sufficient numbers and if the funds are available radio-tracking may be a more suitable method. Because of the problems of the large numbers of fish that need to be marked for such studies, many agencies use echo-sounders (hydroacoustics) to assess fish populations (even at quite low densities). Case Study 4.8 describes how this technique can be integrated with more low-tech methods to study fish populations. Here, short bursts of high frequency sound are directed into the water and the resulting echoes are processed to identify items such as fish that differ in density from the surrounding water. Both density and individual sizes can be monitored in this way. Appropriate techniques are discussed by Ricker (1956), Parker et al. (1992) and Nielsen (1992).

Case Study 4.8 Lake fish populations

Dr Ian J. Winfield is a lake ecologist at the Centre for Ecology and Hydrology in Lancaster, who has studied fundamental and applied aspects of lake fish populations in the UK and elsewhere for over 25 years. Ian's case study describes the use of a destructive low-tech sampling method used in conjunction with a non-destructive high-tech monitoring method to estimate fish numbers in freshwater lakes.

Photo IJW
Fish from a lake in northern Italy
Species visible include pike (*Esox lucius*), black bullhead catfish (*Ameiurus melas*), rudd (*Scardinius erythrophthalmus*) and perch (*Perca fluviatilis*). The black bullhead catfish just below the centre of the photograph is approximately 350 mm in length, but note the range of individual sizes of the other fish.

Model organism and sampling problems faced
Fish populations inhabiting physically restricted habitats such as streams and minor rivers are relatively easily sampled using electrofishing and other shallow-water techniques, but those occurring in lakes present much greater sampling problems. In addition to the obvious complication of a much greater surface area, ranging from several hectares to perhaps tens or even hundreds of square kilometres, lakes also introduce the challenge of considerable depth. Even a small lake is likely to have areas deeper than 5 m and in large lakes maximum depths over 50 m are not uncommon. Add to this that fish have highly developed sensory and locomotory systems, enabling them to both detect sampling equipment and quickly move away from it, and the scientist is left with a considerable sampling challenge. Finally, he or she also has to contend with the complication that other humans have a great interest in fish populations for commercial or recreational reasons and so their capture for whatever reason is tightly controlled by regulatory bodies.

How the problem was overcome
Ian has overcome these sampling problems by combining the low-tech capture technique of gill netting with the high-tech remote-sensing technique of hydroacoustics (echo sounding). Different

forms of gill nets have been used in fisheries for many hundreds of years, but recently scientists across Europe and elsewhere have standardised to a single design that samples fish of all sizes encountered in lakes. By using such equipment, Ian can take a relatively unbiased sample of fish from even the deepest areas of lakes. However, because gill nets work by being set for a period of hours (typically overnight) and catch their targets by entangling their gill covers (hence the name), fish sampled in this way are usually either dead or injured. While Ian makes full use of such specimens by taking them back to the laboratory for detailed examination to determine their individual length, weight, age and other features, this destructive nature means that gill nets must be used with discretion. Furthermore, gill netting cannot directly estimate the number of fish in a lake. Relative or absolute population abundance is of course a critical element of many ecological studies and it is for this reason that Ian also routinely uses hydroacoustics. By emitting a short pulse of high frequency sound (inaudible to both fish and humans) and listening for returning echoes, such equipment can estimate the numbers of fish in a lake with no adverse impacts. In addition, by the detailed analysis of the returning echoes it is also possible to estimate the sizes of fish present and their spatial distributions in the horizontal and vertical dimensions.

Advice for students wanting to study lake fish

Ian's advice for students is that studying lake fish can be very challenging and is not easily undertaken by a student working alone since it requires considerable pre-sampling preparations, including acquiring licences and permissions. Appropriate licences to remove fish for scientific sampling must be obtained before fieldwork starts and typically requires a return of information on the fish actually caught. In addition, further licences are required if the species under study is specifically protected under conservation legislation and the scientist should also consider ethical issues. On the positive side, gill nets are relatively inexpensive and their operation is not particularly difficult, nor is the examination of resulting specimens. However, hydroacoustics is both expensive (in terms of capital outlay) and the analysis of resulting data requires extensive training. Needless to say, both activities also require boat work, which itself requires appropriate training. In short, working on lake fish populations is very much a team rather than a solo activity and the student is encouraged to seek out appropriate teams working in this area. Collaboration is often the key to success.

Amphibians

Photo CPW

Most amphibians are intimately associated with water even if it is only for egg laying. Although some species of newts and frogs live for much of the time in water, others, such as tree frogs, live in rainforests and can lay eggs in little or no water. Although the identification of adult amphibians is normally not too problematic, their larval forms may be difficult to study and often requires the rearing of juveniles to adulthood. Depending on the species this can be quite straightforward, although the correct temperature and appropriate food is important. DNA bar-coding (of adults, larvae and eggs) is being explored as a non-invasive technique to investigate the species diversity of amphibians.

Salamanders will lose their tail if picked up by it and so care needs to be taken in handling such animals. Most amphibians are relatively easy to handle, although some may be toxic and should only be handled wearing gloves, with hands kept away from eyes and mouth and carefully washed after touching the animals. Amphibians can be susceptible to diseases (including fungal infections) that can be readily spread between populations. In fact, one such fungal infection, the chytrid *Batrachochytrium dendrobatidus*, has been implicated in severe declines in populations of amphibians across the Americas and Australia, and is now found in Africa and Europe. Handling amphibians can increase disease transmission and care needs to be taken to avoid keeping too many animals in holding tanks and nets for too long. In addition, any gloves (vinyl only not latex), footwear, containers and other equipment should be disinfected between use at different sites. Animals should not be moved between sites. Further guidelines are given in Beaupre et al. (2004) and ARG-UK (2008). See Case Study 4.9 for a discussion of implementing careful working practice in wild amphibian populations.

Case Study 4.9 Breeding behaviour of neotropical tree frogs

Victoria Ogilvy is a PhD student at the University of Manchester. She is studying the nutritional and lighting requirements of frogs in *ex situ* environments, and the breeding behaviour of wild neotropical tree frogs. This case study describes the problems associated with studying small tree frogs that live in the canopies of tropical forests, as well as highlighting the issues associated with the dangers of researchers being responsible for disease transmission between populations.

Photo VO

A pair of wild *Agalychnis moreletii* frogs spawning at a temporary pond in the Chiquibul Forest, Belize

Model organism and sampling problems faced

The neotropical tree frog, *Agalychnis moreletii*, is a critically endangered species that inhabits moist forest habitats in Central America. Studying the ecology of tree frogs is problematic because they are extremely difficult to locate outside of the breeding season. *A. moreletii* are believed to live predominantly in the tree canopy; however, accessing such areas without disturbing the animals is extremely difficult and potentially dangerous. Furthermore, these frogs are extremely well camouflaged when inactive making them very hard to see, even from very short distances.

Working on such species requires considerable care since populations of amphibians around the world have experienced dramatic declines in recent years. Several species have become extinct, and almost a third are currently threatened with extinction. One of the major reasons for these declines is the spread of infectious diseases, such as ranaviruses and chytridiomycosis (caused by chytrid fungus). These diseases may be easily spread by researchers moving between ponds, particularly if they have been handling frogs. Introduction of these diseases into amphibian habitats can have catastrophic consequences on some species, and there is currently little remedial action that can be taken.

How the problem was overcome

Reproduction in many amphibian species is restricted by water availability. *A. moreletii*, for example, will only breed during the rainy season. After days of heavy rain, hundreds of frogs congregate around temporary or permanent ponds in an attempt to find a mate. These breeding aggregations are often fairly easy to find since many frogs call to attract mates. Consequently, research and monitoring of frogs is often carried out during the rainy season when animals are easy to locate. Recently there have been significant developments in animal tracking devices, including

> radio-trackers and harmonic radars. Miniature devices now exist, and many are suitable for use on relatively small species of frog without causing significant changes in behaviour. Technology such as this enables researchers to more easily study the movement patterns and ecology of frogs outside of the breeding season.
>
> Biosecurity is an important issue to consider when studying amphibians in the wild. Measures taken to prevent the spread of diseases were fairly simple and easy to follow. Victoria used disposable gloves when handling frogs, which were changed between ponds and between species. Any frogs collected, were carried in disposable containers, e.g. zip-seal bags. Victoria also ensured that footwear, vehicle tyres and any equipment coming into contact with the frogs was disinfected with a biodegradable disinfectant when moving between study sites.
>
> Victoria was careful to ensure her safety in the forest at night, and during the rainy season, because she was aware that she may be exposed to many hazards, and careful planning was required in order to minimise the risks involved. For example, heavy rain can lead to flooding and can cause forest terrain to become very unstable, which may increase the occurrence of tree falls. Team leaders must have an action plan for emergencies in the field and all members of any research team must be made aware of, and fully understand, emergency protocols. Researchers should never work in the forest alone at night, and a system should be put in place whereby somebody at the research base knows exactly where work is being carried out and when researchers will return. Radios should always be carried when out in the field, and all members of the research team should know who to contact in case of an emergency.
>
> **Advice to students wishing to study tree frogs**
> Victoria has found that working with tree frogs in the wild can be extremely rewarding. However, students should be aware of the problems and ethical considerations involved. Handling of frogs should be kept to a minimum to reduce the risk of disease transmission and stress to the animal. Students should also be aware that research with frogs is often highly dependent on weather conditions. For example, the location of frogs may be very difficult unless they are in breeding aggregations or males are calling, and many species will not breed without sufficient rain. Waiting for appropriate weather conditions can be very frustrating, and it is therefore always advisable to have an alternative research plan should conditions prevent the planned study from being carried out.

This section covers the major techniques that are employed in monitoring amphibians. Further details of appropriate techniques can be found in Corn and Bury (1990), Heyer et al. (1994), Bennett (1999), Parris (1999) and Gent and Gibson (2003), and details of amphibian biology can be found in Duellman and Trueb (1994).

Direct observation

The most fruitful time of searching for many amphibian species is at night, by torchlight. Warm, damp, wind-free nights can facilitate a reasonable estimation of numbers, and this is particularly suitable for frog and toad populations. Direct counts may be difficult if the pond that is being surveyed is very large. Under these circumstances the pond should be divided up into a grid, working quickly with others to count the numbers in each block (e.g. 5 m × 5 m blocks). Alternatively, canes can be placed around the pond and the numbers estimated in belt transacts around the entire perimeter of the pond. Depending on the species, amphibians may be readily watched in open water and damp bankside vegetation, on trees, and under logs and stones. Surveys during the breeding season can provide opportunities to count breeding pairs.

Indirect methods

The presence of some species can be monitored by observing tracks in mud and other soft substrates. Some frogs and toads can be tracked by their call, both by monitoring the calls made by males (on a ranked scale of abundance, e.g. on a scale from $0 =$ no calls, $1 =$ few calls, $2 =$ many calls with distinguishable individuals, $3 =$ full chorus of many indistinguishable calls). Recordings of such calls can be used to stimulate other males to call and to attract females. An increasing number of guides are being produced that help in the identification of species from their calls (see also an increasing number of sound libraries[7]).

Counting egg masses

Studying amphibian eggs can give an indication of productivity and can be used as an indirect census of the breeding population. Estimating the size of egg masses of amphibians in water bodies can be a relatively simple business as they are often readily identified. Frogs produce egg bundles but toads tend to produce strings of eggs. Newt eggs are a little more difficult to find as they are normally laid singly and not *en masse*. Additionally, newt eggs are attached to plants (often with the leaf wrapped round them for protection) rather than floating on water surface and may be as small as 1 mm across. Newt eggs need to be netted: look for leaves that have been folded over. Try to estimate the numbers, rather than opening all folded leaves, since exposed eggs are vulnerable to predation. Some authors suggest that toad spawn numbers can be estimated in the water without removal, but this depends upon the clarity of the water. However, frogspawn should always be removed because they are not easily counted and many may be submerged. Counting spawn will give an estimated figure of egg production but it is even better to closely observe these in order to estimate the number of fertilised spawn. This is relatively straightforward since healthy spawn becomes pigmented and then divides whereas unfertilised eggs look dull and do not divide. Count the numbers of fertilised eggs and express them as a proportion of the total number.

Capture techniques

Capture of adult or juvenile amphibians by hand or with the assistance of a small net can be very effective, although in some countries some species are protected by law and a licence may be required to catch them. Animals caught in the light of a head torch at

---

[7]https://sounds.bl.uk/Browse.aspx?category=Environment&collection=Amphibians and http://www.cmnh.org/site/ResearchandCollections/VertebrateZoology/Research/IndexFrogCalls.aspx

night will tend to remain still where they can be caught using a slow and steady pick-up rather than a sudden grab. Species that live in burrows can be carefully dug out. Once caught, it may be useful to take a variety of measurements of the animals. At the very least, the snout to vent (cloaca) length should be taken for frogs and toads. Head to tail length is appropriate for tailed species such as newts and salamanders, although here too snout to vent lengths are commonly measured because of the possibility of tail damage and variation within individuals (e.g. through the breeding season).

Where possible, the sex, age grouping and mass of the animal can also be noted. Growth rates can be calculated using their lengths (measured with callipers) or mass, either by regular visits made to study sites to recapture marked individuals, or animals being kept in the laboratory. If the body temperature is required, then measurements should be made as soon as possible after capture, since the body temperature may change rapidly as the animal's environment alters. Temperature measurements can be taken by using a small temperature probe or thermister inserted gently into the cloaca. Regurgitated food and faeces that are produced while the animal is in captivity can be analysed. Some studies have used emetics to stimulate regurgitation for gut analysis, but this is a specialist technique and may require appropriate licences. Gut content analysis from dead specimens (e.g. road kills) can be used to identify the prey; many of the harder parts of insect cuticles for example remain intact in the stomach. Faecal remains can also be used to identify the diet of some amphibians. Plant eaters tend to produce brown or dark green solid faecal mass with some undigested plant remains, in contrast to carnivorous species where insect head capsules and wing-cases may be found.

Sampling adults in water

Catching amphibians while they are in the water is relatively easy using a net (preferably with a net head greater than 30 cm and a handle about 2 m in length). The mesh size should be around 1 mm to prevent injury due to toes becoming caught in the net. It is difficult to use a net to calculate density. The standard method used is to make 1 m sweeps around the entire perimeter of the pond, drawing the net to the shoreline to force escaping animals towards you. You should note down the number of 1 m sweeps made around the pond and the length of the pond perimeter. Seine nets (see p. 179) can also be used to catch amphibians in water. Although netting can catch all types of amphibians it is particularly useful for newts that are difficult to observe. It is probably easier to count the density of frogs and toads without capture.

Counting has the advantage over netting in that it is probably more accurate for frogs and toads and will not damage the emergent vegetation around the pond. However, a series of estimates will be required to track the rise and fall in numbers. If counting breeding pairs, it may be useful to measure the water temperature and the relative humidity of the air as these may correlate with activity. For example, in Britain frogs breed at cooler water temperatures ($\sim 5\,°C$) than toads ($\sim 10\,°C$).

An alternative to netting is to use a bottle trap to catch newts. However, some species are protected and should only be collected under licence. For example, in Britain if great crested newts are found in the pond, a licence must be obtained before traps can be used. A bottle trap is easy to make from a large plastic drinks bottle (Figure 4.53). These traps should be checked every 8–12 hours as occasionally newts lose consciousness in the bottle. If this occurs, place the newt in 1 cm deep cold water for 2–3 hours to revive it. If this happens frequently, reduce the time between checking the bottles. Bottle traps should only be used at night and removed during the day as newts will die if trapped and exposed to sunlight in a confined space. An intensive trapping regime would involve a bottle trap being placed at every metre around the pond. Small fish traps (see p. 179) also work well and, being made of mesh, avoid the problems of lack of oxygen. In the breeding season, large catches can be common in groups such as salamanders because of high activity levels, or pheromones attracting large numbers of individuals into the traps.

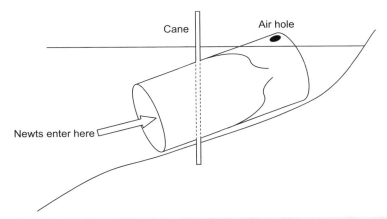

Figure 4.53 Bottle trap for newts
Cut a plastic bottle in half and insert the bottleneck inside the base, and secure with a cane at a pond edge (expose the trap by 2–3 cm with an air-hole clearly above the water level).

Sampling adults on land

Belt transects can be used to search for adults by placing two lengths of tape 1 m apart for a distance of 50–100 m. Walk the belt transect searching for individuals and note their behaviour (e.g. resting under cover) and the habitat (preferably including the vegetation type – for example using the NVC classification see Chapter 2). You can combine this technique with a detailed map of the area to establish the distribution by the habitat of the animals. Alternatively, you can use the habitat map to search for animals in different stands of vegetation. To standardise the search at each sample point, either use a tape to define a 5 m quadrat or set a specified time such as 5 minutes to search for animals in each area. Adjust the search area or time if you find you are sampling too many or too few animals.

A commonly used technique known as drift or ring fencing can be used to collect moving animals (e.g. during migration in the breeding season). This involves digging a 20 cm deep trench within which is erected a 50–60 cm high fence made of a flexible material such as plastic sheeting, window screening or light metal sheeting either placed in straight lines (drift lines) or around the entire perimeter of the pond (ring fences). Avoid mesh-type material as amphibians are good climbers and will easily escape if given a toehold (as a precaution when using ring fences around a pond, bend the top of the fence over towards the pond to prevent escape). Where the material is not rigid enough to stand on its own, it should be supported using posts at regular intervals. For any drift fencing, place buckets dug into the ground at 3–5 m intervals around the inside and outside of the fence and at the ends of the fence line: check these about 3 hours after dusk and soon after dawn for the duration of the survey. To prevent predation and flooding of the traps, either a cover can be placed over the bucket, or side-flap bucket (pail) traps can be used. The latter are lidded buckets with a flap built into the side, buried half way into the ground so that the flap is at ground level (Figure 4.54). Additionally, funnel, hoop or box traps, which use funnel-shaped entrances to trap animals (Figure 4.55), can be set along a drift fence. Drift lines can be set in a variety of patterns to maximise the captures, while ring fences around ponds should be erected about a metre away from the high-water mark (Figure 4.56). Species that can climb (e.g. many frogs) are not efficiently captured using drift fences. Arboreal species can be captured using either drift fencing along branches or traps set along walkways placed in trees. Some studies have used modified drift fences with overhangs at the top where tree frogs may rest and therefore can be captured.

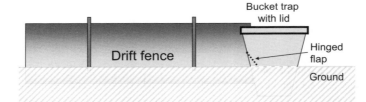

Figure 4.54   Drift fence with side-flap bucket trap
Animals move along the fence, falling into the bucket traps through the hinged flap.

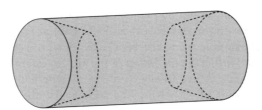

Figure 4.55   Funnel traps for amphibians
The funnel entrances help to retain the catch.

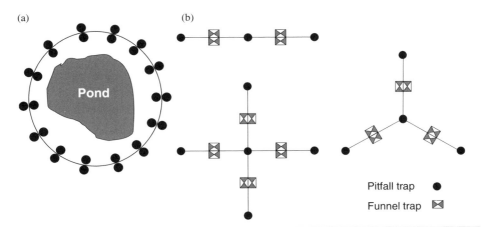

Figure 4.56 Examples of layouts for drift nets
(a) Ring fencing a pond: animals captured on the different sides of the fence can be assumed to be travelling in different directions (into and away from the pond) and thus used to examine movements – animals should therefore be released on the opposite side of the fence to the one in which they were caught; (b) drift nets incorporating pitfall (bucket) traps and funnel traps. Lines indicate drift nets.

Other trapping regimes include the use of refuges that involve a variety of materials being left in the field under or within which amphibians may shelter and subsequently be collected. For example, toads may shelter under pieces of carpet or wood, while 2 cm diameter plastic tubing has been used to collect tree frogs. Cover-boards (simple pieces of plywood) and artificial cover traps (see p. 192) can be successfully used to collect amphibians.

Tadpoles

Unlike the adults, tadpoles are hard to identify and may need to be reared in the laboratory in order to identify them as adults. Tadpoles can be netted or sampled using a 1.5 litre syringe. Ensure that the nozzle is large enough to sample tadpoles but small enough to avoid sucking up juvenile metamorphs.

Juveniles/metamorphs

These are commonly referred to as toadlets, froglets and newtlets. Once juveniles get two pairs of legs, froglets and newtlets tend to leave the pond over a number of weeks whereas toadlets all leave together over a period of a week or so. Once they have left the water, on land they shelter under logs, stones and artificial artefacts. Large buckets can be sunk into the ground using a barrier trap arrangement (see Figures 4.54 and 4.56) to see which direction the juveniles are heading. Remember to include drainage holes to stop flooding and some leaf litter to offer cover in the bottom of the bucket. If combined with a mark–release–recapture study this approach can be very rewarding. Alternatively, you can create artificial cover traps (using split drain pipes, pieces of untreated timber board, or base planks with boards resting on spacers to create wedge-shaped gaps (Figure 4.57) and place them in damp places or in the grass sward, within which juveniles can hide during the day.

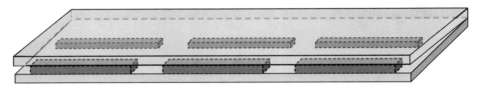

Figure 4.57 Artificial cover trap for amphibians
The trap is set in a suitable habitat and the animals hide in the wedge shaped gaps and can be collected later.

Marking individuals

Marking amphibians is difficult and the use of passive integrated transponder (PIT) tags is one of the best methods, although this is a rather specialised technique and can be very expensive. Dye injected under the skin has been used to mark salamanders. Toe clipping is sometimes used, but is usually rejected on ethical grounds quite apart from potential problems of influencing the animal's behaviour and increasing mortality levels (see May 2004 for a review and further debate in Phillott et al. 2007). Adult and juvenile toads and frogs can be successfully marked using knee tags (small tied tags attached to the leg just above the knee with stretchable thread). Tags can be made from waterproof paper or plastics that can be punched or written on to produce a unique code. Newts can be a little more difficult. Elastic waistbands have been used in a few studies, but these may only survive for a few weeks and are probably not suitable for long-term studies. Increasingly popular is the natural variation approach where individuals are photographed. For example, in many newt species (e.g. great crested newts) there is substantial variation on the bellies of the animals where spots form blotches or separate clusters. The advantage with this technique is that it is non-invasive, but critics suggest that it requires a high degree of skill and familiarity with a population, which takes too much time. **Note:** when using any method that involves a tag, you should check that the tag does not interfere with the body and look out for any signs of chafing when the animal is recaptured. Most amphibians are not adverse to handling, although in dry conditions hands should be moist and animals must be kept in the shade. Woodbury (1956) and Gent and Gibson (2003) give more details on marking amphibians.

Reptiles

Photo CPW

Reptiles can be exciting animals to study and are often of high conservation interest. Although many small diurnal lizards (and nocturnal geckoes) can be relatively easy to find, in many habitats reptiles may be quite difficult to see and/or catch. Some species are protected by law in many countries and some species of snake can be dangerous; therefore, care needs to be taken to understand all of the legal and health and safety issues before attempting sampling. This section discusses the most commonly used sampling techniques for reptiles. Corn and Bury (1990), Bennett (1999) and Gent and Gibson (2003) give further details on techniques used for monitoring reptiles.

Direct observation

Reptiles can be studied while basking either on land or in water. The best conditions for seeing lizards are days when there is little or no wind, and when there is sunshine after rain. Lizards have good hearing and sight and any approaches should be slow and silent. Snakes have relatively poor hearing and are not as sensitive to the wind as lizards, but nevertheless they can detect ground vibrations easily and approaches should be made with caution (especially if they are venomous). Typically, basking snakes can be found more easily in sunshine rather than on overcast days. Tortoises can sometimes be found basking early in the morning and grazing during the rest of the day. Tortoises can be found by searching on a transect basis, although this can be time-consuming. Distance sampling (as used for birds see p. 203) has been employed to estimate tortoise population sizes. Turtles are probably easiest to see when coming ashore for egg laying (although care must be taken to avoid disturbing them at this stage especially when using any sort of illumination at night), or when swimming in shallow seas. Terrapins and crocodilians can sometimes be observed in water, or basking on banksides. The latter are often easier to see at night in bright torchlight.

Indirect methods

In a similar way to amphibians, some species of reptile may be monitored by finding their tracks in soft or loose substrates such as mud or sand. Tracking tunnels may be used to collect reptile tracks in some habitats (see p. 220). The shed skins of some reptiles (e.g. snakes) can indicate the presence of the animals and can sometimes even be identified to species level if found reasonably intact. The dung of some reptiles can be analysed to determine their diet (e.g. to identify the seeds eaten by large tortoises and the arthropods eaten by carnivorous lizards and snakes). Large animals (e.g. large monitor lizards and crocodilians) can be surveyed using camera traps. These can be set up to fire when an animal presses a pressure switch or triggers a baited sensor. Motion sensors and infrared beams have also been used, although some of these are less suitable for reptiles than for warm-blooded animals such as mammals. In all cases, film or digital cameras can be used. Time-lapse video, film and digital cameras have been used that take an image at set intervals throughout the day. These can be adjusted for the frequency of taking the picture and can be replayed quite quickly (in a couple of hours per 24 hour period surveyed). Positioning a camera in an appropriate place known to be frequented by the species will result in a greater success rate of image capture. Suitable situations include at the mouths of burrows, at the sides of regularly used tracks, at watering holes or where bait is left. Like any expensive equipment left in the field, camera traps may attract vandals and thieves, as well as interference and possibly destruction by the animal itself and should be carefully and securely sited (Figure 4.58). Video cameras can also be used to follow an animal's behaviour by attaching it to the individual and recovering it at a later date. This technique has been successfully employed with large tortoises.

Photo CPW

Figure 4.58 Concrete housing for a camera trap
This design was used in a jungle in Central America where poachers were likely to steal the equipment.

Capture techniques

Once captured, the animal can be weighed and measured (snout to vent and/or snout to the end of the tail), have their sex and reproductive condition determined and their parasite load assessed (see Gent and Gibson 2003). If the body temperature is required, then measurements should be made as soon as possible after capture, since the body temperature may change rapidly as the animal's environment alters. Temperature measurements should be taken in the cloaca by gently inserting a thermometer or probe and the ambient temperature should also be taken as a comparison. Faeces from reptiles may contain the remains of hard-bodied animals (e.g. the head capsules and wing-cases of insects). These can sometimes be identified to indicate the type of prey being taken. Flushing the stomach contents using water or manipulating snakes' stomachs to move food back out of the gut are specialist skills and should not be attempted except by those with appropriate training (and licences where necessary). See Case Study 4.10 for a discussion of a research project analysing prey from faecal samples in wild populations of reptiles.

Case Study 4.10 Reptile diet

David S. Brown is a molecular ecologist conducting his PhD at Cardiff University using non-invasive molecular techniques to study the diet of British reptiles. This case study describes the collection of slow worms and smooth snakes and the subsequent examination of their prey using DNA analysis of their faeces.

Photos DSB

Slow worm   **Smooth snake**

Model organism and sampling problems faced

Loss, modification and fragmentation of habitat through agricultural intensification and urbanisation has caused a rapid decline in Britiain's six native reptile species and a better understanding of their dietary requirements is necessary for effective conservation. The focus of this study was primarily on the diet of slow worms (*Anguis fragilis*), one of Britain's most abundant reptiles, and smooth snakes (*Coronella austriaca*), Britain's rarest.

One of initial difficulties in working with wild reptile populations is finding them in the first place, as all species are camouflaged and highly secretive. Once found, handling them has its share of challenges too as reptiles have a range of defensive strategies, from biting to shedding their tail. Often it is necessary to recognise individuals, whether for capture–recapture studies or, in this case, when sampling on multiple occasions in order to avoid pseudoreplication (sampling the same animal more than once) in data collection. All reptiles regularly slough their skin, limiting the usefulness of marking with paints/dyes and tags, and the streamlined shape and behaviour of snakes and slow worms prevents the use of tags altogether. Once reptiles have been found, caught and individuals identified, determination of their diet presents an additional problem. Traditional methods are either highly invasive (forcing the regurgitation of prey or gut dissections) or largely ineffective (identification of prey morphology in faeces), particularly with snakes that digest even the bones of their prey.

How the problem was overcome

With experience, reptiles can be found basking in sunny spots, but this can be time-consuming and makes catching them difficult as they are prone to making a hasty exit at the first sign of movement. They can often be found underneath sunbathed debris (e.g. wood) and, as such, can best be found by laying down artificial refugia (50 cm × 50 cm pieces of corrugated iron or roofing felt) in suitable habitats to attract reptiles to shelter beneath them, making them easy to locate and catch. Reptiles are active between April and October, although the best months for finding them are April, May and September, when temperatures are lower and reptiles need to spend longer to warm up to their active temperature. Between 09:00–11:00 and 16:00–19:00 are ideal times of the day, preferably when the air temperature is between 10–17 °C with hazy sunshine and little or no wind.

The limitations and biases of analysing prey remains in faeces by visual identification were overcome by the detection of prey DNA; a non-invasive approach that has been successfully used in a wide range of taxa. To accomplish this, PCR (polymerase chain reaction) primers were developed to target DNA of specific prey species (where existing primers did not already exist). Extracted faecal DNA was screened in PCR with these primers and the presence of an amplified DNA fragment (of the expected size) indicated the presence of the target species. In this way, many individual samples could be very quickly processed.

Slow worms generally defecate on being handled, which makes collection of faecal material reasonably straightforward. To obtain faeces from snakes, the abdomen can be gently palpated, which stimulates defecation approximately 10–20% of the time. Animals that failed to defecate (~80%) would do so if kept in a ventilated plastic container with water and shelter for a couple of hours. In addition to providing information about diet, molecular analyses allowed for investigation of parasitic prevalence in reptiles and population genetics. Snakes and slow worms were photographed and individually identified by a combination of their snout-to-vent length and natural variation in colouring and patterning of their head and neck, which eliminated pseudoreplication. As well as measuring length, animals were weighed by placing them in a pillow case suspended from a spring balance.

Advice for students wanting to study reptiles

To locate suitable sites for reptile studies, students should contact local amphibian and reptile groups, natural history societies or even county councils. Ecological consultants, who are often involved in the relocation of reptiles for land developers, may also know of suitable sites or may permit students to join them in site surveys/translocations. For areas occupied by smooth snakes or sand lizards, both of which have restricted distributions, the Herpetological Conservation Trust should be contracted, as they manage most local public-access land in these areas. If using artificial refugia, make sure to gain permission from the landowner/manager and to lay refugia out of sight of casual passers by, as reptiles are prone to collection. All UK reptiles are protected under the Wildlife and Countryside Act 1981 making it illegal to kill, harm or injure them or to sell/trade them. A protected species licence must be obtained from Natural England before handling smooth snakes or sand lizards. The Herpetofauna Workers' Manual (Gent and Gibson 2003) is essential reading for any budding herpetologist.

Thanks go to the Natural Environment Research Council (NERC) for research funding, to the Herpetological Conservation Trust for their advice and assistance, and to Prof. Bill Symondson for project supervision.

Hand-Capture

Hand-capture is an option for many species of reptiles. Searching at night looking for eyes shining as the animal is caught in light of a head torch can be productive. In general, reptiles should be slowly and steadily picked up or pinned to the ground at the head end, not suddenly grabbed. Larger animals will need to be held just behind the head and near the base of the tail. It should also be noted that many reptiles (not just venomous snakes) bite : be careful and use gloves as a precaution and, where appropriate, a snake hook. Acrylic tubes can be used to contain venomous snakes, although such animals should only be handled by experts. Hand-capture can range from being relatively easy in the case of tortoises through to very difficult with small fast-moving lizards, and potentially downright dangerous with large crocodiles and venomous snakes. In all cases it is important to know the risks associated with handling the animal both to you as the researcher and to the animal. Many species of lizard will lose their tails if handled roughly, especially if caught by the tail. Putting hands down burrows to try to dislodge reptiles is potentially hazardous and should not be attempted. For some species that hide in burrows, it may be necessary to dig them out; however, tortoises in burrows may be encouraged to emerge by tapping their shell a couple of times with a stick and then withdrawing a little way away. If any venomous snakes are likely to be encountered, it is important to have fast access to appropriate anti-venoms. Venomous species should only be studied under expert supervision. Some small species of reptiles can be found by raking over leaf litter: avoid doing this by hand in case of encountering venomous animals. A useful technique is to cover likely rocks with netting, then to flip the rock over and any reptiles present tend to get caught in the net and can then be collected.

For reptiles seen at a distance, grabbers (Figure 4.59a) or nooses can be used to slowly grab them. Nooses can be made from modified fishing rods (Figure 4.59b). Those used for small lizards should employ thin thread such as dental floss, although those used for snakes and crocodiles should be much more substantial. Carnivorous lizards can sometimes be captured using insects tied to fishing lines and dropped near to the animal's head. Animals in trees can be caught by knocking them off the branches into nets held or placed below. At even greater distances, some researchers use catapults to fire lightweight missiles (e.g. seeds), or fire heavy elastic bands at the tail base to temporally stun the animal enabling it to be quickly picked up. This technique can kill animals that are hit on the head and should only be used by experts. In fact, all of these capture techniques require expert training to prevent damage to the animals.

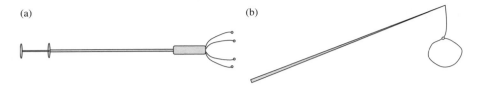

Figure 4.59 Equipment for catching reptiles at a distance
(a) Grabber; (b) noose.
Care should be taken when using either of these techniques (especially the nooses) to avoid damage to the animal. Training and practice are needed to use these techniques effectively. Avoid catching reptiles by the tail since this may be shed. Animals should be placed in soft cloth bags or in buckets or other large containers as soon as possible after capture.

Traps

There is a range of other techniques available to catch reptiles, some of which are similar to those used for amphibians. These include pitfall traps (e.g. side-flap bucket traps – Figure 4.54) with or without drift fencing; nets in trees for arboreal lizards; box, hoop and funnel traps; and the use of refuges.

When using pitfall traps, several buckets sunk in the ground with some leaf litter in the bottom are often adequate. However, these traps should be checked regularly, preferably every 4–6 hours during daylight hours. Where drift fences of plastic or sheet metal are placed between sets of pitfall traps, this can greatly increase the numbers of animals caught. Traps may be baited with food (including wet fruit or tinned fish), although this will tend to only attract omnivorous species. Box, hoop and funnel traps can be used alone or in conjunction with drift fencing.

The use of refuges can be successful. Tin sheeting has also been used to provide basking sites and aid capture of both snakes and lizards. Some researchers have found that by standardising the size of the sheeting, a more objective measure of their density can be established. Other materials (e.g. carpet, heavy cloth and wood) can be colonised by small reptiles. Pipe traps also act as refuges (Figure 4.60).

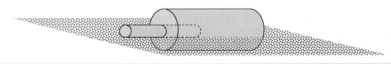

Figure 4.60 Pipe trap
The larger pipe is half buried in the ground and any animal entering the smaller pipe will be caught. These traps should be checked regularly to prevent the animals from suffering.

Sticky traps can capture smaller species. Commercial traps use non-toxic glues that can be removed using vegetable oil. Such traps must be checked regularly and can collect many non-target species as well as debris, especially in windy conditions. Delicate animals may be harmed by such traps and they should not be used for geckoes.

Marking individuals

A variety of marking techniques have been used to estimate population size using the mark–release–recapture technique. Although some authors advocate the removal of lizard toes using nail clippers, more humane methods include clipping scales, or cauterising them with a hand-held soldering iron, or tattooing. In other studies, researchers have made notches in the tail crests or sewn coloured beads into the tail crest (which is difficult, if not impossible, on lizards with slender tails, or those under about 40 mm). Some lizards may retain an elasticated waistband, and knee tags and non-toxic paints may also be useful, although the latter will only be a short-term measure if the skin is likely to be shed in the near future. Be aware that most of these techniques are not permanent, reinforcing the view by some that toe clipping can be justified for long-term studies. Using a technique called spooling, the initial movements following release of animals such as tortoises can be ascertained. Here a spool of thread is attached to the animal's back and as the animal runs the thread is played out and can be followed for a few hundred metres depending on the length of the thread. Telemetry (see p. 108) is also useful when monitoring larger lizards and tortoises. Passive integrated transponder (PIT) tags (see p. 109) can also be useful in marking reptiles. The overriding problem with mark–release–recapture of reptiles is that the recapture rate will probably be low, inviting criticism of any estimation of population size. Woodbury (1956) and Gent and Gibson (2003) give more details on marking reptiles.

Birds

Photo CPW

Birds are one of the best-studied groups of all animals and as a result there are a number of comprehensive guides to their identification and ecology even in areas where otherwise the biodiversity is relatively poorly understood. In many countries, they are also the subject of some of the highest levels of protection, not only for the birds themselves, but also their eggs, nests and habitats. In Britain, the Royal Society for the Protection of Birds (RSPB), the British Trust for Ornithology (BTO) and the Wildfowl and Wetlands Trust (WWT) combined are arguably the most powerful pressure group in wildlife conservation. Such protection often means that birds are not allowed to be captured (or indeed disturbed in any way) except under licence. Such licences may be for particular species or for specific techniques. Working on birds with appropriate licences does mean that researchers have a wider choice of methods available to them. Such techniques include the use of mark–release–recapture of marked birds, radio-tracking, and taking measurements from birds, including identifying the sex (which in some species requires close examination of an individual), breeding condition, moult condition, mass, length of bill, length of tail, and length of wing (Figure 4.61).

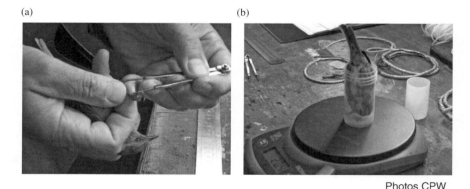
Photos CPW

Figure 4.61 Measuring captured birds
(a) Tarsus length; (b) mass.

Although the majority of birds are relatively easy to identify, some of the commonest species confuse even the most experienced birders (e.g. gulls in flight). Since some birds are obvious and can be of interest to many people, local knowledge can be very useful to researchers. In some places this can take the form of census lists that may have been kept for many years. In other situations local communities (from farmers to indigenous tribes folk) can provide invaluable information about the distribution and habits of a species of interest.

Although the techniques involving the capture of birds are often restricted to researchers who have appropriate training and licences, data that were gathered using such techniques may be available to less experienced researchers. Even if this is not the case, substantial data can be obtained from census methods, although care should be taken to ensure that these follow standard protocols. Gilbert, Gibbons and Evans (1998), Bibby, Jones and Marsden (1999) and Bibby et al. (2000) give more details of these types of techniques for many species of birds. Some groups of birds require more specialised techniques, for example Walsh et al. (1995) give information on monitoring seabirds.

Direct observation

At the very least, studying birds requires a pair of binoculars, but telescopes are also useful, especially for open habitat studies such as wetlands, coasts and observation from hides (refuges from which observers can watch birds without disturbing them). In woodland the majority of your contacts with birds will be by call, so a thorough knowledge of songs and calls is necessary before beginning any such survey. Permanent hides are frequently built in nature reserves and other areas where bird watchers congregate (Figure 4.62). This principle can be used with temporary hides for use during your study. These can either be ready-made tent-like structures, or be home-made from appropriately camouflaged cloth or netting, or even vegetation.

Direct observation is by far the most popular way of studying birds, and is one of the few bird activities that does not usually require a licence (although care should be taken in using binoculars in sensitive areas – e.g. near airports or on military land). The aim of direct counts is to count all individuals in a specified area. Although direct counts are one of the most useful ornithological methods, the approach to counting varies depending on whether birds roost, flock, are migratory or display leking behaviour. This section details some of the commonly used methods that are applicable for a range of bird species. Bibby et al. (2000) give more details on techniques appropriate for special situations, and Gilbert et al. (1998) discuss methods that are appropriate for a range of individual species.

Photo CPW

Figure 4.62 Permanent bird hide

Some species of birds are hidden from view, either because they hide in low vegetation or are only really active in the canopy. Walking through dense vegetation may flush birds out from their hiding places, although care should be taken to avoid causing undue distress and such disturbance may be illegal, especially in the breeding season. For those species that feed or roost in the canopy, where it is difficult to see them from the ground, using natural high ground such as hilltops or artificial structures, including bird observation towers (Figure 4.63) or canopy walkways, can be useful.

Natural aggregations of some species of birds occur, for example at colonies, water sources, leks and clay licks. Concentrations of some species can be high at these sites. Water sources can be a good point at which to observe birds, both those using the water to feed in and those, especially early in the morning, that use the water for drinking and bathing. Leks are gatherings of males of some species of animals (e.g. black grouse) to compete for females. Clay licks are areas that some birds such as parrots frequent, eating the clay because of its mineral content. In addition to natural gatherings, birds can be attracted to bait (e.g. in feeders). Sound can also be used to attract birds. The fact that birds call both for attracting mates and displaying territorial behaviour can be exploited by replaying recordings and monitoring bird activity, including their vocal responses.

Photo EMS Photo CPW

Figure 4.63 Bird observation tower
(a) Observation tower; (b) view over the forest canopy from an observation tower.

A technique called distance sampling is now used extensively to use observations (whether visual or aural) to estimate the densities of birds (and an increasing number of other taxa). This involves a series of observations at points (or along transects) where not only the bird but also the distance from the point (or line) to the bird is recorded. By using the assumption that all observations at the point (or on the line) will be recorded and that those at increasing distances away will be harder to detect, specialist (free) software (DISTANCE[8]) can be employed to calculate densities (Buckland et al. 2001). For this method it is important that the distances are accurately measured, and range-finders can be used to improve this. See Case Study 4.11 for a discussion of implementing this technique on parrots in tropical forests.

Timed species count

This suite of methods yields relative indices of abundance and is useful for preliminary investigations or very diverse communities. The method is simple: walk at a constant speed through a habitat for a set period (∼1 hour) and note the time at which each species was first seen. Observations of the same species later in the same visit are ignored. The time taken to walk the habitat is subdivided into intervals and birds are scored higher the earlier they are seen. By visiting the habitat several times (>10) an average score for each species can be calculated (see Table 4.7 for further details).

---

[8]http://www.ruwpa.st-and.ac.uk/distance/

Case Study 4.11 Counting parrots

Dr Stuart Marsden is a Reader in Conservation Ecology at Manchester Metropolitan University. He has long had a passion for parrots, the habitats in which they live, and especially efforts to minimise the threats they face. This case study examines the issues facing Stuart when surveying for rare species of parrots in tropical forests.

Photo SJM

Female eclectus parrot

Model organism and sampling problems faced

Around half of the world's 350 or so parrot species are threatened with extinction, mainly from habitat loss and capture for the pet trade. A first step is to find out how many there are, and how their numbers differ in different areas or habitats. This is an incredibly challenging thing to do for a number of reasons. Most parrots live in the canopy of tropical forest, a habitat that is difficult to access and move around in. Parrots are surprisingly cryptic, meaning that their detection is not certain even if they are directly above you in a tree. They also fly around a lot, making it easy to double-count them. Some species are relatively easy to count at traditional roost sites so long as a series of evening counts are made (there is a lot of variability between the numbers flying in to roost on different nights). The more usual way to count parrots is to walk transects or conduct point counts within the forest itself.

How the problem was overcome

Whilst some species are relatively easy to count at roost sites, for other species, Stuart's method of choice is distance sampling, either from points or transects, depending on the habitat (Marsden 1999). One huge issue with counting parrots is that they are rare and seldom encountered. Of course, it is the rarest species that are of most interest to conservationists. This rarity means that long field seasons are needed to generate big enough sample sizes (numbers of encounters) to allow precise density estimation. It might take 4 months or more of solid field work to accumulate 50 encounters with a rare species. Transects seem to work better than point counts in very rare species as you get to cover more ground per day – but they come with their own set of problems. With transects, it becomes less easy to guarantee that all parrots directly above you are actually detected – missing these birds underestimates density. It is also more difficult to survey for parrots as you move along a transect (some areas are full of snakes and other hazards) and also it is hard to position transects in random places. Random positioning of survey effort is crucial in parrot surveys as they are patchily distributed. This means that if all the transects or points are in, for example, valley bottoms, then Stuart gets an overly high-density estimate simply because parrots like valley bottoms too!

The next issue is making sure that all parrots close to the observer are detected. This means that transects must be walked very slowly and carefully, or points surveyed for long count periods (at least 6 minutes). Of course, the longer he spends at a point count, then the greater is the probability that a parrot will fly into the plot – these individuals must be excluded from the analysis as they don't belong to the 'snapshot' of birds present when the count period started. There are many other problems of parrot surveying. Time of day, species identification, estimations or measurements of the distance from the observer to the parrot contact, effects of the observer on parrot behaviour (especially in areas where parrots are trapped) can all affect results.

Stuart's problems do not end when he returns to his laboratory with the data. Distance sampling is a tricky method that requires considerable knowledge and skill to analyse the data correctly. The DISTANCE program allows him to calculate parrot densities in different habitat types or areas and to estimate the sizes of parrot populations with estimates of precision, for example standard errors or 95% confidence intervals. Hard work it may be, but these results are incredibly important in guiding habitat management for endangered parrots and assessing the likely threat to populations of trapping for the cage bird trade.

Advice for students wanting to study parrots

Stuart suggests that anyone researching rare species such as parrots needs to be aware of the timescales involved in collecting meaningful data. This involves careful experimental design, an awareness of the risks involved in working in environments such as rainforests, gaining a sound knowledge of the identification of the species and developing appropriate skills in both field work and data analysis. The resulting research can be very rewarding, dealing as it does with species and habitats of some considerable conservation value.

Table 4.7 Example of timed species counts

Using 12 surveys, each of 1 hour (divided into six 10-minute intervals), birds are scored on the basis of when they were first seen. A blue tit seen or heard in the first 10 minutes is scored 6, a robin after 35 minutes is given a score of 3 and a species not seen in a particular survey is scored at zero.

Bird species	Scores for 12 surveys	Average relative index of abundance
Blue tit	6, 6, 6, 6, 6, 6, 5, 6, 6, 6, 4, 6	5.75
Robin	5, 3, 5, 4, 5, 4, 3, 3, 4, 5, 6, 2	4.08
Heron	1, 0, 0, 0, 0, 0, 0, 0, 0, 0, 3, 0	0.33

Common Bird Census/Breeding Bird Survey

One standard method for mapping the territories of birds was the Common Birds Census (CBC – see Box 4.5). The CBC has been used since 1962 but is considered labour-intensive and has been replaced by the Breeding Bird Survey (BBS), which uses only two transects and needs just two visits over the breeding season – see Figure 4.64.

Box 4.5   Common birds census for territory mapping

Although now replaced, some projects may use the large body of data collected using the Common Bird Census (CBC) method and so a knowledge of how the data were collected is useful. The CBC involved:

- Identifying a study area of at least 15–20 ha for closed habitats (e.g. woodlands) and 60–80 ha for open habitats (e.g. grasslands) using a map of a scale of at least 1:25 000.

- Choosing blocks of land rather than strips to avoid edge effects.

- Visiting the area to establish whether it is suitable for the investigation and making a short list of the birds in the area.

- Planning about 10 equally spaced (e.g. every 10 days) morning visits spread across the season. Avoiding dawn (which can be confusing as many birds sing at this time), survey during early morning finishing by midday (woodlands of around 15–20 ha or open habitats of 60–80 ha take about 4 hours to survey – depending on the experience of the observer).

- Producing a habitat map (e.g. using the NVC method – see Chapter 2). Territory mapping required several copies of the habitat map (one for each visit and ultimately one for each species), and assumed a knowledge of the breeding season of the birds in the habitat (e.g. mid-March to mid-June for deciduous woodland).

- Covering the whole area in each visit (using a grid system to ensure systematic coverage – preferably surveying no more than 50 m away from any point on the map during the survey).

- Using a planned route, but changing or reversing it on subsequent visits.

- Using standard species and activity symbols (see Marchant 1983 for further details) to note every behaviour observed on each visit.

- Analysis of the maps was dependent upon the observer correctly identifying individuals without double-recording (consult Bibby et al. (2000) for a discussion of the complexities of such analysis).

- Controlling biases where possible (e.g. time of day and walking speed) and noting those that could not be controlled (e.g. the weather).

See http://www.bto.org/survey/complete/cbc.htm for further details of the use of the CBC in the UK.

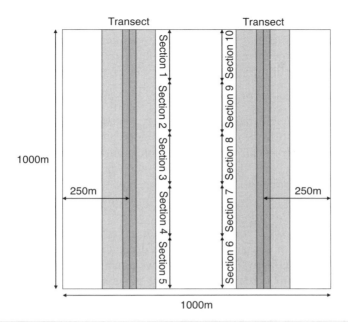

Figure 4.64 Transect layout for Breeding Bird Survey
Birds are counted from 10 sections (200 m long), along two transects, within a 1 km × 1 km square. The transects should be about 500 m apart (and no closer than 200 m at any point) and 250 m from each edge. The actual layout will largely depend on the terrain and the habitat type being surveyed (transects need to be further apart in open areas than they do in woodlands to prevent double counting). Each transect is walked at a steady pace taking about 45 minutes per transect and birds are counted in each zone, separated into those seen (or heard) within three zones: 25 m from the transect (dark grey areas); 100 m from the transect (light grey areas); and more than 100 m from the transect (white areas). Standardised recording for the UK should be done twice at least 4 weeks apart (in early April to mid May and again in mid May to late June) starting before 09:00 and ideally between 06:00 and 07:00. Further information can be obtained from http://www.bto.org/bbs/index.htm.

Point counts

Point counts are made from recording stations that have either been randomly or systematically chosen across a given study site. They can be used to give an index for each species or, if counts have been made within sites of known size, a measure of population density can be obtained. The observer simply arrives quietly at a station and counts the number of birds that are seen and/or heard over a set time period (5–10 minutes is recommended). You should try to ensure that every bird close to the plot's central point is detected (those far away are not so important), and avoid double counting any individual or counting birds that move into the plot once the count period has started as this gives inflated densities. You should not use point counts in bad weather, since this may cause birds to be less detectable.

Point counts can be a useful alternative to transects, since you can stand still and concentrate on identification. Moreover, the points can be easily distributed in discrete habitats without the concerns about edge effects that can bias transects. One serious concern is that of the effect of distance. For example, in a particular habitat at 40 m, observations may become difficult and small birds can be missed, whereas at 80 m nearly all birds could have been overlooked. The best distance for such observations is of course dependent on habitat structure: in open areas (e.g. grasslands) it will be easier to see birds than in more closed environments (e.g. woodlands), making habitat comparisons troublesome. In addition, even when a study is confined to a woodland habitat, the observational distance will vary with season. A distance measure can be incorporated into point counts; however, this requires some skill to implement (see Bibby et al. 2000 for the implications and the many different approaches to this). Various assumptions have been made about point counting, which include that birds do not approach the observer, flee when disturbed, may not behave independently of one-another, are fully detectable close to the central point of the plot (but not further away), and do not move much during recording. Be aware of these issues, and minimise such aspects where possible.

Transect line counts

This method is an alternative to point counts; the major difference is that line counts are best suited to open habitats but need not be confined to observations on foot, as they can equally well be carried out while driving, sailing or flying. It is argued that transects are more accurate than point counts, simply because the observer can potentially come closer to more distant birds (Bibby et al. 2000). Some studies combine the two approaches with transects across environmental gradients being supplemented by point stations at specified distances or within specific habitat types. Usually several transects are established, each at least 200–500 m apart depending on the structure of the habitat. The length of the transect may vary, but 100 m would be considered the minimum appropriate size. Walk, drive, sail or fly each transect at a slow pace recording each species without double counting. A distance measure can be incorporated into line counts: although useful, this requires some skill and training to implement. In this case, an assumption is that every bird that is on the transect line as you pass is detected.

Flush counts

This is similar to point counting in that birds are counted from one position, but here they are disturbed by flapping your arms or making a noise. This method is particularly useful for cryptic game birds and some seabirds. However, disturbing nesting birds is an offence in some countries and the method is, therefore, not used frequently. Furthermore, even for common birds this method may increase the chances of predation by predators that take eggs from unguarded nests. Flush counts should be spread well apart to avoid double counting.

Indirect methods

The tracks of some bird species can be found in soft substrates and in some cases can be used to identify individuals, as can the dropped feathers of some species. Feathers can also be used in studies of migration, where the relative proportions of different stable isotopes of carbon and hydrogen reflect the proportions of such isotopes found in the food consumed by the animal. Since these isotopes vary naturally across different areas, we can use these proportions to identify the location where the animals fed and grew (see Rubenstein and Hobson 2004; Newton 2010, for further details).

Droppings may also be used for the identification, distribution and possibly the abundance of, if not species, groups of species (e.g. wildfowl and game birds) of sufficient size (Figure 4.65). The methods are largely the same as for larger mammals except that transects are rarely used. Brown et al. (2003) provides identification of birds from their droppings for the UK and Europe, whereas Elbroch, Marks and Boretos (2001) cover North America. Be aware that droppings are easily trampled by other birds, that much of the associated survey work is weather dependent, and that this method is unlikely to be successful if the density of individuals is low.

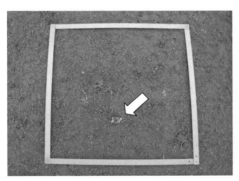

Photos CPW

Figure 4.65 Goose droppings surveyed using a quadrat
To survey bird droppings, first establish several (~20) random quadrats (0.5 m × 0.5 m up to 2 m × 2 m depending on the size and density of droppings) in areas where the birds are known to feed and then clear the droppings from the area. Return a week or so later and count the density of droppings to give an estimated figure of density. The mean number of droppings per unit area per day can be calculated from the number of droppings counted, the number of days between clearing and counting, and the area of each sample plot. If it is possible to estimate the dropping rate of the species (some researchers advocate watching ~20 birds each for a set time, for example 15–30 minutes, to generate a mean dropping rate) then the dropping number can be expressed in terms of 'bird-days'.

Pellets containing undigested remains of food are produced by a wide range of species, including birds of prey, owls, crows, gulls, herons and storks. As well as indicating the presence of the bird, these pellets can be dissected and used to identify the prey, with hard-bodied items often remaining reasonably intact. Pellets can be kept in 70% methyl alcohol to prevent further decomposition. Bang and Dahlstrøm (2006) cover the identification of the pellets from a broad range of species, whereas Yalden (2009) gives details of how to identify the contents of pellets produced by British owl species. Some birds (e.g. seabirds) will regurgitate their food when handled and this can also be used to identify prey items. Other evidence that can indicate the presence of particular species includes the remains of their prey if the type, size and state of the corpse is species specific (as with sparrowhawks).

Counting nests at a distance

This is particularly useful for seabirds (especially gulls, terns and auks), rooks and herons to estimate the colony size of colonial birds. The theory is simple but in practice nest counting is hard because the weight of numbers can confuse and obscure the vision of observers. Furthermore, you have to discriminate between occupied and unoccupied nests to avoid falsely inflating the estimates. The best method is to subdivide the area using small landmarks like rock extrusions, and count nests in each block. Photographs can be used in some circumstances, but these can be expensive if they have to be taken from an aerial position. The number of apparently occupied nests or burrows is half of the actual population size, as birds nest in pairs. **Note:** see Walsh et al. (1995) or Bibby et al. (2000) for more information on specific counting techniques for named species or groups.

Nest searching can sometimes be employed to assess clutch size and survival rates, but disturbance of the nest and/or occupants must be avoided. Mirrors held on a stick can negate the need to climb trees and enable you to get very close to nests in order to check eggs. Nest cameras are an alternative when nest boxes are used or when birds are likely to return to an old nest site. The presence of broken eggshells below nests can indicate successful (or if dead nestlings are found, unsuccessful) fledging.

Bird song

Identifying birds from their song is something of a skill, especially when the bird is not visible. Unless extremely experienced, you are recommended to limit your study to a few bird species or to work within a clearly defined habitat (e.g. deciduous woodland) to limit the number of species possibilities. Many guides to bird song are available with samples of the species' songs and calls as audio-files (e.g. Roche 2003 for Britain and Europe) and there are extensive collections available from the Cornell Laboratory of Ornithology[9] for many areas of North, Central and South America and from the British Library sound archives for British Birds.[10] The study of bird song is particularly useful if

---

[9]http://www.birds.cornell.edu
[10]https://sounds.bl.uk/Browse.aspx?category=Environment&collection=British-wildlife-recordings

the species in question displays territorial behaviour during the breeding season. This includes some ducks, game birds, raptors and most passerines. Currently, research is being undertaken to refine techniques to automatically record and identify birds from their calls (e.g. see Gage, Kasten and Joo 2009).

Capture techniques

Unlike many other groups of animals, the capture of birds is often restricted by law (see Box 4.6). This usually means that only experienced and licensed researchers are able to use many of the techniques described here. This is as a result of the conservation status of many birds in many countries as well as because of pragmatic considerations regarding how delicate birds can be and how easy it can be to damage them either in traps or in the hand. By joining an appropriate group (e.g. a ringing group or wader group), it may be possible to obtain such experience and licences. Being that different species may require different trapping and handling techniques, it is easier to become licensed for working on a single species or species group (e.g. mallards or ducks in general). Commonly used trapping methods are described here so that you have an appreciation of how birds can be captured, measured and marked.

Box 4.6 Restrictions on handling birds

Birds are afforded extremely good protection by both national and international legislation. You are advised to read the RSPB's publications on the restrictions that will apply to trapping or handling of specimens.

For England and Wales: http://www.rspb.org.uk/Images/WBATL_tcm9-132998.pdf

For Scotland: http://www.rspb.org.uk/Images/WbatlScotland_tcm9-202599.pdf

For the regulations concerning the capture of birds in countries other than the UK see:
European bird directive: http://ec.europa.eu/environment/nature/legislation/birdsdirective/index_en.htm
USA: http://www.fws.gov
Canada: http://www.cws-scf.ec.gc.ca/enforce/law_e.cfm
Australia: http://www.environment.gov.au

Of particular legislative relevance in the UK is Schedule 1 of the Wildlife and Countryside Act 1981 that controls activities towards rare breeding species. If you wish to work with a notified species, you can apply for a licence from Natural England or Scottish Natural Heritage (or their equivalent), or through the BTO. Additionally, Annex 1 of the EC Wild Birds Directive ensures habitat protection for wild birds and controls the trapping, killing and selling of birds – consult the RSPB or BTO if in any doubt about these and other issues.

Handling and trapping of **any** wild birds, except for some pest species under certain circumstances, is illegal without a licence issued by the BTO. The process of obtaining a licence can be rather protracted. The licence and associated training can be easier to obtain if a single species study is planned.

Illegal handling, trapping or interference with the nests or eggs of birds can be punishable by a fine.

Mist netting

Mist netting can provide information about population size (e.g. changes in the numbers of adults between years), productivity (e.g. the ratio of adults to juveniles late in the breeding season), survival (e.g. between year trapping rates of ringed birds) and fitness (e.g. observation of parasite load, health, etc.). Mist nets are made of fine mesh and comprise loose netting held between taught strings. They are set up along flight ways and when birds fly into them, they become entangled in the loose netting that acts as a pocket (Figure 4.66). Regular patrolling of the nets is essential to prevent birds from suffering. To standardise the method for

Photos CPW

Figure 4.66 Mist netting
(a) Mist nets set; (b) greenfinch caught in net; (c)–(e) removing bird from net.

capture per unit effort, the same site, length and type of net is used over at least two seasons. Netting must take place at the same time of day for each catch, normally lasting about 6 hours each time. Mist netting can be validated by comparing net catches with point counts (see p. 207) that are taken concurrently. However, the size of the net (e.g. 30 mm or 36 mm) and the position in which it is set will determine the type of catch as species weighing less than 16 g are caught more frequently in 30 mm nets. Thus catches should only be compared between studies when the same size net is used. You should be aware that some species are net shy whereas others, like some passerines and warblers, can be easily trapped. In sites such as tropical forests with high tree canopies, such nets can only be readily placed in the under-storey and hence are biased towards species frequenting this part of the habitat.

Propelled nets

Alternative trapping methods are used to catch birds in flocks at some distance from the researcher (Figure 4.67). These are really only used to trap birds for ringing and cannot yield survival or productivity estimates like mist nets because the capture per unit effort cannot be standardised. Additional licences may be required if trapping methods other than mist nets are used. Clap nets and whoosh nets are sprung traps that are placed over baited areas and are the least powerful of the alternative methods (i.e. do not fire the nets very far from the propulsion zone). Phutt nets use pressurised air propulsion and are more powerful than clap nets. Cannon and rocket nets use explosive propulsion and can target large birds some distance away and are the most powerful of all. Phutt, cannon and rocket nets are only used in areas such as wetlands where visibility is high. In addition to using such nets over bait, decoys can also be used to attract (especially flocking) birds.

Marking individuals

Ringing is one of the most commonly used methods of marking birds. Here specifically sized rings (appropriate for a given species) are placed on the leg just above the foot (Figure 4.68a). These are loose enough so they do not chafe but tight enough not to fall over the foot. In Britain, ringing birds cannot be attempted without a permit. Both metal and coloured plastic rings are used. For small-scale ecological investigations, colour rings are favoured, but metal rings with unique codes are invaluable for migration and/or long-term studies. Colour rings do not require that the bird is recaptured because they are visible at a distance, whereas recapture of birds with metal rings is necessary in order to confirm the identification number. Single colour rings can indicate that the bird has been captured at a

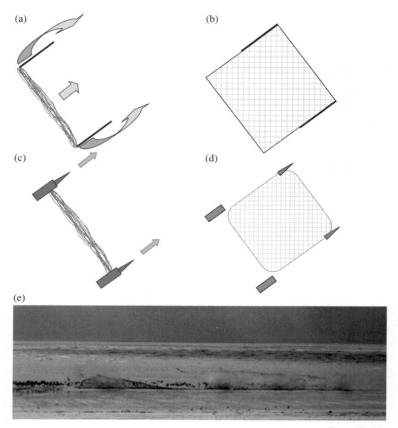

Figure 4.67 Propelled nets
(a) Clap net set; (b) Clap net launched.
The net is fixed at the lower axis and held by elastic or rods attached to elastic stretched along the ground (dark bars) that, when released, flip the net over a flock (whoosh traps work in a similar way to clap nets).
(c) Rocket net set; (d) Rocket net launched.
The net is attached to two or more rockets that, when fired, carry the net over the flock of birds (phutt nets and cannon nets work in a similar way to rocket nets).
(e) Cannon net being fired.

specific time and place. The use of several rings (involving different orders of different colours – see Figure 4.69) allows individual identification of a large number of birds of each recognisable species (over 100 individuals can be recognised using just three rings consisting of different combinations of five colours – double this if both legs are used). Other marking techniques involve the use of neck

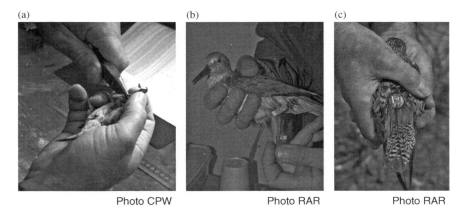

Figure 4.68 Marking birds
(a) Using a standard metal ring; (b) using colour rings and tags; (c) using a radio tag.

rings in species such as geese, tags attached to the leg (Figure 4.68b) or pinned through the wing avoiding the wing bones (also known as patagial tags), dyes (which can be useful on light or white coloured birds, e.g. gulls), tail streamers (pieces of tape that project slightly from the tail) and radio-tracking (Figure 4.68c). See Cottam (1956), Bibby et al. (2000) and Balmer et al. (2008) for further details on marking and ringing birds.

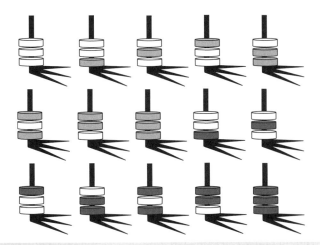

Figure 4.69 Use of colour rings
Here 15 individual birds have been marked using one leg only, and three rings with two colour combinations of three possible colours.

Mammals

Photo CPW

Mammals are often selected for ecological studies because they are attractive and thought to be easily observable. The reality is that mammals may occur at low densities and many are seldom seen, with some notable exceptions such as rabbits and domesticated animals. Working on mammals in your own country can be difficult enough, and even more problems may occur if you intend to work on species overseas (unless this involves working in a zoo or wildlife park). In many countries, certain species of mammal (including bat species) are protected by law and you should check with the appropriate authorities to see if a licence is required for your research. In general it is worth not only reading up on comparable studies before designing your project, but also contacting local mammalogists and trying to make use of local expertise, perhaps by joining an existing large scale project where you could establish a small niche of your own. Several texts cover the methods that can be used to monitor mammals: for example, Barnett and Dutton (1995) deal with small mammals, Barlow (1999) deals with bats, and Wilson et al. (1996) cover standard methods for a variety of mammals worldwide, whilst Harris and Yalden (2008) give details of field signs and descriptions for British mammals.

This section deals with both wild and feral mammals, but excludes sea mammals, although seals can be observed and counted when basking on the shore using some of these methods. Different techniques should be used for different sized animals. Larger mammals over 5 kg are trapped with difficulty and for the purposes of this book the techniques employed for such animals are limited to direct observation. Many small to medium-sized species of mammal (5 g to 5 kg) can be caught using cage traps. Small mammals (under 50 g) tend to be caught in live traps or snap traps depending on whether they are required alive or dead. Bats, being flighted, are another case altogether. Although bats can be captured, the majority of surveys and recommended techniques employ observations.

If mammals are captured, measurements that are taken usually include the sex, mass, length with and without the tail (if it is a long-tailed mammal), forearm, ear and tragus length (for bats), as well as the length of horns/antlers, and the sex and breeding condition where appropriate. Some studies examine regurgitated food with or without the use of emetics, but this may require an appropriate licence. Collecting samples of dung (scats) provides a better source for diet analysis, but be aware of potential health hazards. Use rubber gloves and plastic spoons, and keep each sample, in a freezer, in a separate polythene bag. The gut content of road kills can be a suitable alternative if they have not decayed too far.

Direct observation

The study techniques needed to study larger mammals (e.g. deer, foxes and badgers) are quite different from those of smaller mammals, which can be trapped and handled more easily. Although large mammals can be captured, more usually the favoured method for species like deer is direct counting, which can also be used to record data on behaviour, age and sex composition of herds. One drawback of direct counts is that for many species of deer, for example, woodland is the favoured habitat. This can make observation difficult, leading to biased results. However, deer use the very early morning to graze in open spaces and it is at this time that direct counts are invaluable. Deer are bashful creatures so wear appropriate stalking clothes (i.e. dull colours) and approach the animals upwind, slowly, keeping low to avoid them spotting you so that their behaviour is not influenced by your presence (Figure 4.70). Many reserves have hides that can aid observation, although these are only really useful at dawn and dusk. Large animals with body markings that make them individually identifiable enable life histories to be constructed with some accuracy over a long period of time. The best approach to direct counting of large areas is to divide the area under investigation into a grid system, with many observers counting numbers simultaneously. Care must be taken to avoid putting yourself at risk from large mammals, for example adult male deer can be quite aggressive during the breeding season.

Photos CPW

Figure 4.70 Deer becoming aware of the observer's presence

Direct counts of nocturnal animals such as badgers and foxes are often not feasible and much behaviour may be missed unless night vision equipment is available. However, torches with a red filter have been used with some success in a few studies. If the facilities are available and there are no problems with access to study sites, then some nocturnal mammals such as foxes and hares can be spot-counted by torchlight. The procedure is simple: drive a specified route at night-time in a 4 × 4 truck in an attempt to cover a large area of land. Foxes, on hearing the vehicle, head for cover and at this point individuals may be identified in the headlights of the truck. You should aim to cover significant patches of land on a frequent basis as foxes may occur at low densities.

Camera traps can be used to identify the presence (and activity levels) of mammals that are less frequently seen by direct observation (Figure 4.71). Here cameras are left (usually in secure housings to prevent theft or vandalism – see Figure 4.58) to record any animal that triggers the mechanism. Such triggers can be pressure switches, infrared sensors or motion detectors. Camera traps work best when set up in areas known to be frequented by the species under study or when used in conjunction with appropriate bait. For animals that have unique markings, pairs of cameras can be used to photograph each side of the animal so that individuals can be identified.

Photos PH

Figure 4.71 Animals caught using camera traps in tropical forest
(a) Tapir; (b) jaguar.

Animals such as badgers can be counted with ease if a night-scope or camera is set up near a set, although tracking badgers on their forays is somewhat harder and indirect methods may be favoured. In Britain, badgers are protected in law and it is an offence to injure, kill or possess a badger. Additionally, interference of any kind with the sett is also an offence. A licence may be required for any work on this species that involves such interference, though mapping territories and assessing their densities in different habitats is acceptable. The behaviour of ground-active animals such as baboons, lions and rhinoceros (both in the wild and in zoos or wildlife parks) may be monitored by direct observation, as can that of domesticated animals (e.g. sheep). Arboreal mammals such as monkeys can be surveyed from the ground, or (if canopy-active species) from observation towers or natural high points that overlook the canopy.

As nocturnal creatures, bats can offer exciting opportunities for research. Many species of bats can be observed flying and feeding at night. Bats are often protected in law, may be delicate creatures to handle and can bite the unwary. As such, appropriate training and guidance should be sought before risking disturbing their roosts. See Barlow (1999), Mitchell-Jones and McLeish (2004) and the Bat Conservation Trust (2007) for further details about specialist techniques used to monitor bats. To observe bats mating or giving birth and communal interactions, the bat roost must be located. Some bats have summer and winter roosts that, depending on the species, may vary from gables and trees to caves. For example, the greater horseshoe bat (*Rhinolophus ferrumequinum*) roosts in attics in the summer but chooses mines, cellars and tunnels in winter. Roosts can be classified into four types: hibernacula, transitory, nursery or mating (see Harris and Yalden 2008 for a full account). Entering a roost during daytime with a torch fitted with a red filter to avoid disturbing them will enable you to count them (but needs a licence in the UK). However, like many direct counts there are biases. To counteract inaccuracies make sure that all exit holes have been identified, then use a tally counter and preferably additional help to census the population quickly. It may be better to count bats as they depart to feed around dusk. Obviously this will only give an approximation of the roost size depending on the proportion leaving at any one time and the number of exit holes. If you wish to observe mating behaviour (swarming) in the UK, September is the best time, although some bats continue right through the winter. Be aware that in some countries, bats hibernate over winter.

Indirect methods

Specific tracks and signs left by mammals can be used to identify the presence of particular species. Such field signs can include not only tracks, droppings and hair, but also runways, latrines, nests and burrows, and remnants of their meals (e.g. nuts and bark chewed in a particular way, clipped grasses and food caches). Animals' tracks and signs (including calls) are useful for mapping territories; for example, triangulation mapping to establish the breeding territories of barking foxes, and gnawed hazelnuts to confirm the presence of dormice. In snowy conditions, urine staining helps in mapping territories of foxes for example, but is of limited use for most mammal species. The presence and abundance of field voles, for example, may be assessed by checking quadrats laid in grassland at appropriate densities and spacing, for the occurrence of tunnels, piles of droppings and heaps of grass clippings (evidence of feeding). Strachan (2010) and Harris and Yalden (2008) detail the field signs that can be used in monitoring the presence of British mammals, and Bang and Dahlstrøm (2006) and Elbroch (2003) cover European and North American mammals respectively.

Tracks left in soft substrates (e.g. mud and sand) can be identifiable to species level and are particularly useful for mammals that spend some of their time in or next to water (e.g. water voles, mink and otters), or heavy animals (e.g. deer) that make a deep impression in the soil. The imprints made by these mammals can be very clear, although it can be difficult to tell the age of the imprint. You can standardise the field technique by using a fine rake to 'clear' the soil of any impressions before your survey begins. Then, establish a quadrat or transect in the area that is surveyed regularly. Remember to rake the tracks clear after each field visit. Tracking tunnels can be used in areas where small mammals are active. These can be made out of square section drainpipe with a pad of ink in the centre and strips of paper to receive the inky footprints at each end (Figure 4.72). Such tunnels can be placed on the ground or, to monitor arboreal mammals, tied to branches.

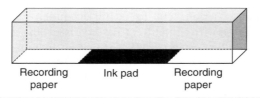

Figure 4.72 Small mammal tracking tunnel

The faeces of mammals can sometimes be recognised as specific to particular species or groups of species. For some species (e.g. hedgehogs), droppings are easily recognisable (see Bang and Dahlstrøm 2006; Chame 2003) and are widely used to analyse their diet, although others (e.g. deer droppings) can be difficult to separate between species, and some (e.g. shrews, bats and rodents) are notoriously difficult to identify to species level unless DNA analysis is applied. Depending on the species, droppings can reveal the diet, age, sexual maturity, seasonal activity and territories of an individual.

For many species, counting droppings is the easiest way of monitoring their presence. Systematic sampling of droppings involves either surveying the number of droppings along a strip transect or within quadrats. Ideally establish a transect where the animals are known to be and then remove all droppings in the defined sampling area. Return 2–4 weeks later and count the droppings to give an estimated figure of density. The most reliable studies compare droppings surveys with another technique (e.g. direct observation or trapping) to give a more accurate figure of density. The length of transects and size of quadrats depend on the species of mammal under investigation (e.g. for deer, a 200 m transect or 10 × 10 m quadrat would be sufficient). You need to take into account whether the animal lives in a herd or is a solitary species. When sampling droppings, be aware of the issues surrounding measuring what constitutes one animal's waste. A pile, known as a 'fumet', is discrete but when the animal begins to move creating a trail of pellets, interpretation can become more difficult. For carnivorous animals, the hard parts of the prey may remain relatively intact in the dung and can be used to identify their food. For some species, there are texts that help in the identification of prey items (e.g. Conroy et al. 2005 for otters).

Badgers and other mustelids are excellent subjects to use in dropping surveys. For example, latrine surveys are now routinely used to establish the territory boundaries of badgers. Food (peanuts and syrup) to which indigestible items (e.g. specially made coloured plastic pellets) have been added is left outside the sett. By baiting a series of setts with differently coloured pellets, the home ranges of the badgers can be mapped by recovering the pellets from the latrines (Figure 4.73). This is a powerful technique that works best in high-density populations and requires about a week of feeding, and then at least as long to survey the latrines.

Photo RS

Figure 4.73 Badger dung pit with bait-marked dung

The occurrence of droppings may be used as an indication of the presence of bats. Counting droppings can also give an indication of bat presence and abundance. Here, the droppings are removed and then surveys carried out some weeks later to estimate the number and extent of the droppings. **Note:** the droppings of different species can be very similar and it may be difficult to identify which species are present. Bat droppings can also be examined to identify the undigested remains of insects and other arthropods eaten (see Shiel et al. 1997 for further details).

For many mammals, an alternative to dropping counts is a hair count using hair left on debris or fences (Figure 4.74) or even collected using hair traps (e.g. using sticky pads in tunnels for small mammals) – see Figure 4.74c. Identifying hair is something of a skill and may be quite time-consuming (texts including Debrot et al. 1982, Teerink 1991, and Cowell and Thomas 1999 provide illustrated keys). With closely related species, the cross-section of a hair may need to be examined, making this method not only time-consuming but laborious. Hair traps for small mammals are a simple affair, using 10 cm lengths of PVC tubing of the correct diameter for the animal under investigation with double-sided sticky tape placed inside the tube on the top and sides. Such traps can be placed at key points on the ground or in trees. For larger mammals, double-sided sticky tape can be nailed to fence posts and trees, but should be protected by some form of rain-guard. It may help to bait sites close to the trap to ensure that the animal rubs against the tape. Alternatively, surveys of barbed wire and fence posts in the area can be revealing. This method is particularly useful if animal runs are prevalent in the area. Make sure that you clean all hair away after each survey to prevent double counting on subsequent surveys. The problem with hair traps is that the number of individuals involved cannot be determined. Hair traps merely indicate the presence/absence of a species and, therefore, may be of limited use in statistical analysis.

Figure 4.74 Sampling mammal hair
(a) Badger hair; (b) sheep wool caught on fencing; (c) a hair trap for pine martens.

Although the calls of mammals can indicate their presence (e.g. with foxes), for most species quantifying the population on the basis of their calls is difficult. Bats are an exception to this and both their use of echolocation and, to a lesser extent, communication calls can be used to identify the species present, determine their distribution and examine their behaviour. Bat detectors (Figure 4.75) have been used in two different but complimentary ways: transect use or triangle walks. For transect surveys, at a set time during the evening, a line transect of between 200 and 400 m in length can be walked, along which the number of bat calls and their location along the transect, together with any micro-habitat changes, are recorded. Transects are particularly useful for species (e.g. Daubenton's bat: *Myotis daubentonii*), which hunt along riverbanks. If a number of transects are planned then make sure that the environmental conditions are broadly similar for each one sampled. There are several types of bat detectors that can detect and, in some cases, record bat calls, and these are compared in Table 4.8.

Photo CPW

Figure 4.75 Bat detector (heterodyne system)

Triangle walks (Figure 4.76) are a standard method used to sample bats, and have been adopted as such by the Bat Conservation Trust in the UK. The triangle design is used to survey within each 1-km² square on a 1 : 25 000 map and may not be suitable for all studies, although is invaluable for long-term research into distributions.

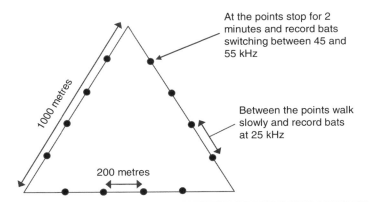

Figure 4.76 Triangle bat walks with frequency settings appropriate for UK bats
See http://www.bats.org.uk/nbmp_tutorials/ for further details of how this approach is used within the Bat Conservation Trust's National Bat Monitoring Programme.

Bat detectors are biased in how they sample bats and some species can be detected from greater distances more easily than others. Furthermore, on evenings when the

Table 4.8 Comparison of bat detector systems

Type of detector	Mechanism	Tuning	Sound analysis	Price
Heterodyne	Filter ultrasound calls with a signal from a high frequency oscillator – creating a low frequency difference sound that we can hear	Tuneable so they can immediately identify bats by their call frequency in the field but need tuning to other frequencies to pick up other bat species	Sounds can be recorded but because the frequency information is not recorded, they cannot be used to create sonograms for sound analysis	Cheap and useful for beginners or when a large number of detectors are needed for surveys with large numbers of researchers
Frequency division	Divide incoming frequencies (usually by 10) to lower them to human range	Detects all frequencies continuously (does not need tuning) and thus does not miss any bats in the field	Sounds can be recorded and used for sound analysis but some information is lost	Cheap to mid-range (cheaper ones do not retain amplitude information)
Time expansion	Digitally record a section of the call, replaying it (usually 10 times) slower so it is audible to humans	Records all frequencies (does not need tuning), but is 'off line' during playback so may miss bats – often combined with heterodyne or frequency division systems to cover this period	Sounds can be recorded and used for sound analysis – produce detailed information that can be used to produce good quality sonograms	Most expensive

relative humidity is much higher than normal, bat detectors may work much less efficiently, making comparisons more difficult over a series of days or weeks. More expensive detectors will tend to have greater detectability. Some bat species cannot be separated by a detector, at which point sonographic analysis from recordings may be the only way to establish identity. If you are working with others, make sure that the detector model is the same for all the researchers in the project. Consider a rotation of detectors across researchers if there is more than one model in use on the project. Social calls are often audible and can also be used to monitor bats and identify the species present (and may need recording for later identification). For UK bats, Harris and Yalden (2008) give more details on calls and echolocation, and Tupinier (1997), Briggs and King (1998) and Russ (1999) provide details of European bat echolocation calls. Sample sequences for many UK species have been collected at Bristol University,[11] and there are a number of sound libraries[12] worldwide that can be useful in helping to identify species that you have recorded, although many such sites require the use of specialist software. The calls of shrews can also be picked up by ultrasonic bat detectors, although (unlike bats) this cannot be used to identify the animals to species.

Capture techniques

Small mammals readily enter traps, of which the most widely used is the Longworth trap (Figures 4.77a and 4.78), although Sherman traps are sometimes used: these have the advantage of being collapsible, but the disadvantage of not having a nest box, so restricting their use in poor weather. The use of the Longworth trap is described in full by Gurnell and Flowerdew (2006), who provide information on trap intensity, location and handling the catch. Longworth traps are made of aluminium and comprise two parts: a tunnel in which the door trigger mechanism is housed and the nest box that attaches to the back of the tunnel when the trap is set. Food, such as wheat or oats, and hay bedding are placed in each trap before the survey begins. Pre-baiting areas before traps are set can be effective in increasing captures, as can carefully positioning food items outside the trap entrance to attract more animals. For quantitative analysis, at least 20 traps should be used for each habitat under investigation (if over 70% of the traps are catching animals then the number of traps should be increased). However, you should avoid using too many traps (e.g. more than 200) as checking traps takes considerable time. Each trap should be spaced well apart: 5 m in grassland, 10–15 m in woodland and 20 m in arable habitats. Traps can be placed in a grid or a line, although Gurnell and Flowerdew (2006) note that

---

[11] http://www.bio.bris.ac.uk/research/bats/calls
[12] For example, http://www.bats.org.uk/pages/bat_sound_library.html and http://www.msb.unm.edu/mammals/batcall/

assessing density is difficult in lines. It is helpful to mark the general location of the traps with a cane, but do not make it too obvious to avoid human interference. It is simple to tell whether a mammal has been caught by a Longworth trap as the trap door has closed shut. Placing the trap in a large polythene bag while breaking open the trap (separating the tunnel from the nest box) will release the animal while isolating it sufficiently to prevent the researcher from being bitten. **Note:** not only can many small mammals bite through plastic bags, you do not always know what you have caught until the trap is emptied: catching a weasel when expecting a field vole can be a surprising experience. Plastic equivalents to Longworth traps are also available (Figure 4.77b). These are lighter and cheaper than their metal counterparts, but are more prone to damage, especially by rodents chewing through them, are less easy to repair, and rain or dew can cause the door to stick.

Photos CPW

Figure 4.77 Small mammal traps
(a) Aluminium Longworth trap; (b) plastic Trip-trap.

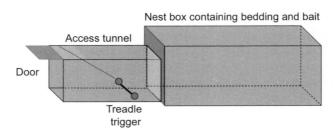

Figure 4.78 Longworth trap for small to medium-sized mammals

Trapping small mammals can be highly successful, but much depends on the siting of the trap. Some small mammals are more active on nights when it is dark, warm and dry, favouring the summer months for survey work. If you plan to survey mammals at other times of the year, you could increase the spacing of traps to counter the decline in activity.

If shrews are caught in Longworth traps, they often perish since if they are deprived from live prey for only a few hours they can die. To avoid this, a 13 mm wide shrew escape hole can be drilled in the Longworth trap to allow them to escape. If shrews are the target animal (for which a licence is required in the UK) they may be caught in Longworth traps baited with blowfly larvae or mealworms, but the traps must be visited every 2–3 hours. The traps can be modified to catch only shrews by placing 'L'-shaped pieces of metal with a 13 mm hole between the treadle and trap door to exclude other species.

Species (e.g. harvest mice, *Micromys minutus*) that spend most of their time in the upper canopy of hedgerow plants, dry reed beds, wheat and oat stems can be difficult to catch. It has been suggested that a tennis ball with a hole cut into the side and attached to a plant stem or to a cane at canopy height in an appropriate habitat can mimic harvest mice nests and may also be used as hair traps (see p. 222). However, some surveys have found this to be relatively unsuccessful in practice, and harvest mice may be best surveyed by checking appropriate vegetation for their characteristic nests (see Harris and Yalden 2008 for details).

Slightly larger animals (e.g. water voles: *Arvicola terrestris* ($=$ *A. amphibius*)) can be captured in a similar trap to Longworth traps, but larger: typical dimensions are 10 cm × 10 cm × 30 cm for the tunnel with a larger nest box attached. Traps are placed next to latrines between April and October baited with 200 g of food (carrots or sliced apple) and with hay bedding in the nest box and should be checked every 4 hours. Remember to secure the trap so that it does not fall into the water with the vole inside. **Note:** since 2008, a licence is required to trap water voles in England and Wales.

Similar (but larger) traps are available to catch rats, made either from metal mesh (e.g. small cage traps such as Havahart traps) or sheet metal (e.g. Sherman traps, which can also be obtained as elongated models to accommodate mammals with very long tails). Other trapping methods for rats include poison, which should be handled carefully, be set in an approved bait container (Figure 4.79) that prevents other animals (including humans) from getting at the poison, and only used by experienced (and where necessary licensed) researchers.

Species (e.g. squirrels) that are particularly cautious about entering closed spaces can be caught in traps with a door at either end, both closing when the mechanism is

Figure 4.79 Poison bait dispenser
Used more in conservation work to remove target animals (e.g. rats) from certain habitats such as islands and near to ground or ledge nesting birds.

triggered. Squirrels are probably best counted indirectly by using their dreys. Squirrel dreys are particularly unmistakeable during autumn after leaf fall (although this technique is less suitable in conifer forest). Furthermore, the animals themselves are also easy to observe at this time before they become torpid during the colder months of winter.

Smaller mustellids (e.g. weasels, *Mustela nivalis;* and stoats, *Mustela erminea*) are among the hardest groups of mammals to trap. You should not approach the study of these animals lightly: their scarcity imposes low sample sizes that are likely to have statistical ramifications. Neither are these animals easy to observe in the wild as they prefer long vegetation: often only their droppings give their presence away. Baited tunnel traps with a see-saw rather than a treadle are sometimes used to catch both stoats and weasels. As the animal enters the tunnel, it crosses the see-saw setting off a mechanism that causes the door to close. Tracking tunnels, made more attractive with the scent of prey or conspecifics, offer an alternative to trapping.

Subterranean mammals (e.g. moles, *Talpa europaea*) can be caught using box or tunnel (Duffus or half-barrel traps) traps, but these have to be dug out each time they have to be checked. Scissor traps have arms that stand proud of the soil and that open once an animal has been caught. Many commercial traps for moles (e.g. scissor traps and barrel traps) are designed to kill them (Figure 4.80), although some may be adapted for live capture, which must then be checked very regularly to prevent caught animals from suffering.

Figure 4.80 Mole traps
(a) Classic scissor trap; (b) Talpex type trap (professional scissor trap) set; (c) Duffus (tunnel or half barrel) trap; (d) these traps are set within the mole runs and kill the animal when sprung.

Bats can be trapped in mist nets (see p. 212) or harp traps (Figure 4.81). Hand nets can also be used, as can nets connected to twin poles (e.g. two fishing rods) that can be flicked through the air as bats fly above. Such techniques are restricted in some countries, including the UK. In Britain, the Mammal Society has designed a ring that can be fitted to the bat, but an additional licence is required for such work and their use usually requires a long-term (~30 year) study. See Case Study 4.12 for a description of studying bats.

Medium-sized mammals can be caught using cage traps, although such traps should be of a size and mesh construction that is appropriate to the animal being trapped. Most cage traps operate by means of a treadle trigger somewhere towards the back

Case Study 4.12 Bat conservation ecology

Matt Zeale is a PhD student at the University of Bristol with over 6 years of experience studying bats in the UK. For the last 3 years Matt has been researching the ecology of the barbastelle, an IUCN threatened species. His case study involves modelling, and molecular and field-based techniques to assess the species' spatial and temporal activity, its foraging behaviour and diet.

Matt locating bats in their day roosts

Assessing the condition of a recently tagged barbastelle
Note the tag antenna at the tail-end and the aluminium ring on the left arm. Rings are used to identify recaptured bats and prevent repeated tagging of individuals (pseudoreplication).

Photos MZ
Two adult barbastelle bats in the hand (between thumb and forefinger), caught in ancient broadleaved woodland using an ultrasonic lure

Model organism and sampling problems faced

Although bats and their associated habitats are protected under UK law, the conservation management of many species continues to be constrained by a significant lack of knowledge regarding their ecological requirements. This is largely due to the difficulties associated with obtaining quantitative datasets from populations in the field. Bats are active fliers, almost exclusively nocturnal and often forage within vegetation, making direct observations of behaviour extremely difficult. Sustained contact with individual bats for extended periods of time is also problematic as nightly forays from roosts may be distant and many species exhibit frequent roost-switching behaviour. Roosts themselves can be difficult to locate, as many species roost in discreet tree cavities that are often inaccessible without climbing equipment. The success of Matt's work depended on his ability to accurately record the movements of bats in the field and to document changes in diet using the least invasive methods as possible. Since direct observations of feeding are often impossible, the determination of bat diets often relies on microscopic examination of prey fragments retained in the faeces. While this method does provide useful information, taxonomic resolution of prey is commonly limited to order or family and biases are inherent due to differential digestion rates of digestion of hard- and soft-bodied prey.

How the problems were overcome

To obtain data on spatial and temporal aspects of bat activity Matt uses lightweight radio-telemetry tags (fitted using degradable glue) to radio-track their precise movements over a period of 2–3 weeks and to locate bats within roosts. In this way he has been able to examine numerous aspects of bat ecology, including foraging behaviour, site fidelity, resource partitioning, habitat preference and roost use. However, such tracking techniques are dependent on researchers first being able to catch bats. Researchers studying rare or sparsely distributed species run the risk of wasting time and effort trying to capture animals in areas where they are not present. Furthermore, bats have highly developed echolocation and many species are extremely sensitive to human disturbance, often making it very difficult to capture and monitor them. Matt overcame these issues in two ways: first, he applied presence-only modelling techniques to identify suitable woodlands for his target species, and, second, he used an ultrasonic lure in conjunction with standard trapping methods to improve catch efficiency within suitable woodlands. For the study of bat diets Matt developed new DNA-based techniques to improve taxonomic resolution of prey and to reduce the biases inherent in traditional methods. Using prey DNA from bat faeces Matt is now able to identify prey to species, offering a much deeper insight into the dietary ecology of insectivorous bats.

Advice for students wanting to study bats

Bats are a remarkably diverse group, found in all but the coldest landscapes, and are among the finest examples of evolutionary design. For these reasons bats offer researchers an abundance of exciting research opportunities spanning many scientific disciplines from field-based practices to laboratory-based molecular work. Integrative approaches to the study of animal ecology are now commonplace and students should be prepared to embrace this practice, learning new techniques and exploring varied scientific fields. This requires collaborations, drawing upon expertise to produce work of the highest standard. Importantly, students wanting to study bats in the field must be prepared to work long and unsociable hours, often becoming nocturnal themselves for months at a time. Researchers studying bat ecology face many sampling issues and the development of innovative methodologies is often required to fill knowledge gaps. Furthermore, conservation legislation demands the close scrutiny of all bat research and in most circumstances considerable experience is required before being eligible for appropriate licences. Therefore, students are encouraged to gain experience as early as possible, either through local bat groups, by volunteering in current research projects or by completing specific licence training.

Figure 4.81 Harp trap to capture bats

of the trap that releases the door mechanism and traps the animal (Figure 4.82). Baited cage traps are usually set in the undergrowth. Cage trapping is especially worthwhile if transmitter collars are to be fitted, the animal marked or the fitness of the animal is to be determined. Unlike the traps used for small mammals, cage traps are not generally used to estimate density. You should consult specialist texts (e.g. Bateman 1979) or experts on the type of trap needed for the species under investigation. Trapping is not always necessary for all species. For example, hedgehogs (*Erinaceus europaeus*) can be located at night by torchlight, and caught quite easily with gloved hands.

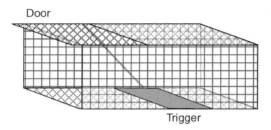

Figure 4.82 Cage trap
The door is closed by the animal stepping onto the trigger plate.

Species such as hares present a special problem for researchers as they are agile, fast and trap-shy. Mountain hares (*Lepus timidus*) have been studied by flush counting in the same way that some ornithologists count birds (see p. 208). The best time to count mountain hares is when they still have their winter white coat but the snow has melted. Because mountain hares occur in open habitats in the uplands of Britain, they are very difficult to catch as they can escape in any direction. The brown hare (*Lepus europaeus*) on the other hand can be successfully driven out of cover by systematic beating towards nets placed at key points in the habitat. This beating method is particularly useful in woodlands, but it is important to ensure that any holes in hedges, or unwanted exit routes, are blocked before the beat begins.

Large cage traps can be used to catch moderately large animals such as racoons, badgers and foxes (Figure 4.83). These are usually baited: prebaiting (where bait is left around the area to be trapped with the cage trap *in situ* but not set) can improve trapping rates.

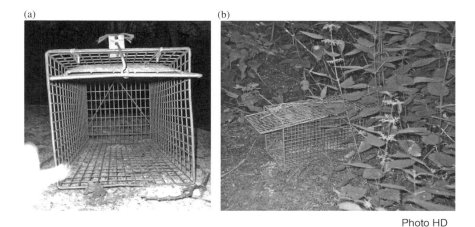

Figure 4.83 Badger trap
(a) Trap from front; (b) trap set in undergrowth.

Marking individuals

Captured mammals can be marked by fur clipping (cutting the fur in patches to represent the individual number of the animal), ear punching in small mammals (which is not recommended from an ethical viewpoint), freeze tattooing and branding (which are specialist techniques that may require a licence to use). Passive integrated transponder (PIT) tags are a modern, but expensive, alternative, although insertion may require veterinary help. Most mammal species have the potential to be radio-tracked (see p. 108), although this is a specialist pursuit and can be expensive. Spooling can be used to track smaller species: by attaching a spool of twine to the back of an animal and following it when it plays out individuals can be tracked for up to a couple of hundred metres. See Taber (1956), Twigg (1975), and Gurnell and Flowerdew (2006) for further details of marking mammals.

5

Analysing and Interpreting Information

We have emphasised throughout this book the importance of planning the data analysis during the inception of the project. A sound research design yielding data that are amenable to analysis is a prerequisite for a successful project. Data analysis can help to identify whether a pilot study is going according to expectation, provides an objective mechanism for the interpretation of differences, relationships and associations between data, and allows complex data to be examined. That said, it is easy to give the data analysis part of a project undue significance: there is a danger that statistical techniques can be used as 'black box' methods without understanding either the basis on which statistics are employed (in terms of any limitations of the technique) or the properties of the data being used. In addition, once statistical 'answers' are obtained, these should be viewed in the context of the system being investigated: for example, finding that there are significantly more species in one habitat type than another may be useful knowledge when managing biodiversity, but is less relevant if the species-poor site contains many nationally rare species whereas the species-rich site is dominated by very common organisms. Thus, we not only need to know what technique is appropriate to use, and how to interpret the results, we also need to discuss such results within the knowledge of the system being investigated. The aim of this chapter is to give an overview of the different approaches to describing and analysing data (see Box 5.1 for a note of caution about the examples we use within this book). For common statistical techniques, you should use an appropriate statistical package or consult the standard text(s) that we cite. For approaches specific to ecology that are not currently grouped in a standard text, we provide our own review and assessment of the most appropriate techniques. Here we point to the more technical literature for those wanting to take their understanding further. In general, we aim here to open up opportunities, to help you to understand what you can do with your data, rather than give you details of how to do it, which is beyond the scope of a book of this size.

Practical Field Ecology: A Project Guide, First Edition.   C. Philip Wheater, James R. Bell and Penny A. Cook.
© 2011 John Wiley & Sons, Ltd. Published 2011 by John Wiley & Sons, Ltd.

> **Box 5.1 A note of caution about the examples used in this chapter**
>
> It is important to be aware that in this book we use a mixture of real and hypothetical data to illustrate how statistical models work. Do **not** draw any conclusions about the ecological implications of any of these test results or cite these test results in any scientific paper or research as findings. Instead, we encourage you to cite this book as a guide. For example, 'Wheater et al. (2011) recommend the use of binomial generalized linear models for analyses of proportion data' **but not** 'Wheater et al. (2011) used binomial generalized linear models to show that blue tits have different relationships with insect biomass depending on habitat'.

In Chapter 1 we introduced the techniques that are commonly used to analyse ecological data. The first of these enables descriptions of the data and includes population estimation, diversity indices and distributions. To take the analysis further and examine differences between samples, relationships between variables and associations between frequency distributions, we require the testing of hypotheses using inferential statistics. More sophisticated analyses can be used to create predictive models using the gathered data to predict outcomes in other, similar situations. Pattern analysis enables complex combinations of sampling units and multiple variables to be examined to identify interrelationships between either the samples, or the variables, or both. There is a variety of software available to run these statistical methods, some of which are quite expensive (and therefore tend to be restricted to use in particular organisations and institutions), although there is an increasing amount available free on the Internet. Box 5.2 gives a brief overview of some of the recommended software appropriate for ecological projects.

> **Box 5.2 Some commonly used statistical software**
>
> Mainstream commercial software includes Genstat, JMP, Matlab, Minitab, SAS, SigmaPlot, SPSS (PASW), Stata, STATISTICA and Systat. Although most institutions have access to one or more of these packages, they are expensive to buy as private, single user licences. A number of packages (including some that are very similar to commercially available software) are available as downloads from the Internet without cost, known as 'open source' or 'freeware'. Be careful to ensure that free downloads are accurate and reliable and remain aware that this type of software is more likely to contain (many more) bugs and niggles than commercially available software. Even though freeware comes at no cost, it is usually distributed with a licence and a disclaimer. Lastly, as a final check it is worthwhile spending time establishing the statistical pedigree of the program developers: check that the method you are aiming to use is adequately described in associated published papers if novel, or that the software comes with formal documentation (all good ones do). Examples include:
>
> - PAST, which is designed for palaeontology, includes many descriptive statistics (including diversity indices), as well as inferential, predictive and pattern analysis techniques (http://folk.uio.no/ohammer/past/).

- PSPP, which is a freeware clone of SPSS that runs most analyses efficiently (http://www.gnu.org/software/pspp/).

- R, which is the most comprehensive and powerful open source environment within which most statistical methods can be run, however, it requires an understanding of statistical programming syntax – i.e. it is not menu-driven (http://www.r-project.org/).

- WinIDAMS, which is a software package for the validation, manipulation and statistical analysis of data developed by the UNESCO Secretariat and its statistical programmers (http://portal.unesco.org/ci/en/ev.php-URL_ID=2070&URL_DO=DO_TOPIC&URL_SECTION=201.html).

- ViSta, which is a visual statistics program (developed at the University of North Carolina) targeted at guiding beginners through complex data (http://www.visualstats.org/).

- Genstat (an otherwise commercial package), which covers most applicable univariate and multivariate statistical analyses, and whose distributors (VSM) provide it free to students and teachers for educational (not research) purposes (http://www.vsni.co.uk/downloads/genstat-teaching/).

There are also MS-Excel add-ins that extend the power of the spreadsheet program including:

- PopTools, which is a CSIRO authored add-in that attempts to avoid using MS-Excel's built-in formulae that are sometimes unreliable (http://www.cse.csiro.au/poptools/).

Multivariate analysis can require more specialist ecological software, although many of the techniques described below and discussed later are increasingly being found in R's statistical environmetrics packages, Vegan and BiodiversityR. We recommend PAST as the program of choice for those who require a windows interface at no cost. A continually updated source of information on multivariate analysis is Mike Palmer's ordination page (http://www.okstate.edu/artsci/botany/ordinate/), which includes a summary of commercially available packages (e.g. Brodgar, Pisces Conservation software, Canoco and PC-Ord) and freeware (e.g. IndVal, and Analysis of Ecological Data – ade4).

More specialist ecological software includes commercially available packages (e.g. from Pisces Conservation – http://www.pisces-conservation.com/) covering diversity analysis (e.g. SDR – Species Diversity and Richness), pattern analysis (e.g. Fuzzy Grouping, and CAP – Community Analysis Package), prediction (e.g. Ecom) and population estimation (e.g. Removal Sampling, Simply Tagging, and DfD – Density from Distance).

Free software for ecological applications includes the standard program DISTANCE (for population estimation using distance methods – http://www.ruwpa.st-and.ac.uk/distance/), MARK, CAPTURE and NOREMARK (population estimation for mark-release-recapture data) from Colorado State University (http://warnercnr.colostate.edu/~gwhite/software.html).

FCStats is a small MS-Excel-based package, written for use with small datasets on ecological field trips in order to bring many commonly used simple statistical techniques together in one package (http://www.sste.mmu.ac.uk/teachers_zone/).

Keys to tests

Choosing a method of data analysis may sometimes seem rather bewildering. Until you have obtained experience in some of the methods, you will have to rely on advice and statistics textbooks. The keys that follow give a guide as to when to use which common test (see Box 5.3 for an explanation of the terms used in the keys). In each case start at the left-hand side of the table and work your way to the right. After the first column there are usually pairs of questions to enable you to proceed (the first question in a pair begins with a capital letter and the second ends with a question mark). At each point you will need to decide on a direction based on some aspect of your data. Following these keys we give a brief description of when and how you might use the techniques. A detailed discussion of statistical analyses is beyond the scope of this book and you should consult an appropriate text if you need more information about a particular technique. Wheater and Cook (2000) provide a useful introduction to project design and statistical analysis and cover many of the most commonly used descriptive and inferential statistics. See Manly (2004), Fielding (2007) and Tabachnick and Fidell (2007) for multivariate analyses. Other books that are useful reference guides, covering a wide range of analyses including more sophisticated techniques (e.g. multivariate methods), are the Electronic Statistics Textbook,[1] Sokal and Rohlf (1995), Legendre and Legendre (1998), Scheiner and Gurevitch (2001), Crawley (2007) and Zar (2009).

Box 5.3 Important terms used in the keys

See also the statistical glossary in Appendix 1

Term	Definition
Binary Data	Data that can be allocated to one of two classes only (e.g. male or female) normally scored as 1 or 0
Binomial data	Data describing the number of 'successes' as a proportion (x/y) of the total (e.g. numbers of healthy plants x in a population y)
Community (multivariate) analyses	Analyses that attempt to simplify the properties of matrix data usually in the context of species and/or site dynamics. At its simplest, two variables may constitute a matrix if, for example, there are multiple sites and species forming a two-way table
Dependent variable	Variable whose values may be determined by another within a model, sometimes referred to as the response variable

---

[1] http://www.statsoft.com/textbook/

Term	Definition
Environmental gradient	In terms of multivariate analyses, an ordering of the subjects (species or sites) along axes may be suggestive of a relationship between it and an independent variable. This inference elucidates a possible underlying gradient in the community that can either be inferred (i.e. indirect gradient analysis) or tested formally (i.e. direct gradient analysis)
Frequency	The number of events within a unit of experimental time or within a set of experimental samples
Independent variable	Variable that may determine the value of another within a model, sometimes referred to as an explanatory variable, treatment or predictor
Interval/ratio data	Measurement data (e.g. age, number of offspring, height, mass, chemical concentration)
Matched (paired) data	Where each data point is not independent of all others, but pairs (or groups) of data points are linked by being taken from the same individual or sampling unit (e.g. an animal's mass before and after diapause, turbidity of a river above and below a sewage outflow, annual herbicide resistance in a series of weed plots over 5 years)
Nominal data	Data on a categorical scale that have no definable order (e.g. woodland, grassland, wetland)
Normal (Gaussian) distribution	Symmetrical bell-shaped curve with the mean value in the middle and most data points around it: a feature of many datasets measured on interval or ratio scales
Ordinal data	Data on a ranked scale (e.g. small, medium, large)
Poisson distribution	A discrete distribution that often describes count data in which large values tend to be rare (e.g. there are many occasions when two magpies have been spotted together but comparatively very few occasions when 30 magpies have congregated). It has a unique property of the variance being equal to the mean
Variable	A characteristic that varies between individuals within a population

Key A – to identify the type of analysis you require	
This key requires you to decide on which of the four broad types of analysis you would like to use	
Describe data, e.g. in terms of average values, graphical displays, or information such as sizes or densities of populations, richness or diversity of items (communities) or spatial/temporal distributions	**Key B**
Test hypotheses, e.g. whether there is a significant difference between samples, or whether there is a significant relationship between variables, or in terms of frequency analysis (significant associations or goodness of fit between frequency distributions)	**Key C**
Predict the values of one variable from a combination of other variables	**Key D**
Examine patterns in communities and generate hypotheses based on those patterns. Typically the data will be organised as a matrix of variables (e.g. species) as columns and individuals (e.g. sites or site-years) as rows. There may also be one or more independent variables measured at the individual level (and recorded in rows)	**Key E**

Key B – to decide on how to describe your data The starting point of this key requires that you decide which of the seven types of analysis you wish to use		
Looking at frequencies, e.g. to see which items were most frequently sampled*		Frequency histograms or frequency tables
Calculating averages and variation within data*	For normally distributed data measured on interval/ratio scales	Means, standard deviations, standard errors, 95% confidence limits
	For non-normal data (e.g. measured on ordinal scales)	Medians and interquartile ranges
Displaying data graphically	To display averages and variation	Bar or point charts for normal data, box and whisker charts for non-normal data
	To display relationships between variables	Scatterplots and regression plots
	To examine comparative frequencies of sampling different items	Pie charts, or clustered or stacked histograms
Identifying clumped, random and regular distributions of items in space and time†		Choice of distribution estimation techniques selected on the basis of whether data are collected in blocks of space (quadrats for example) or time, or by distance sampling methods
Estimating population sizes and densities†		Choice of population estimation techniques selected on the basis of how likely the organisms are to mix thoroughly
Examining the richness or diversity of items (e.g. community structure)†		Choice of diversity and evenness indices selected on the basis of whether simple richness is required or more weight is given to rare or common items
Examining the similarity (or dissimilarity) between communities on the basis of their species (or other component factors)†		Choice of similarity, dissimilarity or distance measure selected on the basis of whether sites are considered in isolated pairs or as part of a larger number of sites, and whether you are interested in species abundance or just presence/absence

*Worked examples can be found in Wheater and Cook (2000).
†Some worked examples can be found in Wheater and Cook (2003).

Key C – Testing hypotheses using basic statistical tests and simple general linear models (e.g. ANOVA and regression).

The starting point of this key (first column) requires a decision about which of the three types of analysis questions (difference in central tendency, relationship between variables, or frequency analysis) you wish to ask. See Wheater and Cook (2000) for worked examples of many of these tests

Differences in central tendency	Two samples	Data in matched pairs	Normally distributed data measured on interval/ratio scales **Paired *t*-test**
			or non-normal or measured on an ordinal scale? **Wilcoxon matched-pairs test**
		or not?	Normally distributed data measured on interval/ratio scales ***t* test**
			or non-normal or measured on an ordinal scale? **Mann–Whitney *U* test**
	or more than two samples?	Data in matched groups	Normally distributed data measured on interval/ratio scales **Repeated measures or nested analysis of variance**
			or non-normal or measured on an ordinal scale? **Friedman's matched group analysis of variance using ranks**
		or not?	Normally distributed data measured on interval/ratio scales **Analysis of variance (one-way, two-way, etc.)**
			or non-normal or measured on an ordinal scale? **Kruskal-Wallis analysis of variance using ranks (one-way, two-way, etc.)**
or relationships between variables	Is a causal relationship suspected		**Linear regression analysis**
	or not?	Are two variables being examined	Normally distributed data measured on interval/ratio scales **Pearson's product moment correlation**
			or non-normal or measured on an ordinal scale? **Spearman's rank correlation**
		or more than two?	**Kendall's coefficient of concordance**
or between associations distributions	Test against an expected distribution		**Chi-square test for goodness of fit**
	or test two or more distributions against each other?		**Chi-square test for association/independence**

Key D – Methods to predict the value of dependent variable(s) from independent variable(s)
Includes general linear models and generalized linear models to predict the value of dependent variable(s) from independent variable(s). General linear models assume a normal distribution, a continuous dependent variable and a least squares approach. Generalized linear models do not necessarily assume a normal distribution or a continuous dependent variable but use a link function to an appropriate error distribution that permits a whole range of possible tests. This key illustrates a useful set of commonly used models for ecologists.

Is the dependent variable continuous	Is the dependent variable measured on an interval/ratio scale	Are all independent variables either measured on binary or interval/ratio scales	Is there one independent variable	**Linear regression analysis**
			or are there more than one independent variables?	**Multiple linear regression analysis**
		or is there a mixture of measurements and nominal independent variables?		**Analysis of covariance (ANCOVA)**
	or is the dependent variable based on counts?[*]	Are predictions linear functions of the dependent variable (i.e. straight line regressions)		**Log-linear generalized linear model (using a Poisson link function)**
		or are predictions non-linear functions of the dependent variable?		**Generalized additive models (GAM) (using a Poisson link function)**
or is the dependent variable binomial, binary or categorical?	**Discriminant function analysis or a generalized linear model such as a binomial regression or a logistic regression (types of generalised linear models that use logit link functions) - if not appropriate then consider some of the basic non-parametric tests such as a chi-square test for association described in Key C**			

[*]Although both multiple linear regression analysis and ANCOVA can also be used to analyse count data (usually following transformation – see the ANCOVA example p. 277), such models are probably not the best ones to use for count data because these models assume a constant variance and normal errors.

KEYS TO TESTS

Key E – Statistical methods to examine the patterns (i.e. the structure of inter-relationships) within a data matrix				
Normally the motivation for examining patterns in the data set is to generate a hypothesis or to classify or group species or sites in terms of some measure. For example, you may have a data matrix comprising sites (rows) that you wish to classify by species (columns). Multivariate techniques can be very flexible with model algorithms tailored to a particular type of data (e.g. Proxscal, Alscal, Indscal, etc. are all forms of multidimensional scaling – MDS). Here we have simplified the vast array of techniques to cover methods commonly used for analysing ecological data. There are three major types of analysis to choose from (see the left-hand column below).				
Do you need a technique that puts species, sites or responses into defined groups and investigate group membership using dendrograms (tree diagrams) – i.e. classification techniques	Is the structure of the relationship one way and either among species or between sites	Are the number of groups unknown		Hierarchical clustering‡ (e.g. single link and complete linkage)
		or are the number of groups known (e.g. you wish to impose 11 groups on the matrix but cannot prescribe how these are allocated),		k-means clustering
		or is the group membership known (e.g. you assign groups according to treatments) and you require a significance test?		Analysis of similarity (AnoSim)(often used in partnership with ordination)
	or is the structure two way and between species and sites concurrently?	Here the group membership is known (e.g. you assign groups according to treatments) and you require a biplot. Further, your data meet assumptions made about linearity		Canonical variates analysis (CVA)$^\#$
or do you need a technique that produces indicator species (i.e. indicator species analysis)	Is the group membership known (e.g. you assign groups according to treatments or by groups derived from a classification technique)			IndVal
	or is group membership unknown and do you require a dendrogram to annotate indicator species?			TWINSPAN
or do you need a technique that explores the dynamics between species and sites and displays them as a biplot? (i.e. ordination techniques)	Are the data two way and measured in ranks, categories or percentages	Are assumptions made about linearity (ideal for percentages where there is a large variation in values e.g. dietary components at the population level)*		Principal components analysis (PCA)$^\#$
		or are there no underlying assumptions (ideal for ranked or category data)*		Correspondence analysis (CA)$^\#$ or Multidimensional scaling (MDS)† (Proxscal algorithm recommended)

	or are the data two way and measured as presence/absence (binary) or as counts (and are underlying environmental gradients hypothesised)?	Are explanatory variables absent*	Detrended correspondence analysis (DCA)#or Principal co-ordinates analysis (PCoA)‡
		or are explanatory variables present†	Canonical correspondence analysis (CCA)#Also consider using DCA and correlating explanatory variables with derived axes scores

*Types of indirect gradient analysis.
†Types of direct gradient analysis.
#Eigenvalue approach.
‡Distanced-based. Note: PCoA is like a hybrid between MDS and PCA in that it attempts to derive distances from dissimilarities (like MDS), but it does so via an eigenvalue approach (like PCA).

Exploring and describing data

Depending on the types of information gathered, there are several ways in which you may wish to describe your data. Summary statistics are useful in order to understand important components of the data (e.g. the range, average value, variability, complexity, etc.) when checking and analysing data, as well as being able to communicate such attributes to others. Counts and measurements (including scores on ranked scales) may need to be summarised either numerically or graphically; distributions (spatially or temporally) can be described in terms of how random they are; population sizes and densities can be estimated (together with standard errors to describe the reliability of the estimates); and community structures can be described in terms of their individual richness or diversity or set in a framework of other communities to examine the similarity (or dissimilarity) between them. Before you describe your information, you need to get an understanding of your data for your own use, including examining the distribution of the data and screening for outliers.

Transforming and screening data

A lot of decisions about how to describe, analyse or present your data depend on whether you have any data that are measured on the interval/ratio scale, and, if so,

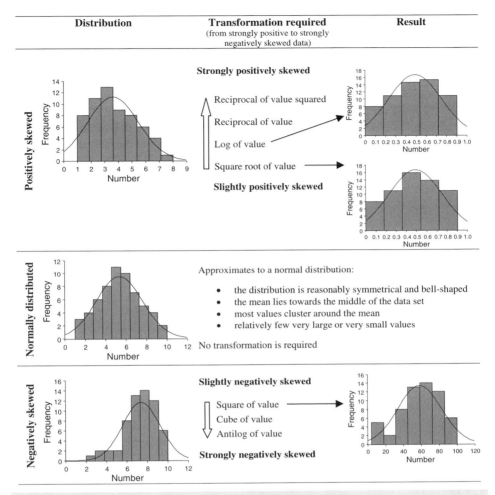

Figure 5.1 Transformations for skewed distributions
The block arrows indicate the range of techniques used for transforming increasingly skewed data.

what the distribution of the data looks like. For example, taking a mean value is only really applicable to situations where the data conform to a normal distribution. Where data are skewed (Figure 5.1), then either they should be mathematically transformed to fit a normal distribution or treated in the same way as ranked data (using the median value rather than the mean). Transformations are mathematical manipulations of all the data points within a variable and there are different methods depending on the type and severity of the problem. For example, slightly positively skewed

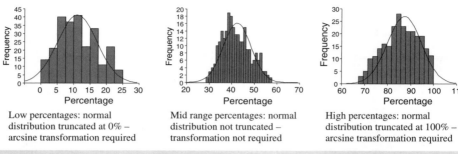

Figure 5.2 Truncation of percentage data

distributions can be transformed using the square root of the value whereas more strongly positively skewed distributions may require the calculation of the reciprocal of the square of the value. Where zero values occur, additional manipulations are required to enable the transformation. For example, where square roots are to be taken, the value of 0.5 is added to each data point before taking the square root, but if a logarithmic transformation is used you should add the value of 1 to each data point before taking the log value. There are other circumstances where transformation is useful, for example where the distribution is truncated (e.g. when measurements are on a percentage scale and most are close to either the 0% end or the 100% end – see Figure 5.2). Wheater and Cook (2000) give worked examples showing how to calculate these transformations.

There are circumstances where you cannot transform your data to conform to a normal distribution and it may be useful to consider using a non-parametric test; see Siegel and Castellan (1988) for a range of non-parametric alternatives to parametric tests. However, in other circumstances, you may obtain a large number of zeros; a situation that you may encounter frequently when counting rare species. Here it would be better to use a test that can deal with a different type of distribution (in this case, this would probably be a Poisson distribution) – see p. 281.

Frequency data can be readily plotted on histograms (similar to those shown in Figures 5.1 and 5.2) and this is important for gaining an understanding of the range and distribution of your dataset (e.g. whether the data are skewed or bimodal – see Figure 5.3). However, you would not usually use such graphs to present the data in a final report. This is because one of the aims of graphing data is to condense information: you will rarely have the luxury of being able to present histograms of every variable.

When inspecting your frequency histograms, you should examine your data to check that there have been no obvious errors in entering the numbers. For example, if the

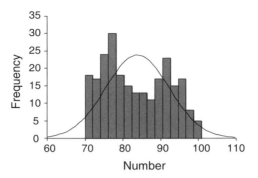

Figure 5.3 Bimodal distribution

maximum value of a measurement is 10, and you plot the distribution and find that a value of 66 has been recorded, then there is an obvious mistake (probably due to you hitting the '6' key twice in error). Of course, there are occasions when you might really have recorded a very high or very low number that looks odd within your dataset. Such values are called outliers and can sometimes influence your data analysis (inflating or decreasing mean values and increasing measures of variation). It may be advisable to delete outliers before analysing the data. If you do remove an outlier, you should always report the fact, and give its value. Similarly, for bivariate data (i.e. where you have two measured variables for each individual) scatterplots should be screened for linearity and outliers.

Graphical display of data

For nominal data, if a graphical display of frequencies is required, this can be in the form of a pie diagram (Figure 5.4), although these are usually more useful for visual presentations than in reports. We do not usually use pie charts in reports, since they are not an efficient way of condensing information and your reader will not want to see pages of pie charts for every variable (multiple pie diagrams should not usually be used). Consider, from Figure 5.4, how much less space would be used by simply stating 'the most common order was Diptera (71 individuals) followed by Trichoptera (44 individuals)', etc. Another drawback of pie charts is that it can be difficult to distinguish between segments of similar size. Pie charts with large numbers of segments can become cluttered and it can be difficult to label segments (especially if the data need to be included) without it becoming untidy.

If multiple samples are compared in terms of the frequency of a nominal variable, then stacked or clustered bar graphs can be used. However, both may also be problematic. It

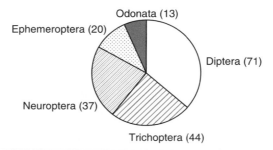

Figure 5.4 Pie diagram of the numbers of invertebrates of common orders found in clean ponds
Note the difficulty of identifying the difference between the slices for Neuroptera and Trichoptera (despite a difference of around 16% of their value) unless the data values are included.

can be difficult to understand size differences between the middle portions of stacked bars, although connecting lines between samples may help in this (Figure 5.5). Clustered bars can become rather thin and difficult to interpret, especially where there are a large number of subunits and/or samples being compared (Figure 5.6). With these sorts of data, it is often best to use a table to display relatively complex comparisons (e.g. Table 5.1).

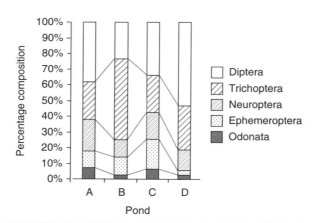

Figure 5.5 Stacked bar graph of the number of invertebrates of common orders found in clean ponds
Note the difficulty of working out the relative sizes of similar sized sections even on adjacent bars unless lines are included (e.g. the values for Ephemeroptera for ponds A and B) and the greater difficulty in distinguishing values that are not on adjacent bars (e.g. the values of Trichoptera for ponds A and C).

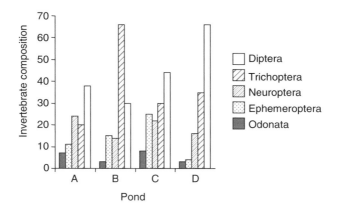

Figure 5.6 Clustered bar graph of the number of invertebrates of common orders found in clean ponds
Note the difficulty of comparing different bars (orders) between different clusters (ponds).

Table 5.1 Abundance of invertebrates in ponds
Numbers of each order collected with percentages in parentheses. Percentages may not sum to 100% due to rounding errors

Taxon	Pond A (%)	Pond B (%)	Pond C (%)	Pond D (%)
Odonata	13 (7)	3 (1)	8 (6)	3 (2)
Ephemeroptera	20 (11)	15 (12)	25 (19)	4 (3)
Neuroptera	37 (20)	14 (11)	22 (17)	16 (13)
Trichoptera	44 (24)	66 (52)	30 (23)	35 (28)
Diptera	71 (38)	30 (23)	44 (34)	66 (53)
Total (100%)	185	128	129	124

Measures of central tendency and sample variability

Where you are interested in a variable comprising data that are normally distributed measurements (whether in their original form or following transformation), the mean should be used as the measure of central tendency (i.e. a measure of the 'middle' of the data that would be expected to be typical of the dataset). This measure is more robust than either the mode (the most frequently occurring value) or the median (the middle ranking value), which are more suitable for nominal and ordinal data respectively. Such statistics should not be used in isolation, and the sample size should always be quoted, as should some measure of the variation in the dataset.

The variability of your data has an impact on how well the measure of central tendency represents the dataset and what sort of statistical tests you should use to analyse them. For normally distributed measurement data, the standard deviation, standard error or 95% confidence intervals should be calculated. The first one is an indication of the range (that is, the mean minus the standard deviation to the mean plus the standard deviation) within which most (~68%) of the data points lie, as illustrated in Figure 5.7. The last two are used more frequently, especially in graphical representations, and measure the reliability of the estimate of the mean as a symmetrical range around the sample mean. In technical terms, based on sampling theory, if you were to take repeated samples and measure the mean each time, the standard error would be the standard deviation of the sampling distribution of the mean, while 95% of calculated 95% confidence intervals would contain the true mean. Means, and their associated measures of variability, can be displayed on point or bar charts, as shown in Figure 5.8. Wheater and Cook (2000) give details about how to calculate these values, or they can be generated by most statistical software or functions within spreadsheets.

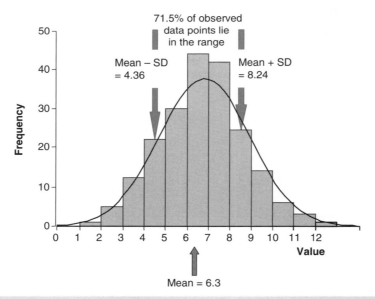

Figure 5.7 The mean and standard deviation plotted on a dataset that approximates to a normal distribution
Note that here 71.5% of data points lie in the range of the mean (= 6.3) minus the standard deviation (= 1.94) to the mean plus the standard deviation (i.e. from 4.36 to 8.24) rather than the theoretical 68.3% that would be seen in a perfect normal distribution.

EXPLORING AND DESCRIBING DATA

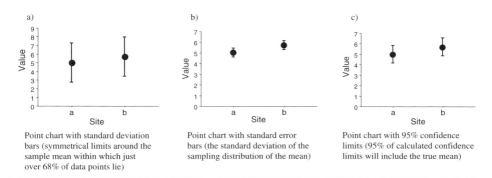

a) Point chart with standard deviation bars (symmetrical limits around the sample mean within which just over 68% of data points lie)

b) Point chart with standard error bars (the standard deviation of the sampling distribution of the mean)

c) Point chart with 95% confidence limits (95% of calculated confidence limits will include the true mean)

Figure 5.8 Comparison of different ways of displaying the variation around the mean using point charts
Using the same data for all three charts and $n = 30$ for both site a and site b.

Medians are used for data measured on ranked scales, or where there are extreme values (outliers) or when non-normal data (that cannot be transformed) are being described. This is simply the mid-point when the variable is ordered from the lowest to highest value. Similarly, the variation can be described using statistics software to define limits taken one quarter along from each end of the data, so obtaining the 25% (quartile 1 or Q1) and 75% (quartile 3 or Q3) points that delimit the 50% of the data in the middle of the distribution. This is called the interquartile range and both this and the median can be plotted on box and whisker plots (Figure 5.9).

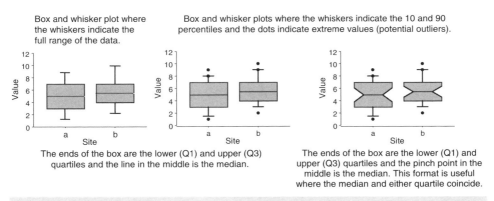

Box and whisker plot where the whiskers indicate the full range of the data.
The ends of the box are the lower (Q1) and upper (Q3) quartiles and the line in the middle is the median.

Box and whisker plots where the whiskers indicate the 10 and 90 percentiles and the dots indicate extreme values (potential outliers).

The ends of the box are the lower (Q1) and upper (Q3) quartiles and the pinch point in the middle is the median. This format is useful where the median and either quartile coincide.

Figure 5.9 Box and whisker plots indicating different ways of displaying median and quartile data

Spatial and temporal distributions

Ecologists have their own specific suite of techniques that are useful to describe, present and display ecological data. One such is the quantification of spatial and temporal distribution of organisms. As discussed in Chapters 3 and 4, individuals may be distributed evenly, randomly or in clumps depending on the scale at which they are measured. For example, greenfly may be clumped on leaves that are randomly distributed on trees which are evenly spaced in plantation woodland. Similarly, the basking activity of reptiles may be clumped in time during the day, but evenly spread across a week. There are two main ways of examining such distributions: by the use of blocks of space (e.g. quadrats) or time, or by the use of intervals between individuals in space (distance) and time (period). The use of quadrats (or other blocks of space or time) is described in Chapter 3, and the use of distances is given in Box 3.3. In both cases, it is possible to test to see whether the distributions are significantly non-random (see Boxes 3.4 and 3.5 respectively).

Generally, analysing data that have strong spatial and/or temporal components is more challenging that most types of statistical modelling. To gain an understanding of the breadth of techniques used beyond what is described here, consult Fortin and Dale (2005), who cite useful statistical techniques that include Spatial Analysis by Distance Indices (SADIE)[2] for regularly and irregularly spaced data and autoregressive models to overcome problems with temporally autocorrelated data (e.g. Autoregressive Integrated Moving Average [ARIMA] models).

Population estimation techniques: densities and population sizes

You may find it useful to measure the densities and/or population sizes of plants and animals. Various methods can be used for this, many of which also calculate standard errors to give a measure of the reliability of the estimates. Following mark–release–recapture of animals, there are several different methods that can be used to estimate the population size, the simplest of which is the Peterson method or Lincoln Index (Box 5.4), which involves only two collections (i.e. the initial collection and one recapture collection). This technique has been found to overestimate population sizes and several modifications have been suggested to the formula to deal with this (Box 5.4). There are two additional assumptions associated with this method:

- the population is closed (that is, that there is no movement into or out of the population). If this is not so, immigration and emigration must be able to be calculated;

- there are no births or deaths in the period between sampling.

The standard error (and hence 95% confidence limits) for the population can also be calculated (Box 5.4).

---

[2] http://www.rothamsted.ac.uk/pie/sadie/SADIE_home_page_1.htm

Box 5.4 The Peterson (Lincoln Index) method of population estimation

The basic formula is:

$$\hat{P} = \frac{(\text{number marked in first sample} \times \text{number captured in second sample})}{\text{number of marks from first sample captured in second sample}} = \frac{n_1 n_2}{m_2}$$

For example, 50 beetles are collected and marked and then released back into the wild. Later a second collection of 100 beetles includes 25 that were previously marked beetles. Since the ratio of the original number marked to the total population should be the same as the ratio of the number of marked animals recaptured to the number caught on the second occasion, the population size can be estimated as:

$$\hat{P} = \frac{(50 \times 100)}{25} = 200$$

Since this has been shown to overestimate the population size, modification formulae include:

$\hat{P} = \dfrac{n_1(n_2+1)}{(m_2+1)}$ used when animals are caught one at a time and returned to the population before capturing the next (thus enabling animals to be caught several times) – also useful for small samples (<20) – Bailey's modification

$$\hat{P} = \frac{50(100+1)}{(25+1)} = 194$$

or

$\hat{P} = \dfrac{(n_1+1)(n_2+1)}{(m_2+1)} - 1$ used when animals are caught and processed as a single sample so each animal is only counted once (i.e. replacement is not taking place) – Chapman's modification

$$\hat{P} = \frac{(50+1)(100+1)}{(25+1)} - 1 = 197$$

By calculating the standard error of the population estimation, 95% confidence limits can be attached to the population estimate:

The population estimate $\pm\,95\%\,\text{CI} = \hat{P} \pm (1.96)\,(\text{SE})$

Where the standard error (SE) depends on the formula used:

For the basic formula:

$$\text{SE} = \sqrt{\frac{(n_1)^2 (n_2)(n_2 - m_2)}{(m_2)^3}}$$

$$\text{SE} = \sqrt{\frac{(50)^2 (100)(100 - 25)}{(25)^3}} = 34.6$$

For the Bailey's modified formula:

$$SE = \sqrt{\frac{(n_1)^2(n_2+1)(n_2-m_2)}{(m_2+1)^2(m_2+2)}}$$

$$SE = \sqrt{\frac{(50)^2(100+1)(100-25)}{(25+1)^2(25+2)}} = 32.2$$

For the Chapman's modified formula:

$$SE = \sqrt{\frac{(n_1+1)(n_2+1)(n_1-m_2)(n_2-m_2)}{(m_2+1)^2(m_2+2)}}$$

$$SE = \sqrt{\frac{(50+1)(100+1)(50-25)(100-25)}{(25+1)^2(25+2)}} = 23.0$$

See Southwood and Henderson (2000) for further details and other variants on this theme.

There are many other methods of estimating population sizes, the choice of which depends on the animals under investigation, the number of recapture events, the likelihood of recapturing the animals and the marking method being used (see Table 5.2 for a summary):

- the Fisher and Ford method is probably better for small samples (especially where the survivorship remains fairly constant during the sampling period);

- the Jolly–Seber method, Bailey's Triple Catch and especially the Manly–Parr method all require large samples, with the first being preferable where you are catching over several sampling occasions.

It can be difficult and logistically expensive to capture (and subsequently recapture) animals. However, if the marks are highly visible on animals that are themselves reasonably obvious, then simply identifying whether animals are marked or not may be a cheaper alternative to recapturing a sample of the population – i.e. using capture–resight techniques. Under these circumstances, the Petersen method can be used to estimate the population (Box 5.4). Such techniques rely on marked animals being from closed populations and being resighted only once. As such, sampling sessions should normally be quite short to prevent double counting of animals. More sophisticated models that can accommodate multiple sightings of marked individuals, multiple sighting occasions, and both closed and open populations can also be applied (see Krebs 1999 or Southwood and Henderson 2000 for further information).

Table 5.2 Summary of commonly used methods of population estimation based on mark–release–recapture techniques

Method	Comments
Bailey's Triple Catch	Extends the Peterson (Lincoln Index) principle by adding a third trapping occasion that enables calculations of the loss and gain of animals over a longer period
Fisher and Ford's method	Uses several trapping occasions and enables estimates of survivorship over the trapping period
Jolly's method (and the similar and independently derived Seber's method)	Uses several trapping occasions and estimates survivorship between adjacent trapping occasions and can accommodate studies where there is both ingress (births and immigration) as well as egress (emigration and deaths – including, for the Jolly method, occasions where deaths occur following capture)
Manly–Parr's method	Another multiple capture technique that is more robust than some other methods in situations where animals are entering the population during the study period and where the mortality rate is probably age dependent (e.g. during the emergence period of some butterflies)
Du Feu's (derived independently and earlier by Craig)	Can be very useful since it uses a single trapping occasion – it operates under the assumption that as each animal is captured, marked and released, it mixes with the remaining population

See Krebs (1999), or Southwood and Henderson (2000) for further details.

Distance sampling is becoming a more widespread technique (having been used by ornithologists for some time). This involves estimating density or abundance by measuring the distances between the sample position and the organism being studied. Since organisms become more difficult to spot with increasing distance from the recorder, there will be fewer recordings at further distances. The technique fits a detection function to the observations to estimate the proportion of organisms that were missed by the survey and hence produces estimates for density and abundance within the surveyed area. See Buckland et al. (2001) and Buckland et al. (2004) for further details.

There is a suite of programs available from Colorado State University[3] to analyse population data, including MARK (for mark-release-recapture data), CAPTURE (for closed populations) and NOREMARK (for resighting data). In addition, the program DISTANCE (for distance sampling data) is available from St Andrews University.[4]

Where animals are found in a closed population and the capture technique involves the removal of those caught (e.g. fish captured and removed from a lake), capture-removal

---

[3] http://welcome.warnercnr.colostate.edu/~gwhite/software.html
[4] http://www.ruwpa.st-and.ac.uk/distance/

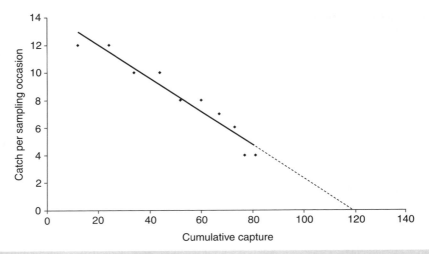

Figure 5.10 Using capture-removal to estimate population sizes
The calculated regression line between catch per unit effort and cumulative capture has been extrapolated onto the x axis to estimate the population at ∼120.

techniques can be used for estimating the population size. Here animals are captured on successive trapping occasions and a regression line calculated between the number caught per sampling occasion and the cumulative catch (see p. 271 on regression). The total population can be estimated by extrapolating the regression line onto the x axis (Figure 5.10).

Richness and diversity

In your research, you may wish to describe species data in terms of the richness of communities. At its simplest this can involve simple reporting of the number of species and the number of individuals found. More sophisticated methods involve the calculation of indices of diversity and evenness. Here the relative proportion that each species makes up of the whole community is used to create an indicator of diversity (usually maximised when there are many species with similar abundances – i.e. no one species dominates). Conversely, indicators of evenness can be produced that describe how balanced the relative proportions of species are (with low evenness being attributed to cases where a single species dominates). Examples of common methods are given in Table 5.3, while more comprehensive discussions can be found in Krebs (1999) and Magurran (2004). Worked examples of the Shannon–Wiener, Simpson, and Berger–Parker diversity indices and the Pielou evenness index can be found in Wheater and Cook (2003). Several of these indices can be calculated using an MS-Excel add-in that is available from Reading University.[5]

---

[5] http://www.reading.ac.uk/SSC/software/diversity/diversity.html

Table 5.3 Common diversity and evenness indices

Index	Type of index	Comments	Calculation
Number of individuals	Individual richness	Simplistic, takes no account of species richness	The total number of individuals found using standardised sampling (N)
Number of species	Species richness	Simplistic, takes no account of relative abundance	The total number of species found using standardised sampling (S)
Shannon–Wiener	Abundance-based diversity index	Assumes all species are represented in the sample. Sensitive to species richness and hence rare species (the potential maximum value increases as the number of species increases). Considered to be flawed, but still popular. (Sometimes incorrectly called Shannon–Weaver)	$H' = -\Sigma p_i \ln p_i$ Increases towards a maximum value of $\ln S$. Rarely exceeds 4 in practice
Simpson	Dominance-based diversity index	Based on the probability that finding a second individual in a sample will be from the same species as the first. Influenced by common species rather than species richness (the addition of small numbers of species makes little difference). Often considered to be a superior alternative to the Shannon–Wiener index	$D = \Sigma p_i^2$ Usually calculated as $1 - D$ or $1/D$ so that as the diversity rises (and dominance falls) the index increases. With $1 - D$, the index has a maximum value of 1, for $1/D$, the maximum is S
Berger–Parker	Dominance-based diversity index	The measure of the proportion of the sample attributed to the most abundant species. Considered to be one of the best indices because of its simplicity	$D = N_{(\max)}/N$ Usually calculated as $1/d$ so that as the diversity rises (and dominance falls), the index increases

(*continued*)

Table 5.3 (*Continued*)

Index	Type of index	Comments	Calculation
Pielou evenness	Evenness index	Based on the Shannon–Wiener diversity index (H'), dividing the value H' by the maximum that one would get if the number of species found was the total number of species in the area. Considered to be flawed but still popular	$J' = H'/\ln S$ Ranges from 0 (not even) to 1 (even)
Simpson's evenness	Evenness index	Based on the Simpson diversity index (D), dividing the value of D by the maximum one that would get if the number of species found was the total number of species in the area	$E_{(1/D)} = (1/D)/S$ Ranges from 0 (not even) to 1 (even)

Where N = the total number of individuals found;
S = the total number of species found;
P_i = the proportion that the i th species forms of the total number of organisms found (N);
ln = the natural log of the value (note for the Shannon–Wiener index, any logarithmic base can be used);
$N_{(max)}$ = the number of individuals found in the most abundant species.

See Legendre and Legendre (1998), and Krebs (1999) for further details.

Similarity, dissimilarity and distance coefficients

There are a large number of ways of measuring similarity (or dissimilarity) based on two distinct approaches. The first of these examines whether two or more communities share species on the basis of the number that are held in common compared to those that are found only in one or the other community (binary approaches). In cases where several communities are being examined in a pairwise fashion, some methods also take account of those species that are not found in either community of a particular pair, but are present in the overall dataset (that is, they include double negatives). The second approach looks at the overlap not only in species but also in the abundance of each species. Table 5.4 gives examples of some of the common methods from both of these approaches, and a more detailed discussion of a range of techniques can be found in Legendre and Legendre (1998) and Krebs (1999). Similarity measures can potentially

Table 5.4 Commonly used similarity measures

Similarity	Type of data used	Comments	Calculation (see table below* for explanation of symbols)
Jaccard index	Presence/absence	One of the simplest formulae, this does not take into account species that are absent from both communities being examined. Can be multiplied by 100 to give a percentage value	$S_J = a/(a + b + c)$ or $S_J = a/(A + B - a)$
Sorensen index (coefficient of community)	Presence/absence	Similar to the Jaccard measure but gives greater weighting to species found in common across the two communities. Can be multiplied by 100 to give a percentage measure	$S_S = 2\,a/(2\,a + b + c)$ or $S_S = 2\,a/(A + B)$
Simple matching coefficient	Presence/absence	Takes into account the species absent from both communities but present elsewhere in the data set	$S_{SM} = \dfrac{(a+d)}{(n)}$
Contingency table (chi-square)	Presence/absence	This uses a 2 × 2 contingency table and a standard chi-square analysis (see p. 275). Expected values should not be less than 5	$X^2 = \dfrac{(n)[\lvert ad - bc\rvert - (n/2)]^2}{(a+c)(b+d)(a+b)(c+d)}$ Where: $\lvert ad - bc\rvert$ = the absolute value of $ad - bc$, i.e. removing any negative sign in the result
Percentage (or proportional) similarity (Renkonen index)	Abundance	Based on an examination of the percentage/proportion that each species contributes to each community. The index is calculated by summing the smallest percentage contribution that each species makes in the pairs of communities being examined	$P = \Sigma$ minimum (p_{Ai}, p_{Bi}) Where: p_{Ai} = percentage of species i in community A p_{Bi} = percentage of species i in community B

(continued)

Table 5.4 (*Continued*)

Similarity	Type of data used	Comments	Calculation (see table below* for explanation of symbols)
Correlation	Abundance	This uses a straightforward correlation analysis (see p. 269) to examine relationships in species abundance between pairs of communities. One advantage is that where species are in similar proportions but different magnitudes, the relationship will be positive. However, correlation is affected by small numbers of species and large numbers (more than 50%) of zero abundances	If there is a linear relationship between the species abundances in the two communities then use Pearson's correlation coefficient, otherwise use Spearman's rank correlation coefficient
Gower's similarity coefficient	Can be based on mixed data types	This can make use of mixed data types including binary (e.g. presence and absence), nominal and/or continuous (abundance) data in the same analysis. For binary variables, comparisons between cases are weighted at 1 where a species is present in both, and zero when a species is absent from either (note that double zeros are ignored unless the data are identified as nominal variables). Missing data can also be accommodated using this method (see Quinn and Keough 2002)	$$S = \frac{\sum w_{ijk} s_{ijk}}{\sum w_{ijk}}$$ Where: s_{ijk} is the contribution provided by the kth variable (i.e. the value of a continuous variable or a score of a binary or nominal variable) w_{ijk} is the weight applied to the kth variable (up to a value of 1 and being given a value of 0 if the comparison is not valid)

*Where the following table represents the presence/absence of species from two communities:

Total no. of species $= a + b + c + d = n$

		Community A (total no. species $= a + c = A$)	
		No. of species present	No. of species absent
Community B (total no. of species $= a + b = B$)	No. of species present	a	b
	No. of species absent	c	d

be used in distance-based ordination approaches (e.g. multidimensional scaling) but must first be transformed into distances (see p. 287). Although classically used to compare sites, similarity measures can also be used to assess the similarity of species in terms of their ecological, behavioural or taxonomic characteristics. However, care should be taken if the characteristics are measured on different scales within the same analysis, since this can artificially weight some variables producing misleading results.

Recording descriptive statistics

Whether you are using summary statistics, diversity indices, population estimates or community similarities, as a rule of thumb, three types of numbers are usually required when recording descriptive statistics:

- the summary statistic (mean, median, percentage, diversity index, index of aggregation, population size, similarity index);

- where possible a measure of the variation, sampling error or other indication of the reliability of the estimate of the statistic, i.e. the standard error (e.g. for means and population estimations), the 95% confidence limits (for means) or the interquartile ranges (for medians);

- where possible the sample size (i.e. the number of samples upon which the statistic is based, the number of individuals and/or species, the number of sites).

When several statistics are being grouped into tables or charts, it is similarly important to report all three of these aspects. For example, where possible, standard error bars or 95% confidence limits should be displayed on point graphs (see Figure 5.8b or c) or bar graphs and the sample size (n) should be recorded in either the title or legend.

Testing hypotheses using basic statistical tests and simple general linear models

Usually the aims of a research project will necessitate moving beyond simple descriptions of different samples and/or variables in order to test defined null hypotheses (see Chapter 1, p. 7, for an introduction to creating null hypotheses). For example, you may wish to determine whether there are differences between samples on the basis of a measured variable. You could of course simply plot the means of the variable (with their standard errors or 95% confidence limits) and use these to discuss the differences observed (Figure 5.11). Because we generally rely on a sample rather than measuring the characteristic of every individual in the population we are studying, these summary

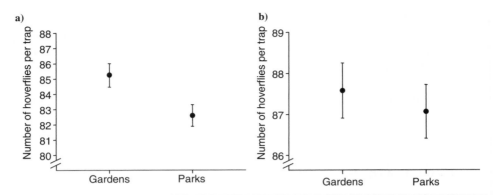

Figure 5.11 Comparison of the central tendency of two samples
(a) Non-overlapping distributions; (b) overlapping distributions.
The points plotted are means, the bars are 95% confidence limits.
Note that where variables are approximately normally distributed, and where the 95% confidence limits do not overlap, then there is a statistically significant difference between the means of the two samples, as in (a). However, just because 95% confidence limits overlap, as in (b), this does not mean that we can say there is no statistically significant difference between the means; we would need to use an inferential test to decide.

statistics (e.g. the mean) are only estimates of the true value. Thus, when we compare two means, we need a measure of likelihood that any apparent difference between them is not simply due to random variation in the sample. This requires an objective method of assessing whether the differences we can see numerically or graphically are indeed statistically significant differences, and the methods used are known as inferential statistics. Such difference testing requires the comparisons of one or more measured (or ranked) variables across two or more samples (the identity of the sample being defined by a nominal variable). So we could compare the heights of plants in fertile soils with those of plants in less fertile soil. Here the height is the measured variable and the category of fertility (fertile or not) is the nominal variable.

The basic principle is that we look at whether the between-group variation (explainable by virtue of being in different groups or samples) is larger than that within groups (inherent variation of each sample). The important components here are that we first set up a null hypothesis (that there is no significant statistical difference between our sample means). We test the null hypothesis using a statistical test that calculates a test statistic. We can use the size of the dataset to arrive at the probability (on a scale of 0 to 1) of obtaining that particular value of the test statistic by chance. Probabilities are sometimes given in percentages (i.e. the range from 0 to 1 is simply multiplied by 100 to give a range of 0% to 100%). If the probability is lower than the critical value (by convention this is usually taken as 0.05), then we can reject our null hypothesis of no significant difference and accept an alternative hypothesis of a significant difference.

We would then look at the means of the data to see which sample is larger than the other. However, if the resultant probability is higher than or equal to the critical value, then the probability that the test statistic was obtained by chance is high and hence we cannot reject the null hypothesis and would therefore state that there is no significant difference between our samples. Note that this critical value of 0.05 is a compromise value. If we set it too high, then we run the risk of rejecting the null hypothesis too readily leading to inappropriately thinking that there is a significant difference, or relationship or association when there is none (this is called a type I error). On the other hand, if we set the critical value too low, then we may accept the null hypothesis too often and think that there is no significant difference, or relationship or association when there should be (type II error).

Similarly, we can look for relationships between variables to see whether as one variable changes, another also does in a systematic and predictable way. This is an appropriate approach if there are two or more measured (or ranked) variables. The final example is when we are interested in whether the frequency of different categories of one variable have an association with categories of another (when both variables comprise nominal data). One example would be to examine whether grasshopper colour (with categories of 'brown' or 'green') is associated with the colour of the substrate (with categories of 'camouflage' or 'contrast'). We shall examine these examples in more detail later (p. 275).

We will summarise some of the commonly used statistical tests later in this chapter, but first there are some general principles that should be borne in mind (see the summary in Figure 5.12). Basically, the first stage in using inferential statistics involves the setting up of a null hypothesis (e.g. there is no significant difference between samples; there is no significant relationship between variables; there is no significant association between frequency distributions). Hypothesis testing examines the likelihood of a null hypothesis being true (and therefore accepted) or, conversely, false (and therefore rejected). All such tests involve the calculation of a test statistic that summarises the comparison (e.g. difference, relationship or association) and has known characteristics depending on the size of the dataset. The size of the dataset and this test statistic can be used to obtain the probability of obtaining the test statistic value by chance (in practice this step is usually done by the statistics software). Depending on the test, the size of the dataset is given as the total number of cases (or the number of cases in each sample, or variable or frequency), or is represented by a related concept, the degrees of freedom. If the resulting probability is below the critical value of 0.05 (5%) then we can reject the null hypothesis of no significant difference, or relationship or association. If we reject the null hypothesis then we need to look at the summary statistics to identify where any significant difference/relationship/association lies. If, on the other hand, the probability is equal to or higher than the critical value of 0.05 (5%) then we cannot reject the null hypothesis and must conclude that there is no significant difference, or relationship or association. Many students wonder why we start with the null hypothesis when we are looking to reject it. It might be helpful to use the analogy of a criminal trial. The

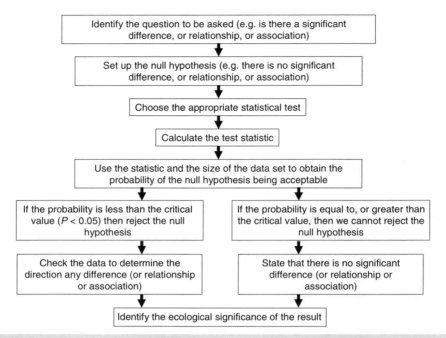

Figure 5.12 Summary of stages in using inferential statistics

defendant is on trail because the prosecuting authorities think that the person is guilty, but the jury are instructed to start from the premise that the defendant is in fact innocent. The defendant should only be found guilty (i.e. the innocent verdict is rejected) if there is evidence of guilt that is 'beyond reasonable doubt' (equivalent to using the critical value as the evidence for significance).

As a rule of thumb, you should state the hypotheses clearly, use appropriate tests, record the test statistic, the size of the dataset either as degrees of freedom or the number in the sample(s), the probability and any important summary statistics (e.g. means or medians) that help the reader to interpret the results. If large numbers of tests are employed, then tabulating such data is a sensible way forward. Note that when calculating a large number of similar statistical tests, we may need to adjust the critical value since the value usually used to reject the null hypothesis ($P < 0.05$) implies that we will accept that one in 20 tests will be significant by chance (see Box 5.5 for a further discussion of this and possible solutions). Where single tests are used, the results can be recorded in the text, thus:

> There were significantly fewer bird species in sparse woodland blocks (mean ± SE = 14.2 ± 0.23, $n = 10$) than in dense woodland blocks (16.1 ± 0.52: $t = 3.12$, df = 20, $P = 0.005$).

Of course such a statement in the results would then be discussed later in the light of the other results generated by the study and information from other works and in terms of the ecological significance of the results. The interpretation of any statistical result should be based on a knowledge of the way in which the results mesh against real life. For example, if a study had found that the number of species was significantly different between habitats, but that it was a result of very costly conservation work that only added a small number of relatively common species to the list, the question would be whether it was of ecological or practical significance.

Box 5.5   Testing for the significance of multiple tests

Note that if many similar statistical tests are employed there are potential problems if the critical value used is the conventional 5% level ($P < 0.05$). The 5% level provides a balance between being too ready to reject the null hypothesis (called a type I error) and being too ready to accept a null hypothesis (a type II error). In fact when we use $P = 0.05$, we are saying that we are happy to accept that we will get a type I error 5% of the time (i.e. 1: 20 tests). So, if we calculate 40 tests, and only get two significant results (where we reject the null hypothesis), then we will not be sure whether these are truly significant or merely a reflection of the probability of gaining test statistics of these values with this size dataset over so many tests. In fact, if the probabilities are close to the critical value, then we would be more likely to believe the latter statement.

Where large numbers of tests are employed, it is possible to simply reduce the critical level (to 0.01 or even 0.001). Other, more sophisticated methods, include the Bonferroni method of using a critical level based on 0.05 divided by the number of tests used. An extension to this (the sequential Bonferroni method) checks each result against a different critical value such that the most significant result is checked against a level of $0.05/n$ (where n is the number of tests being examined), the next most significant against $0.05/(n-1)$, the next most significant against $0.05/(n-2)$, etc. until no more significant values are found (Rice 1989). However, this correction may inflate type II errors and other options have been suggested (e.g. see Sokal and Rohlf 1995). A further solution is to perform a more sophisticated multivariate analysis that takes all such factors into account in one analysis – see p. 276.

Differences between samples

One of the most basic types of question we are likely to ask is whether two or more groups differ in terms of some measured characteristic. For example, we could examine whether there are differences in the height of a particular species of plant (e.g. nettles) growing at the edges of woodlands, and those growing in gaps in the centre of woodlands. Here, a series of patches at the edge and a similar number of centrally sited patches could be surveyed to measure the heights of plants found in each area. The sampling technique (and effort put in) should be as similar as possible, as should all other characteristics (e.g. size of patch, type of woodland, etc.). Assuming that the data are normally distributed, a *t* test would be an appropriate method of analysis for this example.

It is also possible to assess differences between sets of data measured on an ordinal (ranked) scale. For example, we could be interested in potential differences in the amount of light getting through the canopies of deciduous and coniferous woodlands, where the light levels are scored on a ranked scale (e.g. from 1 to 10, where 1 is very dark, and 10 is very bright). If the measured variable is not normally distributed (and cannot be transformed – see p. 244) or is on an ordinal scale, then a Mann–Whitney U test should be used.

For both of these questions, we can describe each individual entered into the analysis in terms of two variables, a nominal variable that identifies the sample (e.g. sample A or sample B, shaded or unshaded) and a measured (interval/ratio) or ordinal variable that is a series of measurements (e.g. numbers of animals, height of plant). The question could be extended to compare more than two groups, for example asking whether the number of decomposers found under logs differs between three tree species (e.g. from oaks, beech and pine trees). Tests for such questions include simple general linear models, including analysis of variance (ANOVA) for data where the measurements are normally distributed (or can be transformed to become normally distributed), and non-parametric analysis of variance using ranks (Kruskal–Wallis analysis of variance using ranks) for non-normal (untransformed) or ranked data.

The statistical tests used to answer these types of difference questions and the statistics that should be recorded when doing each test are listed in Table 5.5. The table goes on to include other tests for designs where two or more samples are being compared using matched data. For example, if you compared the numbers of invertebrates found above and below a sewage outflow into a number of different streams, then the data are matched by stream and we would need a matched-pairs test or repeated measures analysis. More than one factor can be incorporated into the comparison, for example if we were interested in comparing the numbers of birds found in upland and lowland coniferous and deciduous woodlands, we would have two factors to compare: altitude and type of woodland. This could be analysed using a two-way ANOVA, with altitude and type of woodland providing the 'two ways'. There would actually be three analyses taking place simultaneously in looking at the impact on the numbers of birds of (1) altitude, (2) type of woodland and (3) the combination of altitude and type of woodland. This last (new) variable is termed the interaction and looks at whether altitude has the same effect in each woodland type (i.e. comparing upland coniferous against lowland coniferous and upland deciduous against lowland deciduous) and vice versa. In such analyses, it may be necessary to specify which factors are fixed (such as those determined as part of the project design, e.g. site or date) and which are random (i.e. those that are measured, e.g. male/female, adult/juvenile, species a, b or c). Such analyses are discussed in Wheater and Cook (2000).

Bear in mind that the more complicated the question, the more complex the analysis will be, and the more difficult the interpretation. Although the comparison of two samples is easy to interpret (if found to be significantly different, then the sample with the largest mean is larger than the sample with the smaller mean) where more than two

Table 5.5 Statistics that should be reported for difference tests
These are used when one variable is measured (interval/ratio) or ordinal (ranked) and the second variable is nominal (with two or more categories). See Wheater and Cook (2000) for further details.

Statistical test	Test statistic	Size of data set	Probability	Measure to be interpreted if the test result is significant*
t test	t	df	P	Mean values for each sample
Paired t test	t (sometimes labelled t_p)	df	P	Mean difference between sample pairs
Mann–Whitney U test	U or z	n for each sample	P	Median
Wilcoxon matched-pairs test	T	n (the number of non zero differences between pairs)	P	The number of positive differences and the number of negative differences between pairs of values
Analysis of variance	$F_{b,w}$	df for between factors (b) and df within factors (w)[#]	P	The mean values for each factor using a minimum significant difference test (e.g. Tukey's multiple comparison test[‡]) to determine where any differences lie
Repeated measures ANOVA	$F_{b,e}$	df between factors (b) and error df (e)[†]	P	The mean values for each factor - multiple comparison tests can also be used here[‡]
Kruskal–Wallis analysis of variance using ranks	H	df	P	Rank means (arithmetic average of a series of ranked values) for each factor using a minimum significant difference test (e.g. Dunn's multiple comparison test[‡]) to determine where any differences lie
Friedman's matched-group analysis of variance using ranks	F_r	Number of matched groups (g) and the number of samples (k)	P	Rank means of samples using a minimum significant difference tests (e.g. Nemenyi's multiple comparison test[‡]) to determine where any differences lie

(continued)

Table 5.5 (Continued)

Statistical test	Test statistic	Size of data set	Probability	Measure to be interpreted if the test result is significant*
Two-way ANOVA	$F_{b1,e}$ $F_{b2,e}$ $F_{int,e}$	df between factor (b_1, b_2 or the interaction – int) and the error df (e)[‡]	P	First the significance of the interaction term is examined, if significant, then each combination of factors $b_1 \times b_2$ is examined using a minimum significant difference test (e.g. Tukey multiple comparison test[†]) to determine where any significant differences lie. Assuming that the interaction term is not significant, then each factor (b_1 and b_2) is examined separately in the same way as for one-way ANOVA.

*These values should be reported whether or not the test is significant for a number of reasons, including so that other people can use your data (e.g. for subsequent meta–analysis where the results from several independent studies are combined to see if there are any over arching trends).
[#]In some statistics programs these are called residual or error degrees of freedom.
[†]In some statistics programs these are called residual degrees of freedom.
[‡]See Wheater and Cook (2000) for worked examples using multiple comparison tests.

samples are being compared, the interpretation can be more complex. Box 5.6 describes how multiple comparison tests are needed to be able to identify which individual categories are different from the others when more than two categories are being compared.

Box 5.6 Multiple comparison tests

When comparing two means (or medians), if you find a significant difference between them, you can simply look at the means and see which is larger than the other. However, if you find an overall significant difference between more than two means (or medians), then identifying which means differ from the others requires a multiple comparison test. Say we have found a significant difference between three types of woodland (deciduous, mixed and coniferous) in terms of the species diversity and we have looked at the means and found that the largest is the mixed woodland group, followed by the deciduous woodlands and finally the coniferous woodlands are the least diverse, then the significant result could be caused by one of the following scenarios:

- mixed > deciduous > coniferous (i.e. each are different to the next in line);

- [mixed ≈ deciduous] > coniferous (i.e. mixed woodlands and deciduous woodlands are approximately the same but both are bigger than the coniferous woodlands);

- mixed > [deciduous ≈ coniferous] (i.e. the mixed woodlands are bigger than either the deciduous or coniferous woodlands, which are not different from each other);

- mixed > coniferous, mixed ≈ deciduous, deciduous ≈ coniferous (i.e. mixed woodlands are bigger than coniferous ones but not than deciduous, which themselves are not bigger than the coniferous).

There are many multiple comparison tests (most of which calculate a minimum significant difference between pairs of samples) to enable us to identify the most likely pattern. One of the most commonly used when comparing mean values is the Tukey multiple comparison test, which is included in many statistics programs. See Wheater and Cook (2000) for details of how to calculate this and the alternatives that can be used when datasets are not balanced or when ranked data are being compared. Sokal and Rohlf (1995), and Zar (2009) give details of a wide range of alternative tests.

Relationships between variables

A second question that we commonly wish to ask is whether two (or more) measured (interval/ratio) or ordinal (ranked) variables are related. For example, we might be looking at dock plants and wish to know whether the number of insects found on them could be related to the height of the plants. Obviously, for each plant we would collect data on the height and the number of insects. Unless we are looking at a more sophisticated analysis such as multiple regression (p. 276), other factors (e.g. whether the plants are in the sun or shade, or the time of day at which the collections were taken) should be kept as consistent as possible. Now we would have pairs of data that could be analysed using the techniques listed in Table 5.6, which also lists the statistics that should be documented when using these methods. If both variables are measured on interval/ratio scales (e.g. height of plant and number of animals), then we can plot the relationship between them on a scatterplot (Figure 5.13). We can test to see if the data from two such measured variables have a linear relationship using a technique called Pearson's product moment correlation analysis. This method generates a correlation coefficient (r), which can have values between -1 (for a perfect negative relationship) and $+1$ (a perfect positive relationship). Note that a positive relationship means that as one variable increases then so too does the other. A negative relationship is where as one variable increases, the other decreases. This method should only be used if the scatterplot shows that there is an approximately linear (rather than curved) relationship between the two variables, and that there are no extreme cases (known as outliers – p. 247). The

Table 5.6 Statistics that should be reported for relationship tests
These are used when there are two (or more) measured (interval/ratio) or ordinal (ranked) variables. See Wheater and Cook (2000) for further details.

Statistical test	Test statistic	Size of data set	Probability	Measure to be interpreted if the test result is significant*
Pearson's product moment correlation coefficient	r	df	P	The value and sign of the correlation coefficient (r)
Spearman's rank correlation coefficient	r_s	n (the number of data pairs)	P	The value and sign of the correlation coefficient (r_s)
Kendall's coefficient of concordance[#]	W for n below 8 (or X^2 for values of $n > 7$)	k and n when n is below 8 (df is used for $n > 7$)	P	The value of the test statistic, which ranges from 0 (no relationship) to 1 (fully related)
Simple linear regression analysis	$F_{1,w}$	df for the regression line (1) and the df within factors (w)	P	The equation of the regression line and the coefficient of determination (r^2)[†]

*These values should be reported whether or not the test is significant for a number of reasons, including so that other people can use your data (e.g. for subsequent meta-analysis where the results from several independent studies are combined to see if there are any over-arching trends).
[#]Where: $k=$ the number of groups; $n=$ the number in each group.
[†]The coefficient of determination (r^2) has a range between 0 and 1. This value can be multiplied by 100 to produce a percentage scale (R^2).

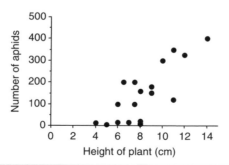

Figure 5.13 Example of a scatterplot
Showing the hypothetical relationship between the height of plants and the number of aphids found on them.

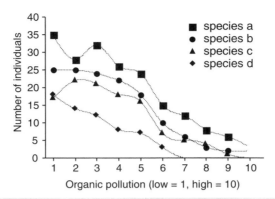

Figure 5.14 Trends of invertebrate numbers with organic pollution

scatterplot should always be inspected even if it is not ultimately used to display your results to your audience.

Similar questions could be asked if either or both of the variables are ranked, for example examining the relationship between the colour of cabbages (ranked on a scale from light to dark green) and the number of aphids found on the plants. The technique used here would be a Spearman's rank correlation coefficient. An extension of this technique (Kendall's coefficient of concordance) can be used to examine potential relationships between more than two measured or ranked variables. For example, we may be interested in whether the numbers of each of four different species of aquatic invertebrate change in some way with changing organic pollution levels (Figure 5.14).

It might be that we wish to extend our analyses of relationships to make predictions regarding the value of one variable from a knowledge of another. For example, we might examine the impact of certain pesticide concentrations on the number of greenfly attacking rose plants. If we discover a straight-line relationship between the two, we could reasonably expect to be able to predict the level of infestation that would be expected with other concentrations of the same pesticide. Note that we can only make predictions for concentrations within our experimental range and should not extrapolate to higher or lower pesticide concentrations. The technique we would use here is called linear regression analysis, which usually requires measurement (interval/ratio) data and can be visualized graphically by calculating a regression line and adding it to the appropriate scatterplot (Figure 5.15). Even though you do not have to do this by hand, it is helpful to understand the principles behind this method. There are several methods of finding the best fit line, one of the most commonly used of which is the least squares approach. Since the best fit line is the one that is the closest possible to all of the points, if the distance from each point to the line is measured and these distances are summed, then the line of best fit is the one that has the smallest sum of such distances. Before summing the distances, they are squared to remove any negative values, hence

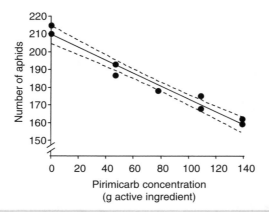

Figure 5.15 Regression line between the number of aphids found at different levels of pirimicarb (pesticide) application
Dotted lines are the 95% confidence intervals. Number of aphids = 209.6 − 0.36 × Pirimicarb concentration ($R^2 = 96.3\%$).

the name the 'least squares' method. Different types of regression analysis need to be used depending on whether the independent variable can be measured without error or is fixed by the experimenter (Model I) or is not fixed by the researcher (Model II). See Zar (2009) for further details regarding linear regression analysis.

A regression equation can be calculated that describes the regression line and can be used to substitute any value of x in order to calculate (predict) the expected value of y. This equation is usually expressed as:

$$y = a + bx'$$

where y is the value of the dependent that you wish to predict; a is the intercept or the point at which the regression line crosses the y axis (i.e. the value of y when $x = 0$); b is the slope of the line; x' is the value of the independent variable for which you wish to predict a value of the dependent variable.

The equation can be associated with a calculation to attach confidence limits to any calculated prediction (the confidence limits will be smaller at values towards the middle of the dataset that has been used to calculate the regression equation and larger towards the extreme values). When regression lines are plotted on appropriate scatterplots, care should be taken to avoid extending the line beyond the range of measured points. In order to be satisfied that regression analysis is appropriate for the data being examined, the distribution of the points above and below the line should not form any sort of pattern (i.e. there should be a similar numbers of points above and below the line at each end and at the middle of the distribution). There are more sophisticated ways of investigating whether the distribution is suitable for regression (see Sokal and Rohlf 1995, or Zar 2009 for further details).

For both regression analyses and Pearson's correlation analysis, we can obtain a measure of the strength of the relationship by calculating the coefficient of determination. This is simply calculated as the square of the correlation coefficient (r^2) and so lies between 0 and 1, where 0 indicates no relationship and 1 a perfect match between the variables. Usually, r^2 is multiplied by 100 to give R^2, which indicates the percentage match between the variables. This R^2 value can be used to annotate the figure, as can the regression formula (Figure 5.15).

If you are having problems with your linear regression model (i.e. you have lots of outliers and a low r^2), you may suspect that it is because the relationship is non-linear (Figure 5.16). For example, if measuring the relationship between population growth and time, there might be periods of growth that barely rise (lower asymptote) followed by an exponential period that eventually reaches saturation (upper asymptote). This form of growth produces S-shaped, sigmoidal curves that are strongly non-linear

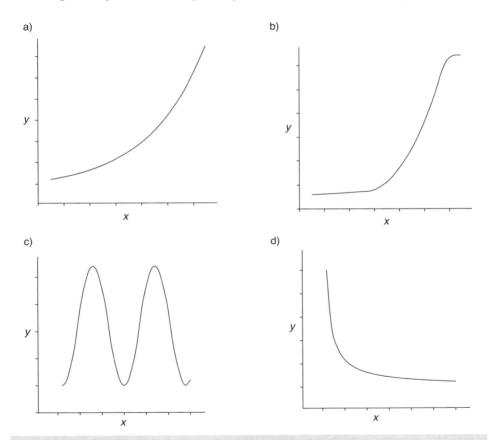

Figure 5.16 Examples of common non-linear graph types in ecology
(a) Polynomial – cubic; (b) sigmoid – Gompertz; (c) polynomial with periodicity (cycling); (d) exponential decay.

(Figure 5.16b). Exploring this statistically is not as hard as may seem, and many standard statistical programs produce a vast array of non-linear models: SigmaPlot[6] is one of the best programs to get you started.

Associations between frequency distributions

A third common type of research question relates to potential associations between (or independence of) the frequencies of categories within two variables. For example, we might be interested in possible associations between animal and plant colouration (e.g. in relation to the incidence of predation of the animals concerned) by examining the frequency of green and brown insects found on green and brown plants. In this case, the frequency in each category (e.g. the number of green insects on green plants, the number of green insects on brown plants, the number of brown insects on brown plants and the number of brown insects on green plants) would be compared using the techniques listed in Table 5.7, which also indicates the statistics that should be recorded when using these methods. Such questions could easily be extended to add more characteristics (increasing either, or both, the number of categories of insect colour and/or plant colour). Such data are usually viewed in a contingency table (Box 5.7). The statistical test (chi-square test also sometimes known as Pearson's chi-square test) examines the cells within the contingency table to see if they differ significantly from the expected values calculated from the total values.

Table 5.7 Statistics that should be reported for tests used to examine associations between two frequency distributions
These are used when there are two nominal variables, each consisting of two or more categories. See Wheater and Cook (2000) for further details.

Statistical test	Test statistic	Size of data set	Probability	Measure to examine if the test is significant*
Chi-square test for association	X^2	df	P	Compare the observed values with the expected ones generated by the test
Chi-square test for goodness of fit	X^2	df	P	Compare the observed (measured) values with those from an expected distribution based on experimental or survey data or from the scientific literature

*These values should be reported whether or not the test is significant for a number of reasons, including so that other people can use your data (e.g. for subsequent meta-analysis where the results from several independent studies are combined to see if there are any over arching trends).

---

[6] http://www.sigmaplot.com/products/sigmaplot/sigmaplot-details.php

Box 5.7 Using a contingency table in frequency analysis

As an example, if we had 68 grasshoppers in total, 31 of which were green and 37 were brown, and if 28 of the grasshoppers were found on green plants, and 40 were found on brown plants, then if grasshopper colour was random with respect to plant colour, we would expect about 13 green grasshoppers to be found on green plants. This expected value is found by first finding the probability of an animal being green ($31/68 = 0.456$) and multiplying this by the probability of an animal being found on a green plant ($28/68 = 0.412$). The resulting probability (0.188) is now multiplied by the total number of animals (68) to get the expected number of green grasshoppers on green plants (12.78). Since in our example we found 25 (see contingency table below), we have observed about twice what we would expect.

		Plant colour		
		Green	Brown	Totals
Insect colour	Green	25 (37%)	6 (9%)	31 (46%)
	Brown	3 (4%)	34 (50%)	37 (54%)
	Totals	28 (41%)	40 (59%)	68 (100%)

A related type of analysis is the examination of the association (or goodness of fit) between observed frequencies and a previously expected outcome. For example, we might be interested in whether we are more likely to find male or female spiders sitting in webs within thistle plants. Since our expectation would be that there are approximately equal numbers of males and females, any deviation from a 50:50 split (the expected ratio) would be of interest.

In both cases, we are looking at the frequencies of nominal (or categorical) data types and the statistics we would record are the test statistic (X^2), the degrees of freedom (df) and the probability (P). If a significant association or goodness of fit is found, then we need to examine the contingency table more closely and for simple tables (e.g. 2×2) we can look at which diagonal has the highest percentages: in the example in Box 5.7, the diagonal containing insects and plants of the same colour has the highest percentage values that, assuming that the chi-square test was significant, would indicate a positive association between insect/plant colour (if the other diagonal had been the highest, it would have indicated a negative association between insect/plant colour). For more complicated situations (more than 2×2 contingency table), the expected and observed valued in each part of the table should be compared (most software has an option to display these) to see where the major levels of difference occur. Objective methods are available to identify the significant components (if this is not obvious). Sokal and Rohlf (1995) gives more discussion about this and other issues relating to this type of analysis.

More advanced general linear models for predictive analysis

Multiple regression

We have already seen a relatively simple predictive analysis in linear regression (see p. 271). Imagine now that we are interested in examining a wide range of environmental variables that could influence the presence or numbers of a particular species or group of species. Here, we could employ a series of correlations (see p. 269) to assess which are related to the dependent variable. However, there is quite likely to be overlap between some of these variables and we should take this into account. Two techniques that will be useful in understanding more complex situations are multiple regression and discriminant function analysis, which will be briefly considered here. You are also encouraged to read the example of the binomial generalized linear model in Box 5.10 to understand how multiple variables can be modelled and their interactions identified.

Multiple linear regression is used to obtain a least squares prediction of a measured dependent variable using two or more independent variables such that the amount of unexplained variation is reduced. There is flexibility in the type of data that can be included as independent variables, which may be either measured or categories. When using categorical variables, these must be binary (i.e. have only two options, e.g. presence/absence or male/female). Dummy variables can be used to create binary variables from those with more categories (Table 5.8).

There are several ways of conducting a multiple regression, including:

- simultaneous use of all the potential predictor variables entered;
- inclusion of the predictor variables one at a time starting with the most significant one (forward stepwise multiple regression);
- deletion of the variables from the model one at a time starting with the least significant one (backward or reverse stepwise multiple regression).

Table 5.8 Using dummy variables
Example of how two dummy variables (large/not large and medium/not medium) can cover three categories (large, medium and small) within a nominal variable. Once you have coded 'small' as being neither large nor medium, we know that the individual must be small without needing a 'small/not small' dummy variable. Thus there is always one fewer dummy variable than there are categories

		Dummy variable	
		Large	Medium
	Large	1	0
Original variable	Medium	0	1
	Small	0	0

In all cases, changing the model is governed by rules set up to maximise the prediction of the model, whilst minimizing the number of variables being used. Using multiple regression we obtain an indication of which of the independent variables are significantly related to the dependent variable (at $P<0.05$) and a regression equation (from which we can predict new values of the independent variable):

$$y = a + b_1 x'_1 + b_2 x'_2 + b_3 x'_3 + \ldots + b_n x'_n$$

where y is the value of the dependent value you wish to predict; a is the intercept (i.e. the value of y when all the independent variables are zero); b_1, b_2, b_3 and b_n are the multipliers of each independent variable; x'_1, x'_2, x'_3 and x'_n are the values of each of the independent variables that are used to predict the value of y.

A word of warning: it is appropriate to mention here the difference between multiple linear regression and generalized linear models. Although both techniques use multiple linear predictors, the regression technique assumes that the residuals are normally distributed, and you need to ensure that your data meet this assumption. In addition, count data may be better modelled as a log-linear generalized linear model rather than using a transformation to satisfy normality. See Sokal and Rohlf (1995) for further details on using multiple regression analysis.

Analysis of covariance and multivariate analysis of variance

Where we wish to see whether a mixture of categorical variables and measured variables have an effect on a measured dependent variable, we can use an analysis of covariance (ANCOVA). This operates in a similar way to ANOVA in comparing the values of a measured variable across categories in one or more nominal variables (known as independent variables). However, here the relationship between the dependent variable and one or more measured variables that may be related (co-vary with) is also taken into account. ANCOVA combines one-way or two-way ANOVA with linear regression and is used when the dependent variable is influenced by a covariate. In practice ANCOVAs adjust the treatment effect size by the inclusion of a continuous covariate. It is easy to use complex analyses and just look at the main effects, but this would be wrong without further model checking. There are two additional assumptions that ANCOVA models demand:

- homogeneity of the regression slopes (HRS) – see Box 5.8 for how this can be an important factor in ANCOVA;
- independence between the covariate and treatment effects.

Perhaps the easiest for ecologists to violate (and ignore) is the HRS. We can easily test this by modelling the interaction between the covariate and all combinations of the independent variables (factors). If we meet the assumptions of HRS then we should not encounter any significant effects for these interactions using a test statistic known as Levene's test. Box 5.8 shows an example of the use of ANCOVA.

Box 5.8 Analysis of covariance

Let us look at an example to see how analysis of covariance works using the numbers of freshwater fish as a dependent variable, and the effects of altitude (upland or lowland) and habitat type (open water or reed bed) as factors (nominal variables). If in a study we were unable to standardize for the size of each water body, and since we do not wish to produce a model that is merely an artefact of size of water body, it is important to include size as a covariate to correct the treatment means. We start by first transforming our fish (dependent) variable to the log series and checking that the new variable (log fish numbers) now follows a normal distribution. We then proceed with the test.

We first use an ANOVA to see what the fixed effects are having on the dependent variable. To summarise the results, altitude, water body type and the interaction between altitude and water body type are all highly significant ($P<0.05$). Looking at the estimated marginal means of log fish number we find the source of the interaction (the lines are crossed) and we see that the means for the upland open water are apparently the highest. Although this is true in this model we have not accounted for the size of the water body in our analyses so those values are vastly inflated (i.e. larger expanse of water, more fish). So we re-do the analysis (again on log fish number), but this time we include water body size as a covariate.

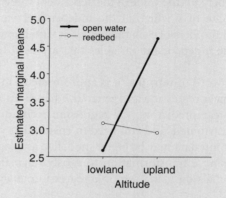

ANOVA estimated marginal means of log fish number

We find that altitude drops out of the model as a non-significant factor ($P=0.216$) changing entirely the estimated means and leaving only habitat type and the covariate size of each water body as significant effects ($P <0.05$) in the model. In this example, we find that there is a significant effect of the habitat*size covariate. This is supported by a statistical test (Levene's test), which shows a significant outcome ($F_{3,50} = 3.912$, $P=0.014$). This means that the groups are not homogeneous and the homogeneity of regression slopes assumption has been violated.

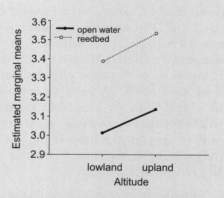

ANCOVA adjusted means showing the overall effect of habitat type using water body size as a covariate

The outcome of the Levene's test indicates that our adjusted means are still not appropriate for the analysis and we should abort the ANCOVA. All is not lost because we may still have the option of a mixed effects model using a restricted maximum likelihood (REML) approach, although this is rather complicated and is beyond the scope of this book (see Crawley 2007 for further details).

More than one dependent variable can be modelled simultaneously using multivariate analysis of variance (MANOVA). This might be a useful technique, for example, when examining whether the numbers of insectivorous and seed-eating birds differ between upland and lowland coniferous and deciduous woodlands. Since the two types of birds might also impact on each other, this technique includes both as dependent variables. Where this type of analysis is combined with ANCOVA, a technique called multivariate analysis of covariance (MANCOVA) is used. So if we add in an interest in the size of the woodlands, we could use a MANCOVA, but the situations in which this type of model are preferred over more elegant models (e.g. generalized linear models or mixed effects models) are few. Zar (2009) gives further details of using ANCOVA and MANOVA, and Scheiner (2001) gives a very thorough worked example of MANOVA in ecology. See Tabachnick and Fidell (2007) for further details of MANCOVA.

Discriminant function analysis

Discriminant function analysis (DFA) is used to examine whether a number of independent variables can discriminate between two or more groups. Effectively it reduces the original variables to a new discriminant score in such a way as to maximise the differences between predetermined groups (i.e. a reduction from many dimensions to a single one). A classification table is used to test the accuracy of assigning groups correctly by comparing the original allocation of sites to groups with that derived using the discriminant functions. The groups can be viewed in a classification table that separates those that are correctly assigned to their group from those that are not. The ideal test of the model is to collect further data and use the best explanatory variables to predict whether a new individual will fit the model. For an example of the use of predictive discriminant function analysis see Box 5.9.

> **Box 5.9 Using classification tables in predictive discriminant function analysis**
>
> Let us go through an example to understand more about how discriminant function analysis works. If we visit a large number of sites (say 1 km × 1 km grid squares) and classify them as either having a badger sett present or not, we might wish to know whether this outcome can be predicted using a collection of habitat characteristics collected at each site (i.e. independent variables). We might also like to know which are the most important variables and which are redundant (i.e. not required in the predictive model). So we proceed with a classification table to derive the test results. The grouping factor is sites with a badger sett or not (coded as a binary variable) and the independent variables are habitat characteristics, e.g. altitude, land use type, etc. Within the analysis a correct classification was assignment of the square to the correct group, while misclassification was assignment to the wrong group. Using this technique we are able to establish that the explanatory variable predicted the sites with setts (71.7%) and sites without (82.7%) very successfully (see table below). In the example shown, the most important variables were sloped land at elevations between 251 m and 300 m, the presence of grassland and deciduous woodland and the absence of urban and suburban land cover types (Wright, Fielding and Wheater 2000).

Classification table

		Predicted group		Number of samples
		Sites with setts	Sites without setts	
Actual Group	Sites with setts	43 (71.7%)	17 (28.3%)	60
	Sites without setts	104 (17.3%)	496 (82.7%)	600

Further statistics can be used to test whether such a classification table (also called a confusion matrix) indicates a classification significantly better than expected by chance. Misclassified data can be further examined to see, for example, whether they are unusual ecologically (e.g. recently used but now disused setts). Note that a much larger sample of non-sett squares was used than for sett squares. This is because those without badger setts would be expected to be much more variable than those with setts within them. See Tabachnick and Fidell (2007) for further details of this technique.

Generalized linear models

Generalized linear models are becoming increasingly popular in ecology, because data often do not conform to the assumptions underlying the linear model (ANOVA and linear regression), in particular that errors should be normally distributed with equal variance. First, we shall discuss generalized linear models and why they are different from general linear models (such as ANOVA and least squares regression) using count data as an example, then move on to a worked example of a binomial regression (Box 5.11), a type of generalized linear model.

In t tests, ANOVAs and least squares regression, the assumption is that the variance is constant and that modelled count data have a normal distribution, which implies no abrupt truncation of the range. However, in reality counts do not exist below zero, the variance of count data tends to increase with the mean, and for small counts the data take only discrete values and cannot be considered continuous. Sometimes, we can log-transform the data to approximate the assumptions of the linear model but the interpretation of transformed count data (for instance dealing with the fact that the log of $17 = 1.23$) is rather clumsy since we are used to handling count data as whole numbers. Furthermore, transformed data must be back-transformed, which may produce predictions outside the valid range of the data. If, however, we attempt to find the correct distribution to model the data, we arrive conceptually at generalized linear models, a much more flexible approach than linear models. There are several components to a generalized linear model that require some understanding:

- The distribution of the dependent variable is generalised to be a member of a group of distributions called the exponential family of distributions. This includes the binomial distribution (often used to model proportions) and the Poisson distribution (often used to model counts). One feature of this family is that the distributions are specified in terms of their expected value.

- The variance is usually a function of the expected value and estimated by the dispersion parameter. For model simplicity (for both binomial and Poisson distributions) the dispersion parameter is often fixed at 1. The dispersion parameter provides information on the necessary scaling of the data, so that the expected pattern of variation is preserved, but is rescaled to match the variation observed in the data.

- The link function is a non-linear function that relates the underlying explanatory variables to the expected value for each observation. For each type of distribution there is one mathematically natural (termed canonical) link function (e.g. logit for binomial, log for Poisson). For binomial data, the link function must map from the proportion scale (0–1) onto an unbounded range; this is achieved by functions such as the logit (Box 5.11). For Poisson data, the link function must map from positive real numbers onto an unbounded range, and the natural log function achieves this simply.

- The linear predictor relates the explanatory variable(s) to the transformed expected value in the form of a linear model.

You will recall that in a simple linear regression, we might use the coefficient of determination (R^2) as a measure of fit. Unfortunately, no such statistic exists for generalized linear models. How then do we know when we have a good model? Surprisingly, this is not as straightforward as you might think. To help us assess whether or not our model is acceptable the following diagnostics are routinely used:

- The deviance compares the predicted value with the transformed value of y, also called the log-likelihood ratio statistic: the lower the deviance ratio the better the fit. It is possible in some programs to use a related log-likelihood ratio statistic called Akaike's Information Criterion (AIC) for the same purpose.

- The residual mean deviance gives a measure of the goodness of fit of a model and should be equal to the value of the dispersion factor expected for the chosen distribution.

- The level of dispersion. If the residual mean deviance is lower or higher than the expected value (1 for binomial and Poisson data) then the model is 'under-dispersed' or 'over-dispersed' respectively and we account for this by estimating the dispersion parameter, from the residual mean deviance. Ignoring an over-dispersed model is serious because it has the effect of underestimating the standard errors of the estimates. Over-dispersion incurs a greater likelihood of detecting a significant effect when none is present, known as a type I error (see p. 263).

- Residuals are plotted as the standardised deviance residuals against normal scores. The reason for this is that we should find that the residuals fall along a straight line, indicating that our assumptions about the data are correct.

- When fitted values are plotted against normal scores, we should find that there is no pattern and therefore no trend if our model is a good one.

- High leverage values are indicative of undesirable outlying data points that are having too much influence on the model fit. Sometimes, removal of those data points from the model is the only cure.

- Wald statistics are an efficient way to make inferences about the independent variables and may indicate if the model is 'over-parameterised' (i.e. too many variables in the model). When Wald statistics indicate that terms are not significant then consider dropping them.

See Crawley (2007) for a discussion of model-checking procedures.

Box 5.10 Generalized linear model: a worked example using a binomial regression

We will consider an example of a research project where 72 blue tit nests across three habitat types (farmland, woodland and gardens) were visited once each to assess the proportion of eggs that have fledged. We suspect that an estimate of insect biomass measured as counts of insects within a 100 m range of each nest will be an important predictor of survivorship simply because adults need food to produce eggs and maintain their young. Before we go any further, there are a few caveats that we must recognise. The experiment, like many in ecology, is unbalanced in that the numbers of nests visited is not equal across habitats (although this is not a problem for a generalized linear model). We also have not measured some potentially important variables associated with 'nest' and 'parents', which are really random variables (i.e. variables that are drawn from a much larger population and are likely to contain some explanatory power) and we are assuming that the proportion of area for foraging is equal and non-overlapping between habitats, which may not be true. Hence, we proceed cautiously with a simple model.

We first fit insect biomass to the proportion of fledged birds, specifying a binomial distribution, a logit link and a parameter estimate fixed at 1. Although significant, this is not a good model because the residual mean deviance is greater than 1 (over dispersed – suggesting an incomplete model) and there is a lot of scatter around the fitted model.

Summary of binomial regression model (y = insect biomass)

Source	df	Deviance	Mean deviance	Deviance ratio	P
Regression	1	36.2	36.203	36.20	<0.001
Residual	70	179.4	2.563		
Total	71	215.6	3.037		

We move on and consider that there are probably different outcomes depending on habitat, so we fit habitat additively as a factor ($y=$ insect biomass + habitat). The residual mean deviance (of 1.190) is closer to the fixed dispersion parameter (i.e. 1), which is encouraging.

Summary of binomial regression model
($y=$ insect biomass + habitat)

Source	df	Deviance	Mean deviance	Deviance ratio	P
Regression	3	134.7	44.903	44.90	<0.001
Residual	68	80.9	1.190		
Total	71	215.6	3.037		

We appear to be heading in the right direction when habitat is added, but the model does not estimate well the effect of insect biomass in gardens because in this technique the slope of the regression models are assumed to be the same. However, on inspection those for farmland and woodland look to be better fits than does the one for gardens.

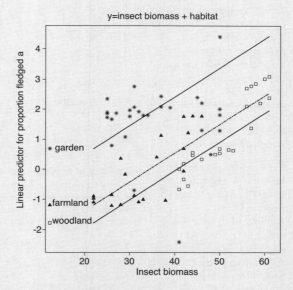

We now move to the maximal model, where we allow insect biomass to interact with habitat and derive different slope parameters using the formula $y=$ insect biomass × habitat. By doing so, all effects will be reported but firstly we see that for the model overall, the residual mean deviance is very agreeable and close to unity (0.9301), and the deviance ratio is also at its lowest (30.85) in the maximal model.

Summary of binomial regression model
($y=$ insect biomass × habitat)

Source	df	Deviance	Mean deviance	Deviance ratio	P
Regression	5	154.2	30.8472	30.85	<0.001
Residual	66	61.4	0.9301		
Total	71	215.6	3.0369		

Secondly, the plot of the model suggests that gardens are less dependent on insect biomass than those wild habitat (they have a much shallower slope) – perhaps supplementary food on bird tables is an explanatory term that we need to consider in future.

We can also inspect in detail the estimates of the parameters. Insect biomass is a significant single parameter, but the habitat parameters should be interpreted a little differently as they are a contrast between farmland (which we have specified as the reference) relative to woodland and gardens.

Gardens have a significantly higher proportion fledging relative to farmland but woodland has a significantly lower proportion, hence the negative t value. Gardens significantly interact with insect biomass but woodlands seem a little different in their interaction. Here, the lack of significance is a suggestion that a model with insect biomass and habitat interacting was not necessarily required for woodland.

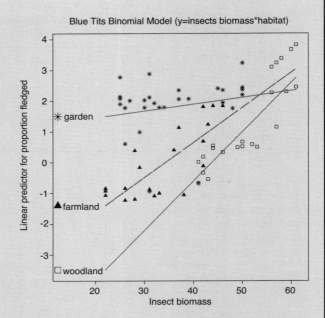

Summary of binomial regression model parameters
(y = insect biomass × habitat)

Parameter	Estimate	SE	t	P
Constant	−3.898	0.779	−5.00	<0.001
insect_biomass	0.1137	0.0228	5.00	<0.001
habitat garden	4.92	1.06	4.64	<0.001
habitat woodland	−3.10	1.46	−2.12	0.034
insect_biomass. habitat garden	−0.0919	0.0303	−3.03	0.002
insect_biomass. habitat woodland	0.0462	0.0342	1.35	0.177

Remember that we are not modelling counts but proportions (i.e. the number of successes). If we wanted to know about the numbers of fledglings by habitat, a log linear model (e.g. Poisson distribution with a log link) could be used to show that there were many more fledglings in woodland than in farmland. See Crawley (2007) for more details about generalized linear models.

Extensions of the generalized linear model

The keyword to all the generalized linear models so far is 'linear' but there is another approach that makes trend lines bend, called generalized additive models (GAMs). In this method, counts are modelled in the same way as a log-linear generalized linear model (i.e. a Poisson distribution with a log link) but with a smoothing function (smoothing splines) on, for example, a year effect that makes the trend bend over time. Box 5.11 shows an example of a log linear GAM. Since the model examines subsequent data points in relation to an index year (usually the first year), problems may occur if there is high uncertainty in counts recorded in the first year. Such uncertainty produces wide confidence intervals that are then inherited along the series producing poor predictions of the mean count. In these circumstances it may be necessary to 'back-fit' the model, starting at the last year rather than the first. See Fewster et al. (2000) and Crawley (2007) for more information about GAM.

Box 5.11 Generalized additive model (GAM)

This analysis uses common rustic moth data collected over 30 years from two sites within the Rothamsted Insect Survey network of light traps. The GAM shows the twists and turns of the moth population relative to the index year (1976). The solid line is the predicted mean count and the dotted lines are 95% confidence intervals. The shape of change is reported using open and filled circles (see below). There is a great deal of variation between the two sites, particularly towards the end of the series, hence the wide confidence intervals.

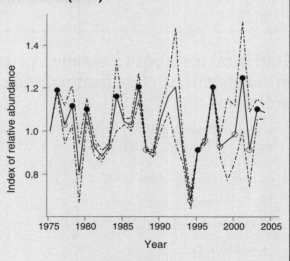

Annual model

The y-axis is an index of abundance that has transformed the counts and begins with the reference year that has been given a value of 1. It is rather difficult to see any pattern in the unsmoothed data, which is known as the 'annual model'.

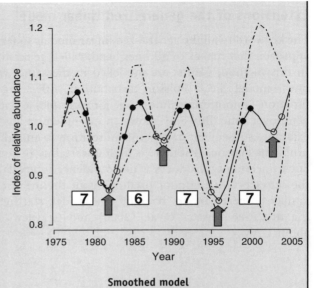

A clear periodicity of between 6 and 7 years can be seen in the smoothed predicted mean count. In a linear model we can only describe what we see using the slope parameter to indicate the *rate* of change. However, for GAMs the slope is continually changing and there is no simple way of reporting the properties of model with a single value. Instead, we report the curvature of the trend at particular time points using a test of the second derivative that measures the *shape* of change. This all happens behind the model that we see and can be confusing because additional information is superimposed on the trend.

Smoothed model

It is a little counter-intuitive but when the predicted mean counts are down-turning (black-filled dots) the shape of the second derivative is like a 'smile' (∪) and falls below a zero mean, but when the shape is upturning (open dots), the shape of the second derivative is like a 'frown' (∩) and above the zero mean. In both cases, the shape of change is bootstrapped to generate approximate confidence intervals and P values.

Statistical methods to examine pattern and structure in communities: classification, indicator species and ordination

Ecological systems are highly complex and it is sometimes necessary to use data reduction techniques to enable us to deal with either a smaller number of variables or to examine patterns in otherwise quite noisy datasets. Multivariate analyses of community patterns fall into three main categories:

- classification (cluster analysis);
- indicator species analysis;
- ordination.

The techniques that are discussed here are by no means exhaustive. Kent and Coker (1995), Legendre and Legendre (1998), Manly (2004), Fielding (2007), Tabachnick and Fidell (2007), Crawley (2007) and Henderson and Seaby (2008) together with a number of websites[7] give more information on ordination and classification.

---

[7] http://www.alanfielding.co.uk/multivar/index.htm and http://ordination.okstate.edu/

Classification

The aim of classification statistics is to partition species or sites into groups. By doing so we are seeking to simplify our understanding of what might be driving group similarities and differences. In a planned experiment where treatments will be known, we may wish to use this information and impose this treatment on our dataset as a grouping variable. Here, the approach might be to test this using methods such as canonical variates analysis (CVA), analysis of similarity (ANOSIM) and ordination. Conversely, many ecological surveys do not used planned experimental designs and consequently the groups are not necessarily known. For these experiments, hierarchical clustering and ordination are ideal because the algorithms will inform us what groups are present.

Classification techniques when the number of groups is known

Partitioning using k-means clustering involves dividing a dataset into a set number (k) of groups or clusters using some criterion (e.g. Euclidean distance between groups – see Box 5.12). It is fairly straightforward in that it works by first allocating a centroid (the central point of a cluster of points) to each of k groups then determines the distance of each object to all the centroids, allocating each object to the most closely aligned centroid, based on the minimum distance within groups. In practice, partitioning methods (e.g. k-means) are of limited use in ecology because they tell us nothing about the dynamics of the clusters, particularly the relationship between them.

Box 5.12   Distance measurements

Distance in the context of classification and ordination methods refers to the distance between data points. Imagine two points on a simple scatterplot, and you can imagine measuring the distance between them. This is relatively easy to envisage even for a three-dimensional scatterplot. In the case of multivariate techniques, we may be talking about many more dimensions. Even though we cannot imagine this, the mathematical principle is the same; any given data point can be defined in terms of its distance from the others.

Both similarity and distance coefficients measure the association between objects (e.g. sites). However, Legendre and Legendre (1998) make a clear distinction between these terms. Similarity coefficients are never metric and cannot be used to derive geometric distances between objects because they do not satisfy the assumptions of the triangle inequality axiom (i.e. the sum of the lengths of any two sides must be greater than or equal to the length of the remaining side). Some similarity coefficients (S) (Table 5.4) can be converted to distance measures (i.e. distance $= 1 - S$) but this is by no means guaranteed (e.g. when there are missing data). Moreover, this conversion will only be meaningful if the coefficient has inherent metric properties that then project symmetry and not distortion (such measures are termed distance measures). Distance measures should ideally be metric in that they satisfy the assumptions of the triangle inequality axiom, but not all measures meet this criterion and are otherwise more properly termed dissimilarities (e.g. Bray-Curtis, which for binary data is a dissimilarity measure of the Sørensen Index). Additionally, not all distance measures are Euclidean but those that are will always be metric. Put simply, if two objects are identical (e.g. two sites have the same species in the case of presence/absence data or the same numbers of the same species in the case of abundance data) then they should be placed on top of each other in a two dimensional biplot, implying that the distances between them must be 0. If you wish to formally test group membership derived from an ordination biplot, it is important to ensure that you use the same distance measure for all tests involved (for example, see Box 5. 13). You are strongly advised to consult Legendre and Legendre (1998) and Fielding (2007) for more information.

The distance measure used depends in part on the project design and in part on the data being compared. Some of the most frequently used distance measures are described below using the example of examining a series of sites on the basis of their component species:

Renkonen Index (based on percentage similarities) involves the calculation of proportions (percentages) of abundances of species and is useful where sampling effort varies across the sites, being that it is less influenced than some other measures by such differences. It is sometimes used in hierarchical cluster analysis and ordination where there are small datasets.

Euclidian distance and squared Euclidean distance are based on calculating the straight-line distance in n-dimensional space between each pair of samples where each dimension represents each variable (species) being used to separate the samples (sites). These geometric distances are at the core of MDS models (see Box 5.17). Squared Euclidean distance emphasises distances that are further apart and is semi-metric and therefore less appropriate for ordination.

City block (or Manhattan) distance is the distance between two points measured along axes that are at right angles to one another (i.e. the base plus the height or more correctly, the abscissa plus the ordinate). It contrasts with the directness of Euclidean distances that take the shortest path between A and B and does not change with axis orientation. City block is sometimes used with the log of relative species abundances.

Mahalanobis generalized distances are used to take into account the correlations between variables that are not necessarily orthogonal (i.e. uncorrelated), and is really only used to look at distances between sites when the group number is known (e.g. canonical variates analysis – p. 289).

Chi-squared distances use relative frequencies but exclude double zeros during the calculation. It has no upper limit, although the value normally falls well below unity. Chi-squared distances are widely used in correspondence analysis and canonical correspondence analysis.

Gower's distance measure enables mixed data types (binary, nominal and continuous) to be used and standardises each variable. Continuous variables are standardised by dividing by its range, whereas for binary and nominal variables a weighting of 1 is allocated when the state of both cases being compared is the same, and 0 if they differ. This measure not only accommodates situations when there are mixed data types, but also when some data are missing. This dissimilarity measure is based on the similarity method described in Table 5.4.

Canonical variates analysis (CVA) can be used to find linear combinations of variates (e.g. species) that maximise the ratio of between-group to within-group variation in order to be able to discriminate between groups (of sites, for example). CVA is used when the groups are known in advance and you want to test membership (see Digby and Kempton 1987 for further details). The analysis results in a biplot (similar to that shown in Figure 5.17) and also provides output in the form the type of a classification table similar to that shown in Box 5.9. In the example in Figure 5.17 the group divisions are stronger

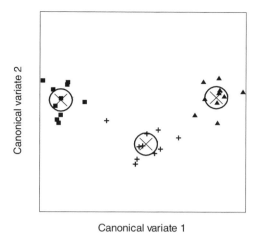

Figure 5.17 A canonical variates analysis (CVA) of spiders across three management treatments
The data are of the occurrence of 10 species of spider across three equal treatment groups (i.e. control and two grass cutting regimes — 3 × 10 sites — a factor variable). Group 1 sites are indicated by crosses, group 2 by filled squares, and group 3 by filled triangles. The centroid of each group is depicted with a cross surrounded by 95% confidence intervals (drawn as a circle). Percentage variation accounted for by axis 1 = 89.27 and axis 2 = 10.73.

on axis 1 (i.e. they show clear separation), with axis 2 having very little explanatory power. The 95% confidence intervals around each group are based on the mean and enclose some, but not all the members of each group. If we wish to extend these to all the members in the group then we calculate the 95% confidence intervals of the population, which will be much wider than those shown. Based on the biplot, we would conclude that membership of the groups on the basis of their characteristics (in this case species composition) is compelling but not without some variation. CVA is perhaps not widely used by ecologists because (as Jongman, ter Braak and Tongeren 1995 have suggested) it is only really of use when the number of sites is much greater than the number of species.

Significance testing for group membership: analysis of similarity (ANOSIM)

Following a technique such as CVA, we may be interested in whether the three groups observed in the biplot are entities in their own right (Figure 5.17), or if they overlap more than the biplot suggests. We can use an analysis of similarity (ANOSIM), a non-parametric test, to examine this more rigorously (Clarke 1993). An ANOSIM requires a distance measure, examples of which are given in Box 5.12. In the example in Box 5.13, we use Euclidean (geometric) distance measure because it is relatively easy to understand. Apart from distance measures, you might think of ANOSIM as equivalent to a

> **Box 5.13 Use of analysis of similarity (ANOSIM)**
>
> ANOSIM produces a test statistic that is calculated as:
>
> $$R = \frac{(\text{mean}(r_b) - \text{mean}(r_w))}{(n \times (n-1)/4)}$$
>
> where:
> r_b is the mean rank of all distances between groups;
> r_w is the mean rank of all distances within groups
>
> For our spider experiment across the three treatments (Figure 5.17), we find strong support for differences between groups, with every comparison yielding significant differences. However, the overall R value for the test is not large ($R = 0.3582$), suggesting that the differences (dissimilarities) between groups, although highly significant ($P < 0.001$), could theoretically be much higher (i.e. the maximum value of R can be 1).
>
> **ANOSIM analysis on spider data across three treatments showing pairwise R values**
> Note the diagonal of the matrix has dashed values because it is not meaningful to test a group against itself.
>
> | | Group 1 | Group 2 | Group 3 |
> |---------|---------|---------|---------|
> | Group 1 | – | | |
> | Group 2 | $R=0.605$ $P<0.001$ | – | |
> | Group 3 | $R=0.374$ $P<0.005$ | $R=0.170$ $P<0.005$ | – |
>
> As mentioned in Box 5.12, we cannot readily compare the output here with that from the CVA in Figure 5.17 since these two techniques have been calculated using different distance measures.

one-way ANOVA where the between-group variation is based on the mean ranks of all the distances between the groups, whereas the within-group variation is based on the mean rank of all distances within the groups. Other techniques that can be used to test groups include two-way ANOSIMs, non-parametric MANOVA and the multiresponse permutation procedure (MRPP). Each has its own merits, depending on the quality of your data. Many of these tests are available in the PAST program (see Box 5.2) and further details can be found in Clarke (1993).

Classification techniques when the number of groups is unknown

Situations where the number of groups is unknown are quite common in ecology. Unlike partitioning methods (e.g. *k*-means clustering), in hierarchical cluster analysis, the number of groups is not predetermined. Examples of hierarchical clustering include agglomerative clustering and divisive clustering. The former technique is more commonly employed in ecological research. In agglomerative clustering, the groups are formed in stages with each level of initial clustering being followed by further groupings, whereas in divisive clustering the total group is successively split into a series of subdivisions, which themselves are further subdivided. Figure 5.18 illustrates partitioning and hierarchical clustering methods. Cluster analysis groups the samples (e.g. sites) together in a graphical form called a dendrogram (tree diagram) with the most closely related ones being linked lower in the series (Figure 5.19). Box 5.14 shows examples of agglomerative clustering methods.

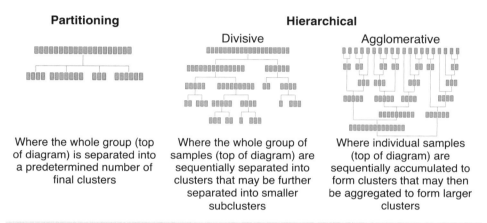

Figure 5.18 Types of cluster analysis

Figure 5.19 Dendrogram following cluster analysis of different habitat types
Data clustered using percentage similarity (Renkonen index) calculated on the basis of their component beetle species.

> **Box 5.14 Examples of agglomerative clustering methods**
>
> The choice of an appropriate clustering method depends on the type of data involved. See Fielding 2007 or http://www.statsoft.com/textbook/cluster-analysis/#d for further details); some commonly used methods are listed below:
>
> **Average linkage** (between-groups linkage) links clusters on the basis of the average distance between items in the different clusters. This includes unweighted pair-groups method average (UPGMA) and weighted pair-group average where the size of the cluster is used to weight the averages.
>
> **Complete linkage** (furthest neighbour) links clusters on the basis of the largest distance between two items in the different clusters.
>
> **Single linkage** (nearest neighbour) links clusters on the basis of the smallest distance between two items in the different clusters.
>
> **Centroid clustering** links clusters on the basis of the distance between the centres (centroids) of different clusters.
>
> **Ward's method** is a statistical (analysis of variance) method of determining the distances between clusters.

Indicator species analysis

Indicator species analysis usually involves the identification of species that characterise groups of sites that can be either known or unknown in advance. There are two ways of generating indicator species lists: hierarchical (using TWINSPAN: Two-Way INdicator SPecies ANalysis) that is employed when groups are not already known; and non-hierarchical (using IndVal: Indicator Value) that is used when groups are known in advance. We start by discussing the properties of IndVal (Dufrêne and Legendre 1997) and then compare the outputs of IndVal against TWINSPAN.

IndVal is a simple tool that uses a data matrix based on species in rows by samples (e.g. site) in columns to compute a percentage indicator value for each species in each habitat group, based upon combining their relative abundance and frequency in groups of samples. The method first defines the degree of habitat specificity measured as the mean abundance of a species across the study sites and then site fidelity measured as the relative frequency of the species across the study sites. These two values are then combined (by taking their product and multiplying it by 100) to give a percentage of the loyalty of a species to a group. A maximum score of 100 indicates that not only are all individuals of a particular species found within a group of sites, but also all sites within that group contain the species. The largest value will be indicative of a species that can causally be associated with a particular habitat type using a technique called the Monte Carlo (MC) randomisation procedure. The MC produces a test result to establish whether the association is significant ($P < 0.05$), concurrently producing the

Table 5.9 A spider indicator species analysis
The respective overall indicator values are expressed as percentages. A breakdown of the indicator values (IndVal$_{ij}$) across the two habitats is also given

Spider species	Habitat indicator	IVmax	Distribution of indicator values (%) across:	
			Grassland	Bare ground
Alopecosa pulverulenta	Grassland	95**	94	2
Agroeca proxima	Grassland	93**	93	1
Coelotes atropos	Bare ground	70**	26	63
Typhocrestus digitatus	Bare ground	65**	0	49

**Statistically significant ($P < 0.01$).

highest indicator value known as IVmax (Table 5.9). IVmax is at its maximum when a species is confined to a single habitat type, is always abundant and never absent.

TWINSPAN is a divisive clustering method that allocates indicator species into groups at the same time as forming divisions (Figure 5.20). It is much more complex and indeed flexible than IndVal, with many options to set minimum group sizes, maximum number of indicators per division, etc. Generally, IndVal is seen as

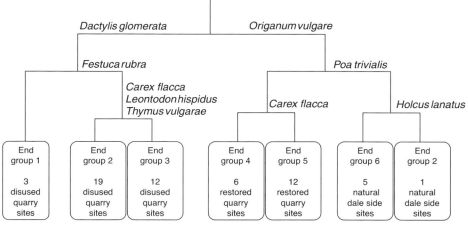

Figure simplified and redrawn from Cullen, Wheater & Dunleavy (1998)

Figure 5.20 TWINSPAN of quarry sites on the basis of their component plant species
Lists of sites that are linked together higher up have the most similar communities. The species names at each separation indicate the indicator species for that division. For further details of this example see Cullen, Wheater and Dunleavy (1998).

statistically robust and TWINSPAN problematic because it sometimes produces unstable solutions (Dufrêne and Legendre 1997). Software is available to run both IndVal[8] and TWINSPAN,[9] The main difference between the two techniques is that IndVal's groups are not forced or imposed but determined in advance. This may be a problem if you do not want to test at the level of treatments, sites or habitat types that are readily understood before analysis begins. In such cases you will need to use hierarchical clustering to determine what the groups are likely to be and then impose these on your IndVal dataset.

Ordination

The purpose of any ordination technique is to produce a summary of matrix-based data, normally presented as a biplot, in as small a number of dimensions as possible. That is to reduce the complexity from many dimensions (variables) to only a few (ideally two). Although the process of reducing the number of dimensions involves the loss of some detail, the aim is to retain the patterns of species assemblages across sites. Ordination may be unconstrained or constrained, depending of whether gradients (e.g. environmental variables) are implied or explicit, known as indirect and direct gradient analysis respectively. In most forms, ordination generates hypotheses but in direct gradient analysis it can be argued that the technique allows for hypothesis testing.

Indirect gradient analysis

Principal components analysis

Principal components analysis (PCA) finds the orthogonal linear combinations of a set of variables and projects them on to a small number of dimensions. Orthogonal variables (or sets of variables) are those that have no relationship with each other: if you were to plot a pair of orthogonal variables against each other on a scatterplot, then the points would be scattered with no evidence of a linear relationship (and the correlation coefficient, r, would be close to zero). It can be thought of as an extension to multiple least squares regression given that PCA seeks to minimise the total residual sums of squares. Ideally, data need to be multivariate normal so consider transforming data using square root or logarithmic transformations as you might do in linear regression. Species matrices are not ideally suited to PCA because of the high number of zeros, which sometimes create serious curvilinear distortions. In these trivial solutions, very distant objects are closer than they should be in PCA space, causing the biplot to 'arch' because the second axis is folding in on itself (producing a horseshoe effect). Here, we give examples of two uses of principal component analysis: data compression (Box 5.15), and to generate biplots in order to understand the structure of our data (Box 5.16).

---

[8] http://biodiversite.wallonie.be/outils/indval/
[9] http://www.ceh.ac.uk/products/software/wintwins.html

Box 5.15 Using principal components analysis for data compression

It is sometimes desirable to compress data from n dimensions to one. For example, if we use a penetrometer (Chapter 2) to measure soil hardness as a surrogate of soil moisture, then we might take five readings at different depths throughout the soil horizon in each of 80 locations in order to gain a proper understanding of the upper soil moisture profile. However, we also wish to use the data to produce a soil moisture profile map. Therefore, we need to consider how we can reduce each sample profile to a singular value that can be interpolated across a surface.

Soil penetrometer readings for each location to a depth of 7.5 cm

	Penetrometer readings at given depths				
	1.5 cm	3 cm	4.5 cm	6 cm	7.5 cm
sample 1	0.8	9.6	19.6	31.4	45.2
sample 2	1.8	9.6	25.8	43.2	50.4
sample 3	0.2	2.8	19.8	43.6	52.4
sample 4	1.6	9.0	20.0	39.0	51.8
sample 5	0.4	7.8	26.6	29.8	42.2
::	::	::	::	::	::
::	::	::	::	::	::
sample 20	4.2	17.6	35.4	61.2	68.2
sample 21	2.0	15.2	37.6	60.2	57.6
sample 22	2.9	24.4	33.4	33.2	36.0
sample 23	1.6	15.8	44.0	45.8	43.8
sample 24	4.0	18.4	34.0	42.0	54.6
etc.	::	::	::	::	::

Using principal components analysis, we can compress this matrix into one variable successfully if the eigenvalues (known as the latent root in PCA) are acceptably high. Statistics programs use the eigenvalues to derive the variation accounted for by each principal component. In our example, this is very high for the first axis (89.59%), and much lower for the second axis (6.75%). For the subsequent axes (3, 4 and 5: there are as many axes as there are variables) the percentage is vanishingly small (<4% in total) since the sum of all the variation is 100%.

Results of principal components analysis comprising eigenvalues (variation accounted for on axes 1 and 2) and the PCA scores

	PCA axis 1	PCA axis 2
Variation accounted for:	89.59%	6.75%
PCA Score (sample 1)	10.064	4.981
PCA Score (sample 2)	22.907	1.64
PCA Score (sample 3)	22.007	10.363
PCA Score (sample 4)	19.446	7.131
PCA Score (sample 5)	9.133	0.037
::	::	::
::	::	::
PCA Score (sample 20)	50.414	3.143
PCA Score (sample 21)	42.637	8.166
PCA Score (sample 22)	10.320	17.111
PCA Score (sample 23)	26.278	17.473
PCA Score (sample 24)	28.761	6.586
etc.	::	::

We conclude that we can use the PCA axis 1 scores as a surrogate for the soil penetrometer readings for each location to produce a soil hardness/moisture map. In this case data compression appears to have worked, but if the eigenvalues were close to being equal for axes 1 and 2 then it would not be valid to use only one of the PCA axes as a surrogate for the original data. These orthogonal factors (or principal components) could also be used as independent variables within other analyses, e.g. ANOVA, regression or GLM.

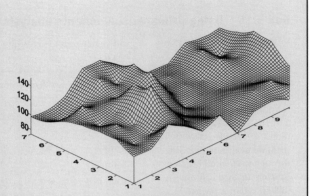

Shaded relief map of soil hardness/moisture produced using interpolated PCA axis scores (z axis) in a field at given xy coordinates
The higher the value on the vertical (z axis) the harder the soil.

Correspondence analysis and detrended correspondence analysis

Correspondence analysis (CA), also known as reciprocal averaging, can be used to investigate patterns within series of sites on the basis of either the presence and absence of species, or using species abundance data. The algorithm attempts to achieve the same outcome as PCA by reducing the number of dimensions while preserving inherent structures. Among ecologists, it is one of the most popular techniques, although it is usually applied in its modified detrended form. Ecologists opt for the detrended form because 'pure' correspondence analysis suffers from quadratic distortions, known as the 'arch effect', analogous to the horseshoe effect seen in PCA. Even though intrinsic scaling of the row and column means it is carried out by the CA algorithm, transformations (log, square root, etc.) are recommended where appropriate (see Jackson 1993). It is not advisable to use CA for species presence and absence data in which there are many rare species, because the method tends to give them too much weight. This is particularly true when significance testing with explanatory variables. However, CA is a useful method for datasets that are based on percentages or relative species frequencies.

Detrended correspondence analysis (DCA) avoids the arch effect by dividing axis 1 into a number of segments. Within each segment, axis 2 scores become detrended when they are adjusted to have an average of zero. It is argued that by using the detrending process, the robustness of the CA algorithm is increased, particularly for extremely large

Box 5.16 Using principal components analysis to produce biplots

The data from the example in Box 5.15 on soil penetrometer readings have been used here to generate simple biplots to understand the structure of our data. Here we have produced a biplot that separates the samples on the basis of the PCA axis scores.

The biplot shows the PCA scores based on soil penetrometer readings to a depth of 7.5 cm. Even though there were five readings per location, we know from Box 5.15 that 97% of the variation was explained by just two underlying factors (PCA axis 1 and PCA axis 2) and therefore can be represented in two dimensions. Each sample is represented by a circle that is projected onto the biplot relative to their coordinates.

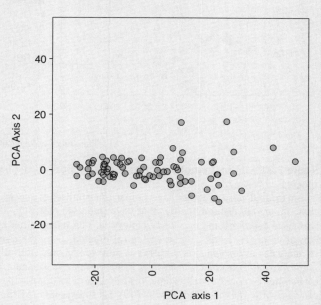

We can superimpose the axes relative to the PCA scores using a method known as rotation. Rotation allows us to interpret graphically how samples fit along the new axes and confirms that much of the variation falls along the first axis (i.e. component) in our example.

This leads us to conclude that that the soil moisture profiles have a fairly simple structure. However, samples 20–23 seem atypical and appear to have very high PCA scores. Looking back at the original data (Box 5.15), we see that these soil profiles are very hard and have little moisture within them.

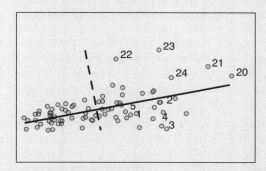

A varimax rotation of the original PCA showing the linear combinations of the original variables as new axes (first and second components) within the set of the 80 individual soil penetrometer readings
The length of the axes estimates the standard deviation. Axis 1 is the solid line and axis 2 is the dashed line.

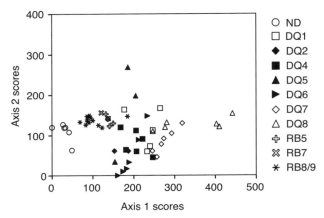

Figure redrawn from Cullen, Wheater & Dunleavy (1998)

Figure 5.21 Ordination of a number of quarry sites on the basis of their component plant species
Points located close together have similar communities of species. ND indicates natural daleside sites, DQ are old disused quarries that have colonised naturally, RB sites are modern quarry faces that have been restored by selective blasting. DCA has a rather nice property in which axis 1 is indicative of the length of the ecological gradient. More precisely, it is scaled as the average standard deviation, which, if it exceeds 4 units (commonly multiplied by 100, as here, to give 400) suggests that two objects have few if any species in common (i.e. there has been complete community turnover). For example, disused quarry 8 (DQ8) is likely to share very few species in common with the natural daleside (ND) because several of the sites are 4 average standard deviations away along axis 1. The number of segments used for detrending was 26. For further details of this example, see Cullen, Wheater and Dunleavy (1998).

datasets (e.g. 300 sites × 450 species). However, the process has also been criticised as being rather *ad hoc* (Jackson and Somers 1991) and is not especially useful in situations where complex ecological gradients are suspected.

In situations where there is a strong environmental gradient DCA may prove very useful. Figure 5.21 shows how a range of quarry environments support different plant communities depending on whether they were restored, or disused or were equivalent natural sites with no quarrying. When using DCA, you are advised to cite the number of segments (26 is the default) used in your analysis until a standardised method of choosing the appropriate number becomes available. Furthermore, you should also be aware that the detrending algorithm modifies the real eigenvalues for axis 2 and beyond so it is very difficult to estimate the variance components. To overcome this, the program PC-Ord[10] now provides a coefficient of determination that can be used as an overall model fit statistic.

---

[10] http://home.centurytel.net/~mjm/pcordwin.htm

Multidimensional scaling and principal co-ordinates analyses

Multidimensional scaling (MDS) and principal co-ordinates analyses (PCoA) are both distanced-based ordination methods that seek to position objects in low dimensional space according to their relative dissimilarities. These dissimilarities are expressed as distances, the default being geometric Euclidean distances (MDS) or squared Euclidean distances (PCoA), but both techniques are very flexible with most distance measures being able to be selected by the user. PCoA can be thought of as more like a hybrid between MDS and PCA in that it attempts to derive distances from dissimilarities (like MDS), but it does so via an eigenvalue approach (like PCA). PCoA is not widely used in ecology, perhaps because it may produce negative eigenvalues; these often result because the distance measure chosen is either semi- or non-metric and neither can be resolved in Euclidean space.

Multidimensional scaling is quite different to PCoA. It attempts to project dissimilarities as distances in Euclidean space, using a measure of stress to indicate the (lack of) fit. There are many different MDS algorithms and rather confusingly they tend to estimate different stress values that cannot easily be compared. However, if you wish to compare models, PROXSCAL will compute various additional stress measures so that you can compare different algorithms. Borg and Groenen (2005) recommend that the stress should *normally* be kept below 0.3. Table 5.10 is a summary of some of the more common algorithms used to measure stress.

Table 5.10 Types of stress measure for computing MDS solutions

Type of stress measure	Algorithm	Method of stress calculation	Programs	General effectiveness
Normalised raw stress (NRS)	PROXSCAL	Iterative majorization	SPSS	Excellent
Delta stress + r^2 to estimate the contribution of each axis to the model*	Taguchi and Oono (2004)	Non-linear optimization	PAST	Excellent but new and not yet tested on 'poor'-quality data
Stress − 1 (square root of NRS)	KRUSKAL	Steepest descent	Statistica, PC-Ord and Primer	Good
S-Stress	ALSCAL	Steepest descent	SAS, SPSS and Genstat	Problematic

*The 'stress' reported in PAST is equivalent to a normalised raw stress value. PAST computes the sum of squares of differences between observed and ordinated ranked distances, over all pairs of data points. The source code and an additional unpublished paper can be found at http://www.granular.com/MDS/src/.
For all other algorithms see Borg and Groenen (2005) for further details.

Box 5.17 Example of distance placement using multidimensional scaling (MDS)

Consider the three lower case letters dna. If we mark points where the letters curve and then measure all distances between points (that is 300 distances between all combinations of the 25 points), we can use MDS to 'map out' those letters from these distances alone (i.e. we use just the distance matrix and not the original x and y coordinates).

Using a non-metric Proxscal model we have been able to recover the letters under low stress. The calculations are relatively straightforward in that the algorithm will rank the relative positions of those points iteratively and in pairs (e.g. v1 to v9 is a lower rank than v1 to v8, which is a lower rank than v1 to v2). As the algorithm progresses, points get dragged by small amounts towards an acceptable 'global' solution.

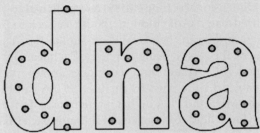

DNA letters with reference points used to derive distances for a lower triangular matrix
For each point, all inter-point distances were derived geometrically

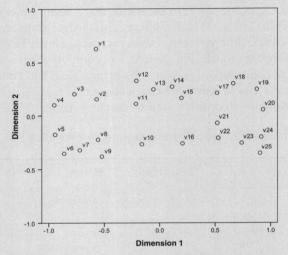

Non-metric Proxscal solution to DNA lettering
Normalised raw stress = 0.00034.

There are many forms of MDS models, but the commonest, and the one we shall concentrate on here is non-metric MDS. Non-metric or ordinal MDS uses monotonic regression (a type of regression where the trend is positive and increases) to compute pairwise distances that approximate to the ranked dissimilarities, rather than their actual distances (i.e. in ratio MDS). Non-metric MDS performs a number of iterations (i.e. keeps calculating distances over and over again) to establish a 'best fit' solution describing the pattern produced at the lowest possible stress (i.e. how far the observed data depart from the expected model). As may now be apparent, non-metric MDS is very useful when scores or ranks are used to describe the relationship between several variables that need not be orthogonal. As an example, consider scores of quality of basking sites for reptiles, based on 10 variable types, such as substrate, vegetation, aspect, etc. One intuitively appealing way of demonstrating the power of a good MDS algorithm is to use it to plot the distances from the triangular matrix of distances between major cities found in many road maps. With this information alone, MDS can precisely place towns and cities proximate to one another or recreate other patterns given distances between points (Box 5.17). MDS has many uses beyond creating landscapes from a small amount of information, including creating functional groups, or predator–prey networks (Bell et al. 2008 and Bell et al. 2010 respectively).

Comparing ordinations and matrix data

There are occasions when we might wish to compare two or more ordinations or data matrices. For example, we might sample the invertebrates at 20 sites along a river bed and then return later and resample the same invertebrate communities. If we ordinate the samples taken on the two separate dates, we can compare the positions of the sites relative to one another using Procrustes analysis. This method measures the fit between the ordinations from the first and second dates to indicate the amount of change between the two sampling occasions. If we keep returning to the river bed and generate new ordinations for each date, we would create a series of ordinations. When you are comparing more than two ordinations, you should use generalized Procrustes analysis. Box 5.18 gives more information regarding such techniques.

Direct gradient analysis

When we not only have a data matrix of sites and species, but we also have a number of environmental variables associated with each sample, we can see whether the site/species composition is linked to one (or more) environmental gradients. If we are interested in how the environmental variables influence the separation of sites on the basis of their component species, we could use an indirect gradient analysis (unconstrained) technique such as DCA and correlate the axis scores with environmental

> ### Box 5.18 Techniques for comparing ordinations and matrix data
>
> Two very useful techniques for comparing ordinations are Procrustes analysis (PA) and generalized Procrustes analysis (GPA). The program Protest enables two matrices to be compared in order to establish whether they are significantly different from one another (see http://labs.eeb.utoronto.ca/jackson/pro1.html on the routine available through the R package). When three or more matrices need to be compared simultaneously, GPA offers a global solution over other methods, which are only partial solutions based on linear models (e.g. Mantel tests). GPA first "skewers" configurations through their centres and then compares the fit of each individual configuration to the group average configuration. To understand these methods further consult Digby and Kempton (1987). Genstat and R provide routines for GPA.

Box 5.19 Example of use of canonical correspondence analysis

Here Dave Brooks (Rothamsted Research) describes how he used CCA to investigate how different types of crop may affect ground beetle communities. Data were derived from a national-scale experiment that collected beetles using pitfall traps from individual fields of break crops across the UK. Break crops are those intermediate crops used in a crop rotation sequence to provide a 'break' from pests associated with the primary crop in the rotation. The total counts of 75 beetle species from each of 251 fields have been transformed using $\log_{10}(n+1)$ and each field represents a sample in the analysis. The analysis examines how the categorical variable (crop type – large inverted triangles) determines ground beetle species responses (small upright triangles). In this analysis, the first and second axes were significant, as were the crops, which were shown to be the most important variables within a much wider set of environmental descriptors. The diagram shows a strong gradient on the first axis representing the contrasting effect of spring crops sown in rows (beet and maize) to the left side of the axis, with winter oilseed rape (right side of the axis). Spring oilseed rape is associated with the second axis and is therefore somewhat uncorrelated with this trend, but still has a notable influence on the beetle communities. In this example there were also strong geographical effects on the beetles as the distribution of the crops also varies across regions of the UK. The model allowed Dave to account for these spatial trends and he was able to conclude that crop can affect the beetle communities.

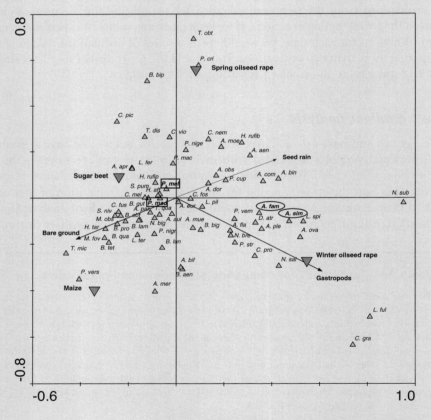

Figure simplified and redrawn from Brooks et al. (2008)

> **Understanding the biplot**
>
> This technique better represents how the relative proportions of species abundances within samples are changing in relation to the environment, rather than the variation in absolute abundance of species across all of the samples. Increasing distances in CCA biplots between samples and species therefore represent increasingly strong shifts in the overall community composition, which reflect distinct changes in the proportional composition of species within samples. To illustrate this point, the two species highlighted in boxes (*Pterostichus madidus* and *P. melanarius*) are often the most numerous species in all samples (i.e. highly dominant), but also have highly significant increases in abundance in the row crops (as measured by simple ANOVA). However, because they also represent a high proportion of total species abundances in the winter crop samples, albeit at far decreased overall abundance compared to the row crops, their species centroids are only displaced a little way from the origin towards the beet and maize crops. In contrast, the two circled species (*Amara familiaris* and *A. similata*) are highly characteristic of the winter crop (and therefore have increased displacement along the first axis). Even though they stand out as being characteristic of the winter crop, their abundances are actually relatively low compared to *P. madidus* and *P. melanarius*. However, their proportions are much higher than in the row crops, where they are often absent.
>
> Although good for highlighting such community dynamics, this technique is only suitable where species are: (a) responding to a limited number of noticeable gradients, for example, between a limited number of influential crop types, as in this example; and (b) when such responses are predominantly unimodal, giving rise to fairly long gradient lengths (typically over 3.5 standard deviation units). Quantitative variables can also be investigated within CCA as long as their metrics do not differ too widely. Here, three such variables (seed rain, bare ground, and gastropods) have been super-imposed to show how they relate to the axes, shown by vectors for their increasing abundance relative to the crops. This has aided the interpretation of how the crops may be influencing the beetles. For example, seed rain from weeds is more numerous in oilseed rape crops and many of the species to the right of the diagram (e.g. the two circled) prefer this food source. Conversely, not surprisingly, there is a higher proportion of bare ground in the row crops and many of the species to the left of the diagram prefer a more open habitat.
>
> See Brooks et al. (2008) for further details of this example.

variables to see whether they are related to any identified site separation. However, direct gradient analysis enables us to draw both of these aspects together into a single analysis. In Box 5.19 we give an example of one such technique, canonical correspondence analysis (CCA). Of the range of programs that will produce such an ordination, one of the best is CANOCO (ter Braak and Smilauer 2002; Leps and Smilauer 2003). Generally, CCA models provide the following:

- Monte Carlo permutation tests to assess the significance of the axes and explanatory variables;
- a method of removal of unimportant variables to get a simple model of community dynamics;
- calculation of the length of the species gradients along the axes to help to decide on the best ordination model for your data.

6
Presenting the Information

The final stage in any project involves communicating the results and conclusions to others, be it through oral presentations, a written report, a thesis or a scientific publication. Whatever the mode of communication, it is important to know who your target audience is and to develop the presentation to be appropriate to the prior knowledge and interests of that audience. Most presentations are restricted in some way by the medium employed. Oral presentations are usually time limited, and require simple visual messages to be imparted rather than dense text. Written reports, theses and journal articles usually have specific guidelines or house styles that must be followed, usually including word count or page length. Notwithstanding these restrictions, there are certain general principles that can usefully be considered for most types of presentation. Here we concentrate on written presentations.

Good presentation and quality control of your work is essential. You will need to evaluate the published work of others, make your own original observations, record and analyse data (if relevant), and formulate logical arguments and conclusions. The quality of the research is almost irrelevant if you are unable to communicate it effectively. The aims of good communication of research are to:

- articulate your hypotheses, experiments and conclusions;
- convince the reader that your results and conclusions are correct;
- set your findings in the context of the existing work in the field.

In order to satisfy these aims, good presentation includes the following:

- careful selection of material for inclusion and/or exclusion;
- structuring for maximum impact and to emphasise the most important results;

Practical Field Ecology: A Project Guide, First Edition.   C. Philip Wheater, James R. Bell and Penny A. Cook.
© 2011 John Wiley & Sons, Ltd. Published 2011 by John Wiley & Sons, Ltd.

- taking care over how the text is written (the writing should be clear and concise) and organised (text and figures should be laid out as attractively as possible);
- use of a consistent style throughout.

What follows is advice on the style and presentation of research reports that should be considered within any corporate structures (e.g. house styles of journals, organisations, courses, etc.).

Structure

Start with a rough outline of what will go in each section and be flexible, as the draft is likely to change as the report is composed. Start your writing early enough to have time for other people (your peers, supervisors, co-authors, etc.) to see drafts of each section. It can be hard to get going when writing a report: the key is to write the easiest part first, and this is very often the methods section. Do not spend time agonising about what to write in a particular sentence. Just write something, come back to it to edit, rewrite and edit again. Your work will benefit by going through this process several times. It is helpful to leave sections for a day or so and re-read them later when you are fresh and distanced from the work.

There are two general structures for reports, depending on whether your work fits more neatly into one story or into discrete subsections that can then be drawn together at the end (Figure 6.1). Model I has a linear structure where you describe the background, detail the issues to be raised, explain the methods you used and the results you obtained, and finally discuss your findings in the context of the wider scientific literature. We shall concentrate on Model I, but it is helpful to briefly mention how Model II differs. Model II is typical of a thesis or major report that has discrete chapters with self-contained sections. Additional to Model I it has both a general introduction and a general discussion. The nature of these two sections can be tricky to get right, but in the general introduction it is advisable to include information on the rationale and scope of the entire thesis or report. The general discussion should bring together a few of the major themes that were discussed within each chapter and demonstrate the impact of the work in the context of current scientific understanding. It is usual to finish by suggesting future work and new hypotheses that are of central interest.

We will now go through each of the sections of the Model I type report. The following gives guidelines for each section in the order of the final report (which may not be the order in which the sections are written). In a report, each main section could have a brief header (e.g. 'Introduction', 'Methods', etc.) to make it easier for the reader to quickly locate a section. Journal articles use a strict template that must be adhered to; if

Model I	Model II
Title	Title
Abstract	Abstract
Acknowledgements	Acknowledgements
Contents	Contents
Introduction	General introduction
Methods	Data Chapter 1
Experiment / survey 1	Introduction
Experiment / survey 2	Methods
etc.	Results
↓	Discussion
	Data Chapter 2
Results	Introduction
Experiment / survey 1	Methods
Experiment / survey 2	Results
etc.	Discussion
↓	etc.
Discussion	↓
Appendices	General discussion
References	Appendices
	References

Figure 6.1 Two formats for research report presentation
Use informative headings rather than 'Experiment/survey 1', 'Data Chapter 2', etc., and additional (informative) subheadings if it helps to structure the work.

you underestimate how exacting journals are in this regard, you will quickly find your work being rejected.

Title

Make your title as interesting and as short as possible. Eliminate phrases such as 'A study of. . . .', 'An investigation into. . . '. For example, the long-winded title:

> An investigation into the factors influencing the distribution of ground-dwelling invertebrates in restored quarries in Derbyshire, UK

could be rewritten as:

> The distribution of ground-dwelling invertebrates in restored quarries

Note that where the location is important to the study then it may be useful to include it in the title; otherwise, it should be left out.

Abstract

The abstract is a summary of your research. It is an important component of the report, being the first part that the reader encounters and its quality can determine whether or not the reader decides to read the full piece of work. Although it is probably best left as one of the final jobs, do not be tempted to leave it until the very last minute and rush it. The abstract must be self-contained, so that the reader can understand what you have done without needing to refer to the text, and should not normally include references unless there are published pieces of work that are fundamental to the aims of your research. Start with a few lines of introduction, setting your work in the context of current knowledge, and then state the aims and objectives. Do not give details of the methods (unless it is a methodological project), but do include a line or two to establish broadly what type of methods were used. Summarise the important results. Finally, end with a couple of sentences covering your main conclusions. Depending on the audience, you may wish to replace the abstract by a section titled Executive Summary, or Key Findings, or Major Recommendations, containing more detail than you would put in an abstract and perhaps formatted as a series of bullet points. Regardless of the type of summary used, this section will be the first (and possibly the only) part of the report that is read. As such, it should be succinct, grab the attention of the reader and provide a hook by which the reader is encouraged to dip further into the report.

Acknowledgements

The acknowledgements section is your opportunity to thank anyone who helped you during your research project. You must acknowledge financial support and facilities, or anyone who has provided you with information or samples. Also thank anyone who helped with the discussion of ideas, read earlier versions of the report, or were generally supportive in either an academic or non-academic context. Any personal communications can be covered here (see p. 319). Although the acknowledgements section is usually placed towards the beginning of a book, thesis or report, in a scientific paper it is more usual to place it at the end, just before the reference list.

Contents

Reports should include a contents section, whereas journal articles do not normally. Many word-processing packages enable this to be produced automatically and using this may encourage you to be more aware of the hierarchy of headings that you use. Contents lists of any figures, tables, boxes, plates, etc. can also be produced and should be placed immediately after the main contents section. In order to make use of a contents section, it is important to number the pages. Again this can be automated using word-processing software and any house style should be adhered to (e.g. putting page numbers at the bottom right of a page). Note that page numbering can also help if others are collaborating or editing your work. Some journals require lines to be numbered as well as pages, but this is removed before final publication. Other house styles include numbering paragraphs or sections for easy reference when the report is likely to be the subject of much discussion.

Introduction

This should introduce your subject area and set your work in context. In it, you should demonstrate a thorough knowledge of the literature and critically appraise the work of others. Wherever you use someone else's work, reference it (see p. 320). The first paragraph should start with the broader subject area; then successive paragraphs should focus in more specifically. As your introduction focuses down, to lead to your exact research question, it is vital that you demonstrate that you have an intimate understanding of the immediately relevant literature, and that you can articulate what is known and what is unknown and, if relevant, any shortfalls in the work of others who have attempted to address similar questions. The final paragraph states the aims and objectives of your study. In the Model II report, the aims at the end of the general introduction will be more general, and those at the end of the introduction to each experimental section will be specific for that section. When phrasing your aims, avoid vague statements that begin with 'To investigate...' or 'To study...'; try to be more specific using 'To determine...' or 'To establish...'. The objectives state how you are going to achieve the aims. For example, if your aim was:

> To determine the colonisation rates of newly created sites by ground-dwelling invertebrates,

then your objectives might be to:

1. survey invertebrate species inhabiting established habitats and new sites using pitfall traps;
2. compare the distribution of invertebrates in the two site types;
3. relate invertebrate distributions to the known ecology of the species concerned.

If, during the course of your research, your aims were altered, your report should reflect these new aims. Also, make sure that the aims reflect the results. If what was originally a minor aim becomes, after the results are compiled, more interesting than initially anticipated (or more interesting than the major aim), then reword your aims to promote the interesting results. The introduction may contain illustrations (tables or figures: see p. 311). If so, this is likely to be background information from external works, the source of which must be referenced. It is increasingly common in papers to include the rationale of the research in the last few sentences of the introduction. Stating your hypotheses is encouraged because it provides scientific focus and these should be written at the very end of the introduction.

Methods

Traditionally, the methods section was called 'Materials and Methods' to emphasise the distinction between the physical objects used (materials) and the approach to sampling (methods). Nowadays this is less common, with many journals preferring simply 'Methods', but the spirit of what this section contains is still the same. Therefore, you need to give a clear a description of what you did and should include sufficient background material (e.g. on techniques) for the reader to understand (and to show that you understand) exactly what you did. To increase clarity and impose a logical sequence to your methods section, try to break the methods into subsections according to topic. For example, 'study site and model organism', 'experimental design' and 'statistical analyses' are a good start.

Where appropriate, the following should be covered:

- The organism(s) studied: common names of species are optional but you should name each species using their scientific binomial names, with an authority and family name. For example, 'The money spider *Tenuiphantes tenuis* (Blackwall 1852) (Linyphiidae) was used to test...'. For further details of how to use species names, authority and family names, see p. 328.

- The location: you need not state a location for a laboratory study but in situations where a field study site was used, give the name of study site, including latitude and longitude and any other significant factors (e.g. mixed broadleaf woodland on clay soils). Do not go into too much detail; a map is advisable for a report but not for a journal publication.

- The experimental design: planned experiments need to be stated formally (e.g. randomised block design) and sampling procedures within that design also included (e.g. '10 replicate quadrats were placed within each of the four treatments and a control and were sampled once per week for 10 weeks').

- Additional information on variable(s) measured, machinery including manufacturer (e.g. Vortis suction sampler, Burkhard Manufacturing, Rickmansworth, UK) and other apparatus. Include concentrations and sources for chemicals used.

- Statistical analyses: this should include a small amount of information on rationale (why you chose the test) and the tests themselves. For example, 'Log-linear models, an ideal test for count data (Crawley 2007), were used to tests for differences in species responses by treatment'.

Generally, there should be sufficient detail included for someone else to repeat the work, but you should assume certain basic knowledge (e.g. being familiar with the operation of a basic piece of equipment). If your method is based on a previously published technique, refer to it here, citing the work in question (see p. 318) and clearly state any modifications or improvements in the technique. If you have chosen a particular technique over another one, or modified a technique, then you may need to justify that here.

Note the distinction between the terms methodology and methods. Methodology is usually used to cover the study of a range of methods within a discipline, and a methodology section would include an analysis of such procedures. Methods, on the other hand, are the processes used to achieve an objective.

Results

Your results should be presented as text, together with appropriate figures and tables. Use the text to describe results, but do not include any reasoning or conclusions since these are points for the discussion. You may, however, interpret major patterns. For example 'the growth of the plants within all treatments followed a sigmoid (Gompertz) function (Figs 1–4)' but not 'the growth of the plants within all treatments followed a sigmoid (Gompertz) function (Figs 1–4) because of the mid-term exponential increase in biomass'. The second example includes too much explanation beyond the simple statement of fact. Figures and tables should contain summaries of your results. Raw data should never be presented in the results section, but if you feel that they may be useful for future work (e.g. species lists), place them in an appendix that must be referred to in the text (Appendix 1, Appendix 2, etc.). Appendices are now commonly used in online journals as supporting material to a published article. They are usually not copy-edited and tend to include a disclaimer stating that they are the sole responsibility of the author. Online appendices are normally inserted at the end of a paper, and the way these are cited varies depending on the journal.

Illustrations (tables, figures, plates, equations, etc.)

You have to decide whether your information is best described in text, or presented in tabular or graphical form (i.e. as a table or a figure). The size and number of tables and figures should be minimised. Make sure that each table, figure and plate (photograph) is referred to at least once in the text, labelling them as three separate numbering series (i.e. Table 1, Table 2, etc., Figure 1, Figure 2, etc., and Plate 1, Plate 2, etc.). The numbering series can either start at the beginning of the report (1, 2, 3), or reflect the number of the section in which they lie (1.1,1.2, 2.1). Place tables, figures and plates near to their first mention with a self-contained caption. Have separate contents lists (placed after the main contents list) for tables, figures and plates.

The reader should be able to understand the results given in tables and figures without having to refer to the text. Similarly, the text must be self-contained and not require reference to the tables and figures, except for the detail. It is preferable to orientate tables and figures vertically so that the final report does not have to be turned on end. If the table or figure is too big to fit in vertically then consider simplifying it. Never present identical datasets twice (e.g. both as a figure and a table). Do not have too many subsections in your results section, since illustrations, including tables and figures, will act to break up the text, and the use of subsections in addition to these may spoil the flow.

If you are presenting sample statistics (e.g. means, standard errors, t test results, etc.), whether in the text, figures or tables, always state what the statistics are and the sample size (where appropriate using the degrees of freedom). See Chapter 5 for information about how to document the statistics.

Tables

Use a table when there are many numerical data to be presented in a logical way in the text, and where the exact numbers are important. The following are guidelines for setting out tables (Table 6.1):

- Normally place the variables you wish to compare (the independent variables) down the left-hand side and the dependent variables across the top in the column headings.

Table 6.1 Mean number of individuals[a] of invertebrate orders found in polluted and clean ponds

	Odonata	Plecoptera	Ephemeroptera	Trichoptera	Hemiptera	Diptera	Neuroptera	Coleoptera
Polluted[b]								
Pond 1	0	0	0	4	1	72	0	1
Pond 2	0	0	1	4	0	41	0	2
Pond 3	0	0	2	6	0	51	0	1
Pond 4	0	1	2	7	2	37	0	1
Pond 5	1	0	3	7	0	28	20	3
Clean								
Pond 1	2	4	4	10	2	16	0	3
Pond 2	1	5	5	9	0	22	0	2
Pond 3	3	2	3	11	1	11	19	2
Pond 4	5	3	3	8	1	9	25	1
Pond 5	2	3	5	9	3	13	0	3

[a]Mean of five 1 litre sediment samples per pond.
[b]Ponds with high levels of nitrates and phosphates.

- Give every table a number and title, and put any explanatory notes underneath the table, with superscript letters (^{a} ^{b} ^{c}), numbers (^{1} ^{2} ^{3}), or symbols (* † ‡) referring to the appropriate part of the table. Put these notes below the table (in a smaller font and single spaced) and make sure it is clear that these belong to the table and not the main text that follows.

- In tables, 0 should indicate a reading of zero, not a missing data point (which are usually annotated by a dash '–').

- Include only the necessary lines on a table: vertical lines are not usually needed and make tables harder to read.

Figures

Figures should only be used for data to which you wish to give prominence and where displaying data as a figure helps to display trends or patterns that illuminate the findings for the reader. Photographs, graphs, diagrams, maps and drawings are all called figures, and should be labelled consecutively throughout the text. Do not give different names to each type of figure (that is, do not use the terms Graph 1.1, Map 1.1, Diagram 3.2, etc., use only Figure 1.1, Figure 2.3 etc.). There is one exception to this: when photographs are interleaved on their own page (rather than embedded in among the text), they may be labelled as 'Plates'. However, even this is becoming less common. Photographs may be useful to help to describe a site or piece of equipment, the species you are working on, or to prove a point (e.g. 'the canal is polluted at site 6, see Figure 6.2').

Photo CPW

Figure 6.2 Pollution in the Forth and Clyde Canal in Glasgow

Table 6.2 summarises some of the uses of some commonly used graphs, which are described in more detail in Chapter 5. It is important to make sure that any figures are correctly labelled (including any units of measurement on the axis labels), that you have made a good use of the scale, that you include error bars where relevant, and that any text and points are sufficiently large to be clear to the reader (Figure 6.3).

Table 6.2 Uses of different types of graphs

Type of graph		Common usage
Histogram*		Display frequencies of a single measured variable (interval/ratio data) for a single sample
Pie chart*		Display frequencies (e.g. as percentages) of several categories of a single sample
Stacked bar chart†		Display frequencies (e.g. as percentages) of several categories of several samples
Clustered (or grouped) bar chart†		Display frequencies (e.g. as percentages) of several categories of several samples
Bar chart using means and error bars		Display central tendency and variation (e.g. mean, standard error, 95% confidence intervals or, more rarely, standard deviation) of a measured variable for several samples
Point chart using means and error bars		Display central tendency and variation (e.g. mean, standard error, 95% confidence intervals or, more rarely, standard deviation) of a measured variable over several samples
Box and whisker chart		Display central tendency and variation (i.e. median, interquartile range, range and outliers) of a ranked (ordinal) variable over several samples
Scatterplot (or scattergraph)		Display the relationship between two measured (interval/ratio) or ranked (ordinal) variables

(Continued)

Table 6.2 (*Continued*)

Type of graph	Common usage
Regression line plot	Display straight line causal relationship between two measured (interval/ratio) variables plus calculated regression line
Dendrogram	Display the similarities (or dissimilarities) between several samples on the basis of their component elements using binary (presence/absence) and/or measured (interval/ratio) data from several variables
Ordination biplot	Display the distances between several samples on the basis of their component elements using binary (presence/absence) and/or measured (interval/ratio) data from several variables

*Not usually used in reports (sometimes useful in presentations – see p. 247).
†Use with care (see p. 248).

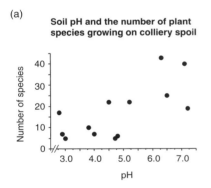

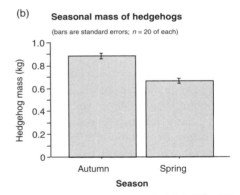

Figure 6.3 Presenting graphs
(a) Scatterplot; (b) bar chart.
Note that in both graphs there is an appropriate title, both axes are labelled, and any units of measurement have been included. Points and error bars are clear, as are the text and numbers. The data have been fitted to the full scale of the graphs: note that in (a) the *x*-axis does not reach zero and hence has been broken to show that it does not reach zero and to avoid a large expanse of the graph without data. This is an ideal way of clearly illustrating the truncation of the scale, but because it is not always produced automatically by statistical programs, it has become less commonplace. Note that in this example, the truncated axis makes the data appear to form a trend going through the origin; since this could be misleading care should be taken to either use truncation symbols (as here) or to extend the *x* axis further past the data points.

If two or more graphs are to be compared, place them near to one another and draw them to the same size and scale. If you take diagrams from another source it is preferable to redraw them so that they take on the 'house style' of your report. Make sure that you give full credit to the source (not to do so would be plagiarism). Do not forget to label axes and include units. Use shading patterns and symbols to distinguish between different series on a figure and black and white illustrations to keep things simple and clear (especially if your work is likely to be photocopied later). Colour is more expensive to produce, can distract the reader and should only be used in exceptional circumstances.

Equations and formulae

If you need to include chemical or mathematical equations and formulae, indent them in the main body of the text and number them for easy reference (e.g. 'as shown in equation 3'). Word-processors have equation editors to help to construct equations. Where several formulae are placed together, wherever possible ensure that the equality signs line up.

Discussion

This is where you interpret your results in the context of the available information. Although you may wish to concentrate on results that are statistically significant, it is worth bearing in mind that non-significant results can also be very interesting, especially if this is following a well-designed study with sufficient replicates. You should make sure that your discussion does not simply rest upon the statistical significance of your results, but extends this to look at the ecological significance. For example, a small but statistically significant decrease in insect diversity may not be particularly interesting ecologically unless the species lost are particularly rare or important (e.g. as food for insectivorous birds or bats). It is important here to use the discussion to highlight those connections between your results and those of other studies. As before, acknowledge any work used in writing this section (see p. 308). Remember that your results rarely prove anything, but they may support or refute predictions. Consider how your work:

- fits with the aims;
- compares to previously published work.

Do not repeat chunks of information from the introduction. If there were problems with the study, discuss them here and suggest how they influenced your results (although try and keep this in a positive context and try to avoid a whole subsection on the limitations of the study unless the format of the publication requires such a section – as in some journals). Suggest how your work highlights avenues for further work. End your discussion with a concluding paragraph reconfirming the main findings and the key messages (including any recommendations) for the reader. In the Model II report, each section's discussion should place the results in the context of current information and the general discussion should draw the themes from all the sections together. For both models, the conclusions may be placed in a separate section rather than being included as the final paragraph of the discussion (or general discussion), but should still be short and concise.

References

The reference list is a list of all the journal articles, books, reports, websites, etc., specifically referred to in your report (see p. 318 for how to refer to the literature). You must have a reference list in your report. A reference list is sometimes confused with a bibliography. A bibliography is list of books or other papers pertaining to a particular topic. They need not have been referred to in the project. For example, a bibliography could be on 'The ice age in Britain' or 'Mammals in North America'. A bibliography may be a useful addition where a series of texts or review articles set the general background for a subject or contain ideas that are generic to the report and are referred to in a general manner, but should not replace a reference list.

As well as listing the references accurately, you must make sure that all the citations in the text are in the list, and that all references in the list are in the text. Carefully read the report, tick every reference in the text while at the same time ticking the equivalent reference in the list. It is important to check that author names and dates tally at the same time, because incorrect citations are misleading.

Citing papers

There are several reasons why we cite the sources of the information that we use in our work. First, it acknowledges the fact that it is someone else's work; to fail to do so would be plagiarism. Plagiarism is a serious issue in academic writing and should be avoided at all costs. Even paraphrasing someone else's work without indicating who wrote it and where suggests that either it is so mainstream as to not need attributing, or more seriously that it was you who thought it up. Second, it allows other people to refer back to that piece of work to extend their understanding and to check that your interpretation of what it says is a reasonable one. To enable this sort of referencing to be useful, then you need to list references in full in the reference list, and cite them properly in the text.

There are two main ways of citing work in the text: the Harvard system and the Vancouver system. The former uses the author and date within the text to identify that a piece of work has been cited (Box 6.1), and the references are always listed in alphabetic order (Box 6.2). In the Vancouver system, an identifier number (in brackets or in a superscript font) within the appropriate part of the text identifies that a citation is being made, and the references are then listed in order of these numbers (i.e. in the order in which they are used in the text). In both systems, house styles may insist on full journal names or allow standard abbreviations (i.e. *J. Ecol.* rather than the *Journal of Ecology*). Such formats should be applied consistently and not be mixed in the same reference list. Most libraries and some Internet sites[1] provide the standard abbreviations for scientific journals. Other styles of referencing exist in science, including those that use footnotes to identify citations within the text, rather than in reference lists.

Box 6.1   Citing works using the Harvard system

There are many variations of the Harvard style and you should use your house style if applicable. The following shows **ONE** variant of the Harvard referencing system.

A paper by a **single author** can be cited in the following ways:

- Many species of invertebrates have been found in grassland areas (Green 2010).

- Green (2010) states that there are many species of invertebrates in grassland areas.

The way in which you cite the paper (i.e. with the author's name in or out of brackets) depends on where you want the emphasis placed within the sentence. In general, if you want to use the citation as supporting evidence it is better to use the first method. In the second example, the use of the author's name at the beginning of a sentence distracts from the main point of the sentence, unless the intention is to emphasise the author's opinion.

A paper by **two or three authors** is cited as (Green, Brown and White 2010) or Green, Brown and White (2010). When a paper has **four or more authors**, the citation is abbreviated in the text as (Black et al. 2010) or Black et al. (2010) and the full list of authors is given in the reference list at the end (see p. 317). However, the full author list for large consortium publications is often not included. It may be included in the reference section using the first few (often the first three) authors only, so:

- Thackeray S. J., Sparks T. H., Frederiksen M., Burthe S., Bacon P. J., Bell J. R., Botham M. S., Brereton T. M., Bright P. W., Carvalho L., Clutton-Brock T., Dawson A., Edwards M., Elliott J. M., Harrington R., Johns D., Jones I. D., Jones J. T., Leech D. I., Roy D. B., Scott W. A., Smith M., Smithers R. J., Winfield I. J. and Wanless S. (2010) Trophic level asynchrony in rates of phenological change for marine, freshwater and terrestrial environments. *Global Change Biology*, **16**: 3304–3313.

would more usually be listed as:

- Thackeray S. J., Sparks T. H., Frederiksen M., et al. (2010) Trophic level asynchrony in rates of phenological change for marine, freshwater and terrestrial environments. *Global Change Biology*, **16**: 3304–3313.

---

[1] http://images.isiknowledge.com/WOK45/help/WOS/A_abrvjt.html

If you wish to cite more than one work by an author (or combination of authors) that were published in a single year, then the citation should be distinguished by letters (a, b, c . . .) e.g. Brown (2010c).

When you wish to cite several sources to support one statement, this is done in chronological order as follows:

- Several experiments have been undertaken on invertebrate colonisation (White et al. 2008; Green and Brown 2009; Green 2010).

If you have seen reference to a work but have not been able to obtain the original work, then it may still be cited in the following way:

- Black (2010) cites White (1990) as demonstrating a relationship between insect diversity and plant diversity.
- White (1990, cited by Black 2010) demonstrated that there was a relationship between insect diversity and plant diversity.
- There is a relationship between insect diversity and plant diversity (White 1990, cited by Black 2010).

Note: the reference list contains Black (2010) not White (1990).

Quotes

If you want to quote text word for word, then the words and punctuation of a direct quote must be exact and placed in quotation marks. When quoting, in addition to the author and date, you also need the page number:

- Black (2010, p.85) concluded that 'herbivory may be important in controlling the spread of the species'.

Personal communications

A personal communication is an unpublished source of information, communicated to you verbally or in a letter, personal email message, etc. It is preferable to use published sources of information rather than personal communications where possible. Where necessary, refer to personal communications as follows:

- Green Mill Lake was restocked with trout and carp in 1996 (A. B. White, pers. comm.).

Personal communications do not appear in the reference list, although you should remember to thank everyone who supplies you with information in the acknowledgements section of your report (see p. 308). In this case it would be helpful to include the affiliation, e.g. 'I would like to thank A. B. White, Site Manager, Green Mill Park for information on the management of the site'.

Box 6.2 Reference lists using the Harvard system

This box gives examples of **ONE** variant of the Harvard referencing style. You should always use the house style where one exists; if preparing a paper for a journal, report or thesis, then you must carefully follow any appropriate instructions.

Journals (single or multiple author)

Name Initials. and Name, Initials. (Date) Title of Article. *Title of Journal*, Volume(Part): Pages.
For example:

- Bell J.R. (2005) The emergence of manipulative experiments in ecological spider research: 1684–1973. *Journal of Arachnology*, **33**: 826–849.

- Wheater, C. P., Cullen, W. R. and Bell, J. R. (2000) Spider communities as tools in monitoring reclaimed limestone quarry landforms. *Landscape Ecology*, **15**: 401–406.

Books and pamphlets
Single author

Name Initials. (Date) *Title*. Series and number if appropriate. Publisher, Place Published.
For example:

- Wheater C. P. (1999) *Urban Habitats*. Routledge, London.

Multiple author

Name Initials., Name, Initials. and Name, Initials. (Date) Title. Series and number if appropriate. Publisher, Place Published.
For example:

- Wheater C. P. and Cook P. A. (2003) *Studying invertebrates*. Naturalists' Handbooks 28. Richmond Publishing Company Ltd., Slough, UK.

Section/chapter in an edited volume

Name Initials. (Date) Title of Section/Chapter. In: Name Initials. (ed.) *Title of Book*. Publisher. Place Published. Page Numbers of Section.
For example:

- Wheater C. P. (2010) Chapter 24. Walls and paved surfaces: urban complexes with limited water and nutrients. In: Douglas I., Goode D., Houck M. and Wang R. (eds). *The Routledge Handbook of Urban Ecology*. Routledge, London. p. 239–251.

Acts of parliament

Author (Date) *Title of Act including the chapter number*. Publisher, Place Published.
For example:

- Great Britain (1981) *The Wildlife and Countryside Act 1981, Chapter 69*, The Stationary Office, London.

> **WWW**
>
> Citing material on the web can be difficult, since web pages do not necessarily include information on the date of publication and the author. The publisher is the organisation responsible for maintaining the site. If the page has a statement to the effect of 'last updated May 2010, for comments on this site please email Zeb Brown, z.brown@smalltown.ac.uk' then these are probably the date and author.
>
> > Name, Initials and Name, Initials (Date). Title of web page [online]. Name and place of publication source, viewed date (day, month and year): URL.
> >
> > For example:
>
> - Wheater C. P. and Cook P. A. (2010) FCStats v1.1i. Manchester Metropolitan University, School of Science and the Environment, viewed 1 January 2011: http://www.sste.mmu.ac.uk/teachers_zone/
>
> Note that web references, especially those only available on-line, are often listed with their digital object identifier (DOI) before they are published and their volume and page numbers are known.
>
> > For example:
>
> - Thackeray S. J., Sparks T. H., Frederiksen M., et al. (2010) Trophic level asynchrony in rates of phenological change for marine, freshwater and terrestrial environments. *Global Change Biology*, DOI: 10.1111/j.1365– 2486.2010.02165.x

Appendices

Include in appendices details that are not central to the main text of the report. Materials appropriate for an appendix might include species lists, calibration curves, data analysis algorithms and maps. Make sure that each appendix is referred to at least once in the text.

Writing style

This section provides some general guidance to scientific writing and should be used in conjunction with any house styles that are associated with your report. In addition to a respected dictionary and thesaurus (e.g. those published by Collins, Oxford, Chambers or Webster), you may find it useful to consult appropriate books on grammar (e.g. Jackson 2005), punctuation (e.g. Truss 2003), and spelling and word usage (e.g. Bryson 2009). Other texts on writing for students may also prove useful, examples include Barrass (2005) and Creme and Lea (2008), whereas Day (2006) covers writing for scientific publication.

Each paragraph should cover an idea, and usually the main idea of a paragraph is introduced in the first sentence. Start a new paragraph for a new idea. Vary sentence length since occasional use of short sentences has more impact.

Avoid metaphors, similes, slang, acronyms and other figures of speech that are common in informal writings; these may not be translatable (remember that scientific writing should communicate ideas to overseas workers as well as to native English speakers). One of the most important tips for good writing style is to avoid jargon. Jargon includes both technical terms and the pretentious use of long words and convoluted sentences. The use of technical terms may be unavoidable, but at least make sure you define them carefully in plain English. Pretentious use of long words should be eliminated. Avoid sentence constructions such as:

> surveys of the vegetation were carried out using random quadrats.

Verbs such as 'to carry out' (along with to achieve, to perform, to accomplish, to undertake) are usually unnecessary. Imagine how ridiculous language like this would be in everyday speech: 'the shopping was accomplished on Saturday' instead of 'I went shopping on Saturday'. Similarly in science writing use:

> the vegetation was surveyed using random quadrats

or

> random quadrats were used to survey the vegetation.

Also watch out for the verbs to occur, to obtain, to be noticed, etc., that are also used to contort science writing. For example, change:

> a higher percentage vegetation cover occurred at lower altitudes

to

> percentage vegetation cover was higher at lower altitudes.

Simplify and increase the impact of your writing by leaving out unnecessary words (Table 6.3).

The following list of general points is a useful checklist when writing your report:

- Check for subject/verb agreement, for example:

  > the greatest decrease in invertebrate numbers **occurs** during ....

  the verb refers to the decrease, which is singular. Also, be aware that the word 'data' is plural: use 'data were' rather than 'data was'.

- The use of the word 'this' or 'these' can be ambiguous. In the example below:

  > Rough grassland areas include scrub communities of hawthorn or gorse, and ferns, including bracken. This provides a wealth of opportunities for wildlife.

Table 6.3 Examples of words used unnecessarily when qualifying terms

Instead of ...	use ...	Instead of ...	use ...
absolutely correct	correct	not actually correct	incorrect
an actual experiment	an experiment	obviously true	true
by means of	by (or using)	positive developments	developments
circular in shape	circular	quite improbable	improbable
completely submerged	submerged	quite steep	steep
definite proof	proof	really developed	developed
facing up to	facing	solid evidence	evidence
few in number	few	the actual value	the value
hairy in appearance	hairy	the smallest minimum	the minimum
highly necessary	necessary	this moment in time	currently
highly problematic	problematic	totally justified	justified
highly relevant	relevant	totally new	new
it really was	it was	totally unique	unique
large in size	large	valid question	question
mossy in character	mossy	very steep	steep

If the intention is to say that together the communities (hawthorn, gorse and bracken) provide a wealth of opportunities, it would be better phrased:

> Rough grassland areas include scrub communities of hawthorn or gorse, and ferns, including bracken. These habitats provide a wealth ...

If, on the other hand, the author intends to say that bracken provides a wealth of opportunities, it would be better to say that

> Rough grassland areas include scrub communities of hawthorn or gorse, and ferns, including bracken. Bracken provides a wealth ...

If there is any possibility of ambiguity, then it is preferable to repeat the noun.

- Avoid double negatives ('not unlikely' should be 'likely', 'not unjustifiable' is 'justifiable').
- Use gender-neutral language: language is considered sexist if it applies only one sex when both are intended. Instead of using 'he' or 'his', 'man' or 'men', reword the sentence to use 'people' or 'humans', or if it is unavoidable, use 'he or she'.

- Use standard abbreviations and spell them out in full in the first instance (consider including a glossary of terms if you include many abbreviations).

- Avoid capitals for accepted common names of organisms. Objects or processes should neither be capitalised nor put in quotation marks.

- Check for careless mistakes introduced through word-processing (such as duplicate paragraphs and inaccurate cross-referencing).

Tense

Most of your work should be in the past tense. This is because you are reporting things you have done and things you have found. However, in the introduction and discussion, established facts or general theories should be stated in the present tense. Present tense should also be used to refer to tables or figures ('Table 2.3 shows that ...' or 'the results from ... are illustrated in Figure 3.5').

Passive tense

In the past it has been traditional for scientists to write in the passive tense (i.e. avoiding the use of 'I' and 'we'), possibly because such writing was thought to be more objective. However, the use of the passive tense leads to stilted writing, and it is now becoming more common for scientific papers to be written in the active tense. A mixture of both probably provides the best balance: use personal pronouns when you want impact, for example in stating the important results: 'I found ...' (active tense), rather than 'it was found that' (passive tense). Use the passive tense in your methods ('the machine was calibrated...', 'the site was surveyed') to avoid the rather repetitive I did this, then I did that, etc.

Numbers

There are several conventions that are associated with the use of numbers in scientific writing:

- use SI units for all measurements (see Table 6.4);
- do not add an 's' after a unit to make it plural: e.g. 50 m means 50 metres, whilst 50 ms means 50 milliseconds;
- spell out numbers from one to nine **unless** it is a measurement with an SI unit (e.g. 7 m);
- spell out any number at the beginning of a sentence (or reword the sentence so that the number does not come first);

Table 6.4 SI units of measurement
To standardise the units in which measurements are made, an international system has been agreed (Système International d'Unités, abbreviated to SI units). The standard units for major types of measurement are shown, together with the common prefixes indicating multiples of SI units (in thousands).

Measurement	SI unit (symbol)	Example of use
Amount of substance	mole (mol)	1 mol
Length	metre (m)	1.5 m
Mass	kilogram (kg)	0.25 kg
Temperature	Kelvin (K)	273.1 K
Time	second (s)	45 s

Common prefixes

Amount	Prefix	Abbreviation
10^{-9}	nano	n
10^{-6}	micro	μ
10^{-3}	milli	m
10^{3}	kilo	k
10^{6}	mega	M
10^{9}	giga	G

- spell out numbers if successive numbers refer to different things (e.g. 'fifteen 10 m transects');
- if you spell numbers between 21 and 99, hyphenate them (e.g. forty-five);
- avoid billion, trillion etc. as these can cause confusion between American and English usage;
- do not leave naked decimal points. Instead of .01, use 0.01.

Abbreviations

Abbreviations include contractions, acronyms and initializations, and are generally defined as shortenings of words. Contractions are usually words that have been shortened in the middle (e.g. Dr or can't). In formal writing, including research reports, colloquial contractions such as can't and didn't should not be used and such words should be spelt out fully. Acronyms are words formed from the use of initial letters to represent longer terms. These are used for technical equipment and processes (e.g. Laser: Light Amplification by Stimulated Emission of Radiation), or organisations (e.g. NATO: North Atlantic Treaty Organisation). Some terms are acronyms made up of more than just the initial letters of the term (e.g. sonar: SOund NAvigation and Ranging). It is certainly the case that using abbreviations can reduce the density of text where terms are occurring over and over again. For example, the use of NVC for the National Vegetation Classification. It is important that acronyms are clarified at least the first time they are used (unless they have come into such common usage that they have become terms in their own right – for example, laser), preferably the first time in each section, and ideally (and especially where there are a large number of acronyms used) listed and defined in a glossary. Some conventions regarding the use of abbreviations are listed in Table 6.5.

Table 6.5 Conventions for the use of abbreviations

Abbreviation	Convention
Contractions where the beginning of the word is abbreviated, e.g. 'plane	Strictly speaking you should use the complete word (e.g. aeroplane) but certain words have acquired almost full status in their own right and so are often written as the abbreviation and without the leading apostrophe
Contractions where the centre of a word or number of words is abbreviated and the word gains a different pronunciation, e.g. can't, don't	The apostrophe should be used here to indicate the missing letters (this type of contraction is not usual in report writing)
Contractions where the centre of a word is abbreviated and the word is pronounced the same as previously, e.g. Dr, Mr	UK writers do not use a full stop at the end of such an abbreviation, whereas American writers do
Contractions where the end of the word is abbreviated, e.g. Prof.	A full stop is used here to indicate the missing letters (note that the full stop should not be used in such cases when they are turned into plurals, hence Profs). Where such abbreviations are used at the end of a sentence, only a single full stop should be used.
Capital letters in acronyms	Some writers prefer to use capital letters for all acronyms, whilst others simply use (at most) the initial letter when the term has become common usage as a word (e.g. SONAR, or Sonar, or sonar, RADAR, or Radar or radar – RAdio Detection And Ranging)

Punctuation

There are several types of punctuation mark in English, of which five are associated with:

- The flow of reading using full stops (periods) and commas. The former indicates the end of a sentence and the latter the separation of phrases within sentences. Commas can either be in pairs placed around a phrase within a sentence, or used as a single separator between phrases. Note that commas can also be used:
    - to replace 'and' in a list of adjectives (e.g. the 'light, brown soil' instead of the 'light and brown soil', i.e. soil that is both light in texture/weight and brown in colour – note that the 'light brown soil' is another thing all together referring to soil that is a light shade of brown);
    - after words such as 'however', 'nevertheless' and 'moreover' at the beginning of a sentence;
    - to separate names and positions of people when they are being addressed: compare 'it's false Dr Black' with 'it's false, Dr Black'.
- The links between items using semi-colons, colons and dashes. Semi-colons are used where two phrases are strongly linked. The colon points forward, indicating that there is more to come, whereas the dash points backwards, reflecting previous statements.
- The separation of lists using a colon just before the list. Either commas or semi-colons can be used between items within the list. The latter are usually more useful since you may need to use commas within descriptions of items within the list and this can be confusing if commas are also used between items.
- The inclusion of additional information using pairs of parentheses (round brackets), pairs of commas, and pairs of dashes (in order of increasing emphasis). These separate off words or phrases that are not part of the main point of the sentence, including phrases that explain, expand or modify the meaning of the sentence.
- The tone of the sentence using question marks and exclamation marks; the use of the latter being unusual in scientific writing.

Other punctuation marks that are commonly used in scientific writing include:

- Quotation marks that are used to quote from people and sources in an exact way (rather than paraphrasing the information), or to provide emphasis, or to identify the item or phrase being discussed. Single quotes (' ') are usually used rather than double ones (" "). When a set of quotes occurs within another, use single quotation marks for the outer set and double for the inner. In Britain (but not in the USA), other punctuation marks are usually placed within the quote if they are part of the phrase (so if a quotation ends with a question mark this is placed before the closing quotation mark, while if the question mark belongs to the whole sentence and not just to the quotation then it should be placed after the closing quote).

- Apostrophes, which are used either to denote to whom or what something belongs, for example the student's (belonging to one student), or the students' (belonging to several students), or in contractions, where two words have been informally squashed into one, for example don't, won't, etc. Contractions should not usually be employed in scientific writing. Never use an apostrophe when making a plural. Although you should be careful to avoid the grocers' apostrophe (e.g. tomato's), there are occasions where its use in forming plurals is acceptable (e.g. do's and don'ts) because without it, the word is ambiguous (e.g. dos). Note that it should be CFCs and DVDs, **not** CFC's and DVD's.

Choice of font

It is good practice to use a single, standard font of appropriate point size throughout any piece of scientific writing; the recommendations[2] for enhancing accessibility for people with dyslexia are useful guidelines for clear, easy to read text for all readers. The use of underlined, bold and/or italic fonts should be avoided unless it is necessary to draw the reader's attention to a particular word or phrase. Quotation marks are more often used to identify the item under discussion where this may be ambiguous. Bold text is often used where a term is subsequently defined in a glossary (although this may also be better achieved by providing the definition in a footnote or in parentheses immediately after the first use of the word). Latin and foreign words that have not entered normal usage in English may be put in italics, while those in common usage should not (so we use *inter alia* but per capita – see Table 6.6), although there may be a house style that you should follow. If in doubt, use a good-quality dictionary.

The scientific names of fauna and flora should be in italics. The first time you mention a species, it should be written in full, followed by the authority. The authority is the name of the person who originally described the species. The authority's name is not italicised. If the authority is well known (e.g. if they have named many species), an abbreviation may be used (e.g. L. for Linnaeus, F. for Fabricius). An authority in parentheses indicates that the original species name has been altered (e.g. if the organism has subsequently been placed in a different genus). In the absence of parentheses, the original name given to the organism is still in use today. The authority may be followed by the date when the species was first described, although this is frequently omitted. You should also give the common name (if it has one) the first time a species is referred to. Thereafter, refer to the species by either its common name, or shortened scientific name. Thus after the first mention, the small white butterfly *Pieris*

---

[2] http://www.bdadyslexia.org.uk/about-dyslexia/further-information/dyslexia-style-guide.html

Table 6.6 Examples of Latin and foreign words and their emphasis

Not usually italicised	Usually italicised
a priori	*ad hoc*
et al.	*en masse*
etc.	*en route*
e.g.	*ex situ*
i.e.	*in situ*
vice versa	*inter alia*
per annum	
per cent	

rapae L. would become *P. rapae*. Where an unknown species is identified only to genus, then the genus name is followed by sp. for species (e.g. *Pieris* sp.). Where more than one species of that genus is mentioned, the genus name is followed by spp. (which stands for species pluralis), e.g. *Pieris* spp.

Common mistakes

Proofread your report carefully for spelling mistakes. Even spell-checkers on word-processors will not pick up words that are wrong but spelt correctly, for example 'in' instead of 'is', 'ant' instead of 'and'. Also be aware of the differences between American and English spellings (behavior/behaviour, color/colour, gray/grey). Box 6.3 lists several words that are often misused.

Box 6.3   Commonly misused words

Affect

In science writing, this is usually the verb:

'Pollution affects the dissolved oxygen concentration.'

Effect

In science writing, this is usually the noun:

'Pollution has an effect on the dissolved oxygen concentration.'

or 'the greenhouse effect'

Amount

Mass or volume

Number

Items that have been counted

Compliment

To imply flattery

Complement

The finishing touches to something, fitting together, completing

It is composed of
'the microscope lens turret is composed of three stages'
not: 'the microscope lens turret is comprised of three stages'

It comprises
'This report comprises three sections'
not: 'This report is comprised of three sections'

Its
Belonging to it

It's
Contraction for 'it is'. Never use 'it's' in report writing

Lead
The element (Pb) and the verb to lead

Led
The past tense of the verb to lead. If it sounds like led, write led (unless it is the metal)

Less
In quantity
'site A had less pollution'

Fewer
In number
'site A had fewer pollution incidents'
not 'site A had less pollution incidents'

Lose
To misplace

Loose
Not secure

Passed
Often means accepted or carried:
'a law is passed'

Past
Just gone by, referring to time:
'in the past'

Practice
This is the noun:
'the practice of dumping waste is avoidable'
American English uses the spelling of practice for both the noun and the verb (and similarly for license)

Practise
This is the verb:
'it is important to practise the technique before collecting data'
To remind yourself of the English spelling for this and similar words, substitute the words advice (noun) and advise (verb)

Principle
A truth, fundamental basis of something:
'The principles of the investigation were'

Principal
Main, the most important, the highest in rank, the foremost:
'The study is made up of five principal sections'

Which
Introduces a non-essential clause:
'the rivers, which were polluted, were included'
(implies that all rivers were polluted)

That
Introduces an essential clause:
'the rivers that were polluted were included'
(implies that only the polluted rivers were included)

Computer files

The importance of saving your work regularly and making regular backups cannot be emphasised too much. At the end of each work session, make a backup of both data and word-processing files. Keep the originals and the backups in different places, for instance one set at home and one at work or on an on-line data storage site. Do not rely on a single computer's hard drive, it can become corrupted or be stolen. Rename your files occasionally (giving each generation a consecutive number) and keep old versions of your work in case both your main copy and current backups are lost. Save your backups to data pens regularly and keep the old ones. Use the most up-to-date virus scanner that you can obtain to check your data pen especially after it has been used in another machine (e.g. on a university network).

Save your work before you try a word-processing command with which you are not familiar. In this way, if anything on the file changes and you cannot get it back, you can exit the program without saving and re-open your original document. If you do make a mistake, do not panic, use the undo function (this is usually obvious through the menu, but pressing the 'control' and 'z' keys together will do this in most programs).

Incorporating several figures (and especially high-definition photographs) can make files large and unmanageable. If your computer is short on memory, it may be easier to keep figures as separate files and slot them into the text just prior to the final edit of the report. Similarly, keeping separate sections in separate files will reduce the size of each file and make loading and saving (especially autosaving) faster processes.

Summary

In summary, taking care throughout the presentation of your work will enhance the extent to which you get your message across effectively. Targeting those you expect to read the report and following any guidance (including house styles), a consistent approach throughout, and rigorous quality control (including checking all references) will avoid distractions for the reader. So, sound project design and preparation, thoughtful development of aims, objectives and hypotheses, the effective use of appropriate methods, targeted analyses and high-quality presentation should result in a readable and useful research report that will be read and used instead of being left to languish on a shelf – enjoy!

References

Aldiss D. T. (2007) *A safe system of fieldwork.* NERC Guidance Note. Natural Environment Research Council, Swindon. http://www.nerc.ac.uk/about/work/policy/safety/documents/guidance_fieldwork.pdf (accessed October 2010).

Allen S. E. (1989) *Chemical analysis of ecological materials.* 2nd edition. Blackwell Scientific Publishers, Oxford.

ARG-UK (2008) *Amphibian disease precautions: a guide for UK fieldworkers.* Version 1. ARG-UK Advice Note 4. Amphibian and Reptiles groups of the UK, Dorset. http://www.arg-uk.org.uk/Downloads/ARGUKAdviceNote4.pdf (accessed October 2010).

ASAB (2006) Guidelines for the treatment of animals in behavioural research and teaching. *Animal Behaviour*, 71: 245–253. http://asab.nottingham.ac.uk/downloads/guidelines2006.pdf (accessed October 2010).

Backiel T. and Welcomme R. L. (eds) (1980) *Guidelines for sampling fish in inland water.* European Inland Fisheries Advisory Commission Technical Paper No. 33. Food and Agriculture Organization of the United Nations, Rome. http://www.fao.org/docrep/003/AA044E/AA044E00.htm#TOC (accessed October 2010).

Balmer D., Coiffait L., Clark J. and Robinson R. (2008) *Bird ringing: a concise guide.* BTO, Thetford.

Bang P. and Dahlstrøm P. (2006) *Animal tracks and signs.* Oxford University Press, Oxford.

Barlow K. (1999) *Expedition field techniques: bats.* Royal Geographical Society, London. http://www.rgs.org/NR/rdonlyres/2C503639-2BCE-4580-AA66-8725DA4D412A/0/BatManualUpdated.pdf (accessed October 2010).

Barnett A. and Dutton J. (1995) *Expedition field techniques: small mammals (excluding bats)* Royal Geographical Society, London. http://www.rgs.org/NR/rdonlyres/C6CC4037-B52E-4E3C-A0B3-E0FBECD8E07F/0/SmallMammalsupdated.pdf (accessed October 2010).

Barrass R. (2005) *Students must write.* 3rd edition. Routledge, London.

Barrow C. (2004) 10. Risk assessment and crisis management. In: Winser S. (ed.) *The expedition handbook.* The Royal Geographical Society, Profile Books, London. pp. 106–116. http://www.rgs.org/NR/rdonlyres/24084ADC-C738-464C-A4D4-91F6CB4ED23E/0/HBSREracm.pdf (accessed October 2010).

Bat Conservation Trust (2007) *Bat surveys – good practice guide.* Bat Conservation Trust, London. http://www.bats.org.uk/download_info.php?id=379&file=BCT_Survey_Guidelines_web_final_version.pdf&referer=http%3A%2F%2Fwww.bats.org.uk%2Fpages%2Fprofessional_guidance.html (accessed October 2010).

Bateman J. (1979) *Trapping: a practical guide.* David and Charles, Newton Abbot.

Practical Field Ecology: A Project Guide, First Edition.   C. Philip Wheater, James R. Bell and Penny A. Cook.
© 2011 John Wiley & Sons, Ltd. Published 2011 by John Wiley & Sons, Ltd.

REFERENCES

BCMF (1996) *Techniques and procedures for collecting, preserving, processing, and storing botanical specimens*. Res. Br., British Columbia Ministry of Forests, Victoria, BC. Working Paper 18/1996. http://www.for.gov.bc.ca/hfd/pubs/docs/wp/Wp18.pdf (accessed October 2010).

Beaupre S. J., Jacobson E. R., Lillywhite H. B. and Zamudio K. (2004) *Guidelines for use of live amphibians and reptiles in field and laboratory research*. 2nd edition. Herpetological Animal Care and Use Committee of the American Society of Ichthyologists and Herpetologists, Lawrence, KS. http://www.asih.org/files/hacc-final.pdf (accessed October 2010).

Bell J. R., King R. A., Bohan D. A. and Symondson W. O. C. (2010) Spatial co-occurrence networks predict the feeding histories of polyphagous arthropod predators at field scales. *Ecography*, **33**: 64–72. http://www3.interscience.wiley.com/cgi-bin/fulltext/123309567/PDFSTART (accessed October 2010).

Bell J. R., Mead A., Skirvin D. J., Sunderland K. D., Fenlon J. and Symondson W. O. C. (2008) Predator functional groups vs. classical taxonomy: does implicit ecological information improve model prediction?*Bulletin of Entomological Research*, **98**: 587–597.

Bennett D. (1999) *Expedition field techniques: reptiles and amphibians*. Royal Geographical Society, London. http://www.rgs.org/NR/rdonlyres/9E3CF152-C817-470C-AC0C-C8FE82A8CBAD/0/Reptilesupdated.pdf (accessed October 2010).

Bibby C. J., Burgess N. D., Hill D. A. and Mustoe S. (2000) *Bird census techniques*. 2nd edition. Academic Press, London.

Bibby C., Jones M. and Marsden S. (1999) *Expedition field techniques: bird surveys*. Royal Geographical Society, London. http://www.rgs.org/NR/rdonlyres/E9386FEB-F085-47DA-8F9D-74E6F9633743/0/BirdSurveysupdated.pdf (accessed October 2010).

Bonar S. A., Hubert W. A., and Willis D. W., (eds) (2009) *Standard methods for sampling North American freshwater fishes*. The American Fisheries Society, Bethesda, MD.

Borg I. and Groenen P. (2005) *Modern multidimensional scaling*. 2nd edition. Springer, London.

Bosshard W., (ed.) (1986) *Kronenbilder mit Nadel- und Blattverlustprozenten*. Eidgenössische Anstalt für das forstliche Versuchswesen, Birmensdorf.

Briggs B. and King D. (1998) *The bat detective: a field guide to bat detection*. Batbox Ltd, Conservancy Trust, London.

Brodie J. (1985) *Grassland studies*. George Allen and Unwin, London.

Brooks D. R., Perry J. N., Clark S. J., Heard M. S., Firbank L. G., Holdgate R., Shortall C. R., Skellern M. P. and Woiwod I. P. (2008) National-scale metacommunity dynamics of carabid beetles in UK farmland. *Journal of Animal Ecology*, **77**(2): 265–274. http://www3.interscience.wiley.com/cgi-bin/fulltext/119391907/PDFSTART (accessed October 2010).

Brown R., Ferguson J., Lawrence M. and Lees D. (2003) *The tracks and signs of the birds of Britain and Europe: an identification guide*. 2nd edition. Christopher Helm, London.

Bryson B. (2009) *Bryson's dictionary: for writers and editors*. Black Swan, London.

Buckland S. T., Anderson D. R., Burnham K. P., Laake J. L., Borchers D. L. and Thomas L. (2001) *Introduction to distance sampling: estimating abundance of biological populations*. Oxford University Press, Oxford.

Buckland S. T., Anderson D. R., Burnham K. P., Laake J. L., Borchers D. L., and Thomas L. (eds) (2004) *Advanced distance sampling*. Oxford University Press, Oxford.

Burrows M. T., Harvey R. and Robb L. (2008) Wave exposure indices from digital coastlines and the prediction of rocky shore community structure. *Marine Ecology Progress Series*, **353**: 1–12.
http://www.int-res.com/articles/feature/m353p001.pdf (accessed October 2010).

Cadima E. L., Caramelo A. M., Afonso-Dias M., Conte de Barros P., Tandstad M. O. and de Leiva-Moreno J. I. (2005) *Sampling methods applied to fisheries science: a manual.* FAO Fisheries Technical Paper No. 434. Food and Agriculture Organization of the United Nations, Rome.
http://www.fao.org/docrep/009/a0198e/a0198e00.htm (accessed October 2010).

Campbell J. B. (2007) *An introduction to remote sensing.* Taylor and Francis, London.

CCAC (2003) *Canadian Council on Animal Care guidelines on: the care and use of wildlife.* Canadian Council on Animal Care, Ottawa.
http://www.ccac.ca/en/CCAC_Programs/Guidelines_Policies/GDLINES/Wildlife/Wildlife.pdf (accessed October 2010).

Chadd R. (2010) Assessment of aquatic invertebrates. In: Hurford C., Schneider M., and Cowx I. G., (eds) *Conservation monitoring in freshwater habitats: practical guide and case studies.* Springer, London. pp. 63–72.

Chadd R. and Extence C. (2004) The conservation of freshwater macroinvertebrate populations: a community-based classification scheme. *Aquatic Conservation: Marine and Freshwater Ecosystems*, **14**: 597–624.

Chame M. (2003) Terrestrial mammal feces: a morphometric summary and description. *Memórias do Instituto Oswaldo Cruz [online]* **98**(Suppl. 1): 71–94.
http://www.scielo.br/scielo.php?pid=S0074-02762003000900014 &script=sci_arttext&tlng=en (accessed October 2010).

Clarke K. R. (1993) Non-parametric multivariate analyses of changes in community structure. *Australian Journal of Biology*, **18**: 117–143.

Coad B. W. (1998) *Expedition field techniques: fishes.* Royal Geographical Society, London.
http://www.rgs.org/NR/rdonlyres/914F3A5F-9675-4AD3-88C6-B8E384766E86/0/FISHESupdated.pdf (accessed October 2010).

Colwell R. K., Mao C. X. and Chang J. (2004). Interpolating, extrapolating, and comparing incidence-based species accumulation curves. *Ecology*, **85**: 2717–2727.

Connor D. W., Allen J. H., Golding N., Howell K.L., Lieberknecht L. M., Northen K. O. and Reker J. B. (2004) *The marine habitat classification for Britain and Ireland.* Version 04. 05. JNCC, Peterborough.
http://www.jncc.gov.uk/pdf/04_05_introduction.pdf (accessed October 2010).

Conroy J. W. H., Watt J., Webb J. B. and Jones A. (2005) *A guide to the identification of prey remains in otter spraint.* 3rd edition. The Mammal Society, London.

Corn P. S. and Bury R. B. (1990) *Sampling methods for terrestrial amphibians and reptiles. Sampling procedures for Pacific Northwest vertebrates.* US Department of Agriculture Forest Service, Portland, Oregon.
http://www.fs.fed.us/pnw/pubs/pnw_gtr256.pdf (accessed October 2010).

Cottam C. (1956) Uses of marking animals in ecological studies: marking birds for scientific purposes. *Ecology*, **37**: 675–681.

Cowell D., and Thomas G. (1999) A key to the guard hairs of British canids and mustelids. *British Wildlife*, **11**: 118–120.

Crawley M. J. (2007) *The R book.* John Wiley and Sons Ltd, Chichester.

Creme P. and Lea M. R. (2008) *Writing at university.* Open University Press, Buckingham.

REFERENCES

Cullen W. R., Wheater C. P. and Dunleavy P. J. (1998) Establishment of species-rich vegetation on reclaimed limestone quarry faces in Derbyshire, UK. *Biological Conservation*, **84**: 25–33.

Dance A. (2008) What lies beneath. *Nature*, **455**(9): 724–725.

Davies J., Baxter J., Bradley M., Connor D., Khan J., Murray E., Sanderson W., Turnbull C. and Vincent M. (2001) *Marine monitoring handbook.* JNCC, Peterborough.
http://www.jncc.gov.uk/PDF/MMH-mmh_0601.pdf (accessed October 2010).

Davy-Bowker J., Clarke R., Corbin T., Vincent H., Pretty J., Hawczak A., Blackburn J., Murphy J. and Jones I. (2008) *River invertebrate classification Tool.* Final Report Project WFD72C, SNIFFER, Edinburgh.
http://www.sniffer.org.uk/Webcontrol/Secure/ClientSpecific/ResourceManagement/UploadedFiles/WFD72C%20FINAL%20REPORT%20with%20security.pdf (accessed October 2010).

Davy-Bowker J., Murphy J. F., Rutt G. P., Steel J. E. C. and Furse M. T. (2005) The development and testing of a macroinvertebrate biotic index for detecting the impact of acidity on streams. *Archiv für Hydrobiologie*, **163**(3): 383–403.

Day R. A. (2006) *How to write and publish a scientific paper.* 6th edition. Cambridge University Press, Cambridge.

Debrot S., Fivaz G., Mermod C. and Weber J. M. (1982) *Atlas des poils de mammiferes d'Europe.* Universitè de Neuchâtel, Switzerland.

Digby P. G. N. and Kempton R. A. (1987) *Population and community biology series: multivariate analysis of ecological communities.* Chapman and Hall, London.

Duellman W. E. and Trueb L. (1994) *Biology of amphibians.* The John Hopkins University Press, Baltimore, MD.

Dufrêne M. and Legendre P. (1997) Species assemblages and indicator species: the need for a flexible asymmetrical approach. *Ecological Monographs*, **67**: 345–366.

Edwards C. A. and Bohlen P. J. (1996) *Biology and ecology of earthworms.* 3rd edition. Chapman and Hall, London.

Elbroch M. (2003) *Mammal tracks and signs: a guide to North American species.* Stackpole Books, Mechanicsburg, PA.

Elbroch M., Marks E. and Boretos C. D. (2001) *Bird tracks and sign: a guide to North American species.* Stackpole Books, Mechanicsburg, PA.

Elzinga C. L., Salzer D.W. and Willoughby J. W. (1998) *Measuring and monitoring plant populations.* US Department of the Interior, Bureau of Land Management, Colorado.
http://www.blm.gov/nstc/library/pdf/MeasAndMon.pdf (accessed October 2010).

English S., Wilkinson C and Baker V., (eds) (1997) *Survey manual for tropical marine resources.* Australian Institute of Marine Science, Townsville.

Everett B. S. and Skrondal A. (2010) *The Cambridge dictionary of statistics.* 4th edition. Cambridge University Press, Cambridge.

Extence C. A., Balbi D. M. and Chadd R. P. (1999) River flow indexing using British benthic macroinvertebrates: a framework for setting hydroecological objectives. *Regulated Rivers: Research and Management*, **15**(6): 545–574.

Fay N. and de Berker N. (1997) *Veteran Trees Initiative: Specialist Survey Method.* English Nature, Peterborough.

Fewster R. M., Buckland S. T., Siriwardena G. M., Baillie S. R. and Wilson J. D. (2000) Analysis of population trends for farmland birds using generalized additive models. *Ecology*, **81**: 1970–1984.

Fielding A. H. (2007) *Cluster and classification techniques for the biosciences.* Cambridge University Press, Cambridge.

Fielding A. H. and Howarth P. F. (1999) *Upland habitats.* Routledge, London.

Forsythe T. G. (2000) *Common ground beetles.* Naturalists' Handbooks 8. Richmond Publishing Company Ltd, Slough.

Fortin M-J. and Dale M. R. T. (2005) *Spatial analysis: a guide for ecologists.* Cambridge University Press, Cambridge.

Fry R. and Waring P. (2001) *A guide to moth traps and their use.* The Amateur Entomologist Series 24, Amateur Entomologists' Society, London.

Gabriel O., Lange K., Dahm E. and Wendt T. (2005) *Von Brandt's Fish catching methods of the world.* 4th edition. Blackwell Publishing, Oxford.

Gage S. H., Kasten E. P. and Joo W. (2009) *Deployment and application of environmental acoustics for monitoring bird pathways in wind resource areas.* Technical report MSU-REAL-09-1, Remote Environmental Assessment Laboratory, Michigan State University, East Lansing, MI. http://www.real.msu.edu/~kasten/publications/online/2009-windresourcearea.pdf (accessed October 2010).

Gange A. C. (2005) Sampling insects from roots. In: Leather S. (ed.) *Insect sampling in forest ecosystems.* Methods in Ecology. Blackwell Publishing, Oxford. pp. 16–36.

Gee J. H. R. (1986) *Freshwater studies.* George Allen and Unwin, London.

Gent T. and Gibson S. (eds) (2003) *Herpetofauna workers handbook.* JNCC, Peterborough.

Gibb T. and Oseto C. (2006) *Arthropod collection and identification: laboratory and field techniques.* Academic Press, London.

Gibson D. J. (2002) *Methods in comparative plant population ecology.* Oxford University Press, Oxford.

Gilbert F. S. (1993) *Hoverflies.* Naturalists' Handbooks 5. Richmond Publishing Company, Slough.

Gilbert G., Gibbons D. W. and Evans J. (1998) *Bird monitoring methods: a manual of techniques for key UK species.* RSPB, Sandy.

Gordon N. D., McMahon T. A., Finlayson B. L., Gippel C. J. and Nathan R. J. (2004) *Stream hydrology: an introduction for ecologists.* 2nd edition. John Wiley and Sons, London.

Grime J.P., Hodgson J. G. and Hunt R. (1988) *Comparative plant ecology: a functional approach to common British species.* Unwin Hyman, London.

Gurnell J. and Flowerdew J.R. (2006) *Live trapping of small mammals: a practical guide.* 4th edition. Occasional Publications 3, Mammal Society, London.

Hagler J. R. and Jackson C. G. (2001) Methods for marking insects: current techniques and future prospects. *Annual Review of Entomology,* **46**: 511–543.

Harris S. and Yalden D. W. (2008) *Mammals of the British Isles: handbook.* 4th edition. The Mammal Society, London.

Hauer R. F. and Lamberti G. A. (2006) *Methods in stream ecology.* 2nd edition. Academic Press, London.

REFERENCES

Hawksworth D. L. and Rose F. (1976) *Lichens as pollution monitors.* Studies in Biology 66. Edward Arnold, London.

Hayek L-A. C. and Buzas M. A. (1998) *Surveying natural populations.* Columbia University Press, New York.

Hayward P. J. (1988) *Animals on seaweed.* Naturalists' Handbooks 9. Richmond Publishing Company Ltd, Slough.

Hayward P. J. (1994) *Animals of sandy shores.* Naturalists' Handbooks 21. Richmond Publishing Company, Slough.

Hellawell J. M. (1986) *Biological indicators of freshwater pollution and environmental management.* Elsevier Applied Science Publishers, London.

Henderson P. A. and Seaby R. M. H. (2008) *A practical handbook for multivariate methods.* Pisces Conservation, Hants.

Heyer W. R., Donelly M. A., McDiarmid R. W., Hayek L-A. C. and Foster M. S. (1994) *Measuring and monitoring biological diversity: standard methods for amphibians.* The Smithsonian Institute, Smithsonian Institution Press, Washington.

Heywood I., Cornelius S. and Carver S. (2006) *An introduction to geographical information systems.* Prentice Hall, London.

Hill M. O., Mountford J.O., Roy D. B. and Bunce R. G. H. (1999) *ECOFACT Research Report Volume 2a Technical Annex - Ellenberg's indicator values for British Plants.* Institute of Terrestrial Ecology, Huntingdon.

Hodgetts N. (1995) Bog moss in Britain – the identification and role of *Sphagnum. British Wildlife,* **7**: 9–17.

Hurlbert S. H. (1984) Pseudoreplication and the design of ecological field experiments. *Ecological Monographs,* **54**: 187–211.

Hurst C., Crawford R., Garland J., Lipson D., Mills A. and Stetzenback L. (2007) *Manual of environmental microbiology.* John Wiley and Sons Ltd, London.

Jackson D. A. (1993) Multivariate-analysis of benthic invertebrate communities - the implication of choosing particular data standardizations, measures of association, and ordination methods. *Hydrobiologia,* **268**: 9–26.

Jackson D. A. and Somers K. M. (1991) Putting things in order: the ups and downs of detrended correspondence analysis. *American Naturalist,* **137**: 704–712.
http://labs.eeb.utoronto.ca/jackson/amnat137.pdf (accessed October 2010).

Jackson H. (2005) *Good grammar for students.* Sage Publications, London.

Jenkins M. (1983) *Seashore studies.* George Allen and Unwin, London.

JNCC (2003) *Handbook for Phase 1 habitat survey: a technique for environmental audit.* Joint Nature Conservation Committee, Peterborough.
http://www.jncc.gov.uk/pdf/JNCC%20A4%20Handbook%20for%20Phase%201%20habitat%20survey%20April%202008.pdf (accessed October 2010).

Jones J. C. and Reynolds J. D. (1996) Environmental variables. In: Sutherland W. J. (ed.) *Ecological census techniques: a handbook.* Cambridge University Press, Cambridge. pp. 281–316

Jongman R. H. G, ter Braak C. J. F and Tongeren O. F. R. (1995) *Data analysis in community and landscape ecology.* Cambridge University Press, Cambridge.

Kent M. and Coker P. (1995) *Vegetation description and analysis: a practical approach.* John Wiley and Sons Ltd, London.

Krebs C. J. (1999) *Ecological methodology.* 2nd edition. Addison Wesley Longman, Harlow.

Lawrence A. P. and Bowers M. A. (2002) A test of the 'hot' mustard extraction method of sampling earthworms. *Soil Biology and Biochemistry*, **34**: 549–552.

Leather S. (ed.) (2005) *Insect sampling in forest ecosystems.* Methods in Ecology. Blackwell Publishing, Oxford.

Legendre P. and Legendre L. (1998) *Numerical ecology.* 2nd edition. Elsevier Science, Amsterdam.

Lehner P. N. (1996) *Handbook of ethological methods.* 2nd edition. Cambridge University Press, Cambridge.

Leps J. and Smilauer P. (2003) *Multivariate analysis of ecological data using CANOCO.* Cambridge University Press, Cambridge.

Lincoln R. J. and Sheals J. G. (1979) *Invertebrate animals: collection and preservation.* British Museum (Natural History). Cambridge University Press, Cambridge.

Luff M. L. (2007) *The Carabidae (ground beetles) of Britain and Ireland.* Handbooks for the Identification of British Insects, Volume 4, Part 2. Royal Entomological Society, London.

McGavin G.C. (1997) *Expedition field techniques: insects and other terrestrial arthropods.* Royal Geographical Society, London. http://www.rgs.org/NR/rdonlyres/D7012894-AC14-4B7F-AF57-5EDA36E85E8E/0/Insectsmanualupdated.pdf (accessed October 2010).

Mackereth F. J. H, Heron J. and Talling J. F. (1989) *Water analysis.* FBA Scientific Publication No. 36, Freshwater Biological Association, Ambleside.

Mackie E. D. and Matthews R. W. (2008) *Field guide to timber measurement.* Forestry Commission, Edinburgh.

Magurran A. E. (2004) *Measuring biological diversity.* Blackwell Publishing, Oxford.

Mahoney R. (1966) *Laboratory techniques in zoology.* Butterworths, London.

Maier R. M., Pepper I. L. and Gerba C. P. (2009) *Environmental microbiology.* 2nd edition. Academic Press, London.

Majerus M. and Kearns P. (1989) *Ladybirds.* Naturalists' Handbooks 10. Richmond Publishing Company Ltd, Slough.

Manly B. F. J. (2004) *Multivariate statistical methods: a primer.* 3rd edition. Chapman and Hall/CRC, London.

Marchant J. H. (1983) *BTO common bird census instructions.* BTO, Thetford.

Marsden S. J. (1999) Estimation of parrot and hornbill densities using a point count distance sampling method. *Ibis*, **141**: 377–390.

Martin P. and Bateson P. (2007) *Measuring behaviour.* 3rd edition. Cambridge University Press, Cambridge.

Mason C. F. (1996) *Biology of freshwater pollution.* 3rd edition. Longman, Harlow.

Matthews R. W. and Mackie E. D. (2006) *Forest mensuration: a handbook for practitioners.* Forestry Commission, Surrey.

May R. M. (2004) Ethics and amphibians. *Nature*, **431**: 403.

Mech L. D. (2002) *A critique of wildlife radio-tracking and its use in national parks.* A report to the US National Parks Service, Biological Resources Division, U.S. Geological Survey, Northern Prairie Wildlife Research Center, Jamestown, ND.
http://www.npwrc.usgs.gov/resource/wildlife/radiotrk/radiotrk.pdf (accessed October 2010).

MELP (1998a) *Live animal capture and handling guidelines for wild mammals, birds, amphibians and reptiles.* Standards for Components of British Columbia's Biodiversity, No.3, Version 2.0. Ministry of Environment, Lands and Parks Resources Inventory Branch, British Columbia.
http://archive.ilmb.gov.bc.ca/risc/pubs/tebiodiv/capt/assets/capt.pdf (accessed October 2010).

MELP (1998b) *Wildlife radio-telemetry.* Standards for Components of British Columbia's Biodiversity No. 5. Version 2.0. Ministry of Environment, Lands and Parks Resources Inventory Branch for the Terrestrial Ecosystems Task Force Resources Inventory Committee, British Columbia.
http://www.ericlwalters.org/telemetry.pdf (accessed October 2010).

Millar I. M., Uys V. M. and Urban R. P. (2000) *Collecting and preserving insects and arachnids: a manual for entomology and arachnology.* ARC Plant Protection Research Institute, Pretoria.
http://www.spc.int/PPS/SAFRINET/inse-scr.pdf (accessed October 2010).

Minteer B. A. and Collins J. P. (2005a) Ecological Ethics: Building a New Tool Kit for Ecologists and Biodiversity Managers. *Conservation Biology*, **19**(6): 1803–1812.

Minteer B. A. and Collins J. P. (2005b) Why we need an 'ecological ethics'. *Frontiers in Ecology and the Environment*, **3**(6): 332–337.
http://academic.regis.edu/ckleier/Conservation%20Biology/Why%20we%20need%20ecological%20ethics.pdf (accessed October 2010).

Minteer B. A. and Collins J. P. (2008) From environmental to ecological ethics: toward a practical ethics for ecologists and conservationists. *Science and Engineering Ethics*, **14**(4): 483–501.

Mitchell-Jones A. J. and McLeish A. P. (2004) *The bat workers manual.* 3rd edition. JNCC, Peterborough.
http://www.jncc.gov.uk/page-2861#download (accessed October 2010).

Mueller G. M., Bills G. F. and Foster M. S. (eds) (2004) *Biodiversity of fungi: inventory and monitoring methods.* Elsevier Academic Press, London.

Murray D. L. and Fuller M. R. (2000) A critical review of the effects of marking on the biology of vertebrates. In: Boitani L. and Fuller T. K. (eds) *Research techniques in animal ecology: controversies and consequences.* Methods and case studies in conservation science. Columbia University Press, New York. pp. 15–64.

New T. R. (1998) *Invertebrate surveys for conservation.* Oxford University Press, Oxford.

Newton I. (2010) *Bird migration.* New Naturalist 113. Collins, London.

Nichols D. (1999) *Safety in biological fieldwork.* The Institute of Biology, London.

Nielsen L. A. (1992) *Methods of marking fish and shellfish.* Special Publication No. 23, The American Fisheries Society, Bethesda, MD.

Nietfeld M. T., Barrett M. W. and Silvy N. (1996) Wildlife marking techniques. In: Bookhout T. A. (ed.), *Research and management techniques for wildlife and habitats.* Wildlife Society, Bethesda, MD. pp. 140–168.

Oldroyd H. (1970) *Collecting, preserving and studying insects.* Hutchinson Scientific and Technical, London.

Ozanne C. (2005) Techniques and methods for sampling canopy insects. In: Leather S. R. (ed.) *Insect sampling in forest ecosystems*. Methods in Ecology. Blackwell Publishing, Oxford. pp. 146–167.

Parker N. C., Giorgi A. E., Heidinger R. C., Jester D. B., Prince E. D. and Winans G. A. (eds) (1992) *Fish-marking techniques*. Symposium 7,The American Fisheries Society, Bethesda, MD.

Parris K. M. (1999) Review: amphibian surveys in forests and woodlands. *Contemporary Herpetology*, 1: 1–14. http://www.contemporaryherpetology.org/ch/1999/1/CH_1999_1.pdf (accessed October 2010).

Parris K. M., McCall S. C., McCarthy M. A., Minteer B. A., Steele K., Bekessy S. and Medvecky F. (2010) Assessing ethical trade-offs in ecological field studies. *Journal of Applied Ecology*, 47: 227–234. http://www3.interscience.wiley.com/cgi-bin/fulltext/123224678/PDFSTART (accessed October 2010).

Peterken G. F. (1993) *Woodland conservation and management*. Springer, New York.

Phillott A. D., Skerratt L. F., McDonald K. R., Lemckert F. L. and Hines H.B. (2007) Toe-clipping as an acceptable method of identifying individual anurans in mark recapture studies. *Herpetological Review*, 38(3): 305–308.

Pollard E. and Yates T. J. (1993) *Monitoring butterflies for ecology and conservation*. Chapman and Hall, London.

Pollock K. H., Nichols J. D., Brownie C. and Hines J. E. (1990) Statistical inference for capture-recapture experiments. *Wildlife Monographs*, 107: 1–97.

Price E. A. C. (2003) *Lowland grassland and heathland habitats*. Routledge, London.

Quinn G. P. and Keough M. J. (2002) *Experimental design and data analysis for biologists*. Cambridge University Press, Cambridge.

Ragge D. R. and Reynolds W. J. (1998) *A sound guide to the grasshoppers and crickets of western Europe*. Harley, Colchester.

Read H.J. and Frater M. (1999) *Woodland habitats*. Routledge, London.

Reed B. T. and Jennings M. (2007) Promoting consideration of the ethical aspects of animal use and implementation of the 3Rs. *Proc. 6th World Congress on Alternatives and Animal Use in the Life Sciences, August 21-25, 2007, Tokyo, Japan. AATEX 14*, Special Issue, 131-135 http://content.www.rspca.org.uk/cmsprd/Satellite?blobcol=urldata&blobheader=application%2Fpdf&blobkey=id&blobnocache=false&blobtable= MungoBlobs&blobwhere=1232990798336&ssbinary=true (accessed October 2010)

Rice W. R. (1989) Analysing tables of statistical tests. *Evolution*, 4(1): 223–225.

Richardson D. H. S. (1992) *Pollution monitoring with lichens*. Naturalists' Handbooks 19. The Richmond Publishing Company, Slough.

Ricker W. E. (1956) Uses of marking animals in ecological studies: the marking of fish. *Ecology*, 37: 666–670.

Roche J-C. (2003) *Bird songs and calls of Britain and Europe*. WildsoundsNorfolk.

Rodwell J. S. (1991a) *British plant communities. Volume 1: woodlands and scrub*. Cambridge University Press, Cambridge.

Rodwell J. S. (1991b) *British plant communities. Volume 2: mires and heaths*. Cambridge University Press, Cambridge.

Rodwell J. S. (1992) *British plant communities. Volume 3: grasslands and montane communities*. Cambridge University Press, Cambridge.

Rodwell J. S. (1995) *British plant communities. Volume 4: aquatic communities, swamps and tall-herb fens.* Cambridge University Press, Cambridge.

Rodwell J. S. (2000) *British plant communities. Volume 5: maritime and weed communities and vegetation of open habitats.* Cambridge University Press, Cambridge.

Rodwell J. S. (2006) *National vegetation classification: users' handbook.* Joint Nature Conservation Committee, Peterborough.
http://www.jncc.gov.uk/pdf/pub06_NVCusershandbook2006.pdf (accessed October 2010).

Roloff A. (1985) The classification of damage in the beech. A proposal for a standardised nationwide classification of the beech into four categories of damage based on terrestrial photographs. *Der Forst- und Holzwirt* **40**(5): 25–34.

RSPCA and LASA (2010) *Guiding principles on good practice for ethical review processes.* A report by the RSPCA Research Animals Department and LASA Education, Training and Ethics Section. (M. Jennings ed.).
http://www.lasa.co.uk/GP%20ERP%20July%202010%20print%20FINAL.pdf (accessed October 2010).

Rubenstein D. R. and Hobson K. A. (2004) From birds to butterflies: animal movement patterns and stable isotopes. *Trends in Ecology and Evolution,* **19**(5): 256–263.

Russ J. (1999) *The bats of Britain and Ireland: echolocation calls, sound analysis and species identification.* Alana Books, Bishop's Castle.

Scheiner S. M. (2001) MANOVA: Multiple response variables and multispecies interactions. In: Scheiner S. M. and Gurevitch J. (eds) *Design and analysis of ecological experiments.* 2nd edition. Oxford University Press, Oxford. pp. 99–115.

Scheiner S. M. and Gurevitch J. (eds.) (2001) *Design and analysis of ecological experiments.* 2nd edition. Oxford University Press, Oxford.

Shiel C., McAney C., Sullivan C. and Farley J. (1997) *Identification of arthropod fragments in bat droppings.* Occasional Publication No. 17. The Mammal Society, London.

Siegel S. and Castellan N. J. (1988) *Nonparametric statistics for the behavioural sciences.* 2nd edition. McGraw-Hill, New York.

Skinner G. J. and Allen G. W. (1996) *Ants.* Naturalists' Handbooks 24. Richmond Publishing Company Ltd, Slough.

Smith D. (1984) *Urban ecology.* George Allen and Unwin, London.

Sokal R. R. and Rohlf F. J. (1995) *Biometry: the principles and practice of statistics in biological research.* 3rd edition. W. H. Freeman and Company, London.

Southwood T. R. E. and Henderson P. A. (2000) *Ecological methods.* 3rd edition. Blackwell Science, Oxford.

Sparre P. and Venema S. C. (1998) *Introduction to tropical fish stock assessment: Part 1 Manual.* FAO Fisheries Technical Paper 306/1 Rev. 2. FAO, Rome.
http://www.fao.org/docrep/W5449E/w5449e01.htm (accessed October 2010).

Speight M. C. D. (1986) *Criteria for the selection of insects to be used as bio-indicators in nature conservation research.* Velthuis H. H. W. (ed.) Proceedings of the 3rd European Congress of Entomology, Amsterdam 1986, 485–488.

Stednick J. D. (1991) *Wildland water quality sampling and analysis.* Academic Press, San Diego, CA.

Stewart A. J. and Wright A. F. (1995) A new inexpensive suction apparatus for sampling arthropods in grassland. *Ecological Entomology,* **20**: 98–102.

Stonehouse B. (1978) *Animal marking: recognition marking of animals in research.* RSPCA Wildlife Advisory Committee, University Park Press, Baltimore, MD.

Stork N. E., Adis J. and Didham R.K. (eds) (1997) *Canopy arthropods.* Chapman and Hall, London.

Strachan R. (2010) *Mammal detective.* Whittet, Linton.

Stubbs A. E. and Falk S. (1996) *British hoverflies.* British Entomological and Natural History Society, Reading.

Sutherland W. J. (ed.) (1996) *Ecological census techniques: a handbook.* Cambridge University Press, Cambridge.

Tabachnick B. G. and Fidell L. S. (2007) *Using multivariate statistics.* 5th edition. Pearson, London.

Taber R. D. (1956) Uses of marking animals in ecological studies: marking mammals; standard methods and new developments. *Ecology*, **37**: 681–685.

Taguchi Y.H. and Oono Y. (2004) Nonmetric multidimensional scaling as a data-mining tool: new algorithm and new targets. Geometrical Structures of Phase Space, Multidimensional Chaos, Special Volume of *Advances in Chemical Physics*, **130**: 315–351.

Teerink B. J. (1991) *Hair of west European mammals: atlas and identification key.* Cambridge University Press, Cambridge.

ter Braak C. J. F. and Smilauer P. (2002) *CANOCO reference manual and CanoDraw for Windows user's guide: software for canonical community ordination.* Version 4.5. Microcomputer Power, Ithaca, NY.

Trudgill S. (1989) Soil types: a field identification guide. (FSC Publication 196). *Field Studies*, **7**(2): 337–363.

Truss L. (2003) *Eats, shoots and leaves.* Profile Books, London.

Tupinier Y. (1997) *European bats: their world of sound.* Société Linnéenne de Lyon, Lyon.

Twigg G. I. (1975) Marking mammals. *Mammal Review*, **5**: 101–116.

Unwin D. M. (1978) Simple techniques for microclimate measurement. *Journal of Biological Education*, **12**: 179–189.

Unwin D. M. (1980) *Micromeasurement for ecologists.* Academic Press, London.

Unwin D. M. and Corbet S. A. (1991) *Insects, plants and microclimate.* Naturalists' Handbooks 15. The Richmond Publishing Company, Slough.

Van Belle G. (2002) *Statistical rules of thumb.* John Wiley and Sons Inc., New York.

Vink C. J., Thomas S. M., Paquin P., Hayashi C. Y. and Hedin M. (2005) The effects of preservatives and temperatures on arachnid DNA. *Invertebrate Systematics*, **19**: 99–104.

Walsh P. M., Halley D. J., Harris M. P., del Nevo A., Sim I. M. W. and Tasker M. L. (1995) *Seabird Monitoring Handbook for Britain and Ireland.* JNCC, Peterborough. http://www.jncc.gov.uk/PDF/pub95_SeabirdHandbook.pdf (accessed October 2010).

Waring P. (1994) Moth traps and their use. *British Wildlife*, **5**: 137–148.

Wasklewicz T., Staley D., Mihir M. and Seruntine L. (2007) Virtual recording of lichen species: integrating terrestrial laser scanning and GIS techniques. *Physical Geography*, **28**(2): 183–192.

Watts S. and Halliwell L. (1996) *Essential environmental science.* Routledge, London.

Wheater C. P. (1999) *Urban habitats*. Routledge, London.

Wheater C. P. and Cook P. A. (2000) *Using statistics to understand the environment*. Routledge, London.

Wheater C. P. and Cook P. A. (2003) *Studying invertebrates*. Naturalists' Handbooks 28. The Richmond Publishing Company, Slough.

Wheater C.P. and Read H. J. (1996) *Animals under Logs and Stones*. Naturalists' Handbooks 22. Richmond Publishing Company, Slough.

White G. C. and Garrott R. A. (1990) *Analysis of wildlife radio-tracking data*. Academic Press, London.

White J. (1998) *Estimating the age of large and veteran trees in Britain*. Forestry Commission Information Note 12. Forestry Commission, Edinburgh. http://www.forestry.gov.uk/PDF/fcin12.pdf/$FILE/fcin12.pdf (accessed October 2010).

Wilkinson D. M. (1991) Can photographic methods be used for measuring the light attenuation characteristics of trees in leaf? *Landscape and Urban Planning*, **20**(4): 347–349.

Williams S. T. (2007) Safe and legal shipment of tissue samples: does it affect DNA quality? *Journal of Molluscan Studies*, **73**: 416–418.

Wilson D. E., Cole F. R., Nichols J. D., Rudran R. and Foster M. S. (1996) *Measuring and monitoring biological diversity: standard methods for mammals*. Smithsonian Institution Press, Washington.

Winser S. (ed.) (2004) *The expedition handbook*. The Royal Geographical Society, Profile Books, London. http://www.rgs.org/OurWork/Publications/EAC+publications/Expedition+Handbook/Expedition+Handbook.htm (accessed October 2010).

Woodbury A. M. (1956) Use of marking animals in ecological studies: marking amphibians and reptiles. *Ecology*, **37**: 670–674.

Woodbury A. M., Ricker W. E., Cottam C., Taber R. D. and Pendleton R. C. (1956) Symposium: uses of marking animals in ecological studies. *Ecology*, **37**: 665–688.

Wright A., Fielding A. H. and Wheater C. P. (2000) Predicting the distribution of Eurasian badger (*Meles meles*) setts over an urbanized landscape: a GIS approach. *Photogrammetric Engineering and Remote Sensing*, **66**(4): 423–428.

Wright J. F., Sutcliffe D. W. and Furse M. T. (2000) *Assessing the biological quality of fresh waters: RIVPACS and other techniques*. The Freshwater Biological Association, Ambleside.

Yalden D. W. (2009) *The analysis of owl pellets*. 3rd edition. The mammal Society, London.

Young M. (2005) Insects in flight. In: Leather S. (ed.)*Insect sampling in forest ecosystems*. Methods in Ecology. Blackwell Publishing, Oxford. pp. 116–145.

Yates D. and McKennan G. (1988) Solar architecture and light attenuation by trees: conflict or compromise? *Landscape Research*, **13**(1): 19–23.

Zar J. H. (2009) *Biostatistical analysis*. Prentice Hall International Inc., London.

Appendix 1 Glossary of Statistical Terms

Statistical term (alternative names)	Definition
Analysis of variance (ANOVA)	A family of statistical tests that examines the ratios between different sources of the variance (between and within groups). Includes one-way (where the effect of a single independent categorical variable on a measured dependent (response) variable is examined), as well as two-way versions (where the test simultaneously examines the effect of two independent variables and their interaction on a dependent variable) as well as parametric and nonparametric versions
Binary data	Data that can be allocated to one of two classes only (e.g. male or female), normally scored as 1 or 0
Binomial data	Data describing the number of 'successes' as a proportion (x/y) of the total (e.g. numbers of healthy plants x in a population y)
Classification	Techniques including cluster analysis and discriminant function analysis that enable matrix data (usually sites against species) to be grouped
Community (multivariate) analyses	Analyses that attempt to simplify the properties of matrix data usually in the context of species and/or site dynamics. At its simplest, two variables may constitute a matrix if, for example, there are multiple sites and species forming a two-way table
Confidence interval	The range that contains the value of interest (e.g. mean, regression line) at a particular level of probability, thus 95% of calculated 95% confidence intervals will contain the true mean (or regression line, etc.)

Practical Field Ecology: A Project Guide, First Edition.   C. Philip Wheater, James R. Bell and Penny A. Cook.
© 2011 John Wiley & Sons, Ltd. Published 2011 by John Wiley & Sons, Ltd.

APPENDIX 1: GLOSSARY OF STATISTICAL TERMS

Statistical term (alternative names)	Definition
Confidence limits	The end points of the confidence interval
Contingency table (cross-tabulation)	Table summarising two frequency distributions, one laid out along the row axis and the other along the column axis, with the table cells containing the frequencies (or percentages) held in common by the intersection of each row and column
Correlation analysis	Statistical test for relationships between two measured (or ranked) variables where cause and effect are not implied
Critical value	Arbitrary probability level, at values less than which an event is said to be significantly unlikely to occur (usually taken as 0.05)
Degrees of freedom	The number of data points in a sample that are free to vary once a given parameter has been calculated. Calculated as the number of data points minus a value based on the number of parameters (e.g. the mean) that have been estimated from the sample
Dependent (response) variable	Variable whose values are determined or hypothesised to be determined, at least in part, by the values of another variable
Distance measure	A type of dissimilarity measure which satisfies the triangle inequality axiom (i.e. where the dissimilarity between two points is less than or equal to the sum of their dissimilarities to a third point)
Distance-based technique	Techniques that use the distances calculated between each of a series of sites (for example) based on their component species to create biplots (xy graphs) that visualise all pairwise relationships simultaneously
Eigenvalue (latent roots, characteristic roots)	A measure of the strength of a contributing axis within a multivariate model
Eigenvalue-based technique	Multivariate methods that attempt to maximise the variance between sites (or other factors) based on their component species (or other variables), captured along a series of axes, starting with the highest values for axes 1

APPENDIX 1: GLOSSARY OF STATISTICAL TERMS

Statistical term (alternative names)	Definition
Environmental gradient	In terms of multivariate analyses, an ordering of the subjects (species or sites) along axes that may be suggestive of a relationship between it and an explanatory variable. This inference elucidates a possible underlying gradient in the community that can either be inferred (i.e. indirect gradient analysis) or tested formally (i.e. direct gradient analysis)
Factor	Usually a nominal variable (often used in ANOVA to describe the independent or treatment variable)
Fixed variable	Measurement, classification or score whose value is set as part of the experimental or survey design
Frequency	The number of events within a unit of experimental time or within a set of experimental samples
Frequency distribution	Summary of a variable (usually a measured variable) that shows the values of that variable (x axis) against the frequency that each value or range of values occurs (y axis)
Heterogeneity	Variation, often used in the context of within samples
Homogeneity	Equality, often used in the context of between samples
Hypothesis	Prediction about the differences between samples, or relationships or associations between variables. The null hypothesis predicts that samples do not differ, or are not related or associated. This is the hypothesis at the start of a statistical test, and is rejected in favour of the alternative hypothesis (that samples differ, or variables are related or associated) if the probability is less than the critical value
Independence of data	Condition where the measured value (or score) of an individual does not affect the value (or score) of any other individual, either within or between samples. A condition for many statistical tests (except where matched or paired tests are being employed)
Independent (explanatory, predictor or treatment) variable	Variable whose values are hypothesised to determine the values of another variable, while being unaffected in return
Interval data	Data on a measurement scale without an absolute zero point, so that data points can be distinguished from each other in terms of magnitude and directional differences, but ratios between points are not possible

Statistical term (alternative names)	Definition
Interpolate	Estimate a statistic or data point from a knowledge of the data often by calculating from the adjacent values
Matched (paired or grouped) data	Where each data point is not independent of all others, but pairs (or groups) of data points are linked by being taken from the same individual or sampling unit
Metric data	Data on interval or ratio scales
Multivariate analysis	Family of analysis in which a large number of independent and/or dependent variables are analysed at the same time (examples are principal component analysis, factor analysis, multiple correlations, multiple regression and logistic regression, analysis of covariance, and multivariate analysis of variance)
Nominal (categorical) data	Data on a categorical scale where values can be allocated to mutually exclusive categories and where categories have no magnitude or directional differences
Non-metric data	Data on ordinal or nominal scales
Non-parametric tests	Statistical tests that do not require the data to be normally distributed, suitable for data measured at least on an ordinal scale (with the exception of tests of association/independence or goodness of fit, which use nominal data)
Normal (Gaussian) distribution	Frequency distribution with a symmetrical, bell-shaped frequency curve with the mean value in the middle and most data points around it and that has defined mathematical properties (e.g. that 68.27% of data points lie within the mean ± the standard deviation): a feature of many variables measured on interval or ratio scales
Ordinal (ranked) data	Data on a ranked scale where data points can be distinguished from each other in terms of directional differences, but not of magnitude
Ordination	Techniques that reduce the number of dimensions (variables) in matrix data (often sites against species) usually enabling a visual perspective to be plotted on a biplot (xy plot)

APPENDIX 1: GLOSSARY OF STATISTICAL TERMS

Statistical term (alternative names)	Definition
Parameter	Measurement derived from a population (e.g. population mean or population standard deviation), values of which are usually estimated by taking a random sample of the population
Parametric tests	Statistical tests that require the data to be normally distributed
Poisson distribution	A discrete distribution that often describes count data in which large values tend to be rare. It has a unique property of the variance being equal to the mean
Population	Collection of all possible items (individual sample units) from which samples are taken
Predictive model	Use of the values of one or more independent variables to predict the values of a dependent variable
Probability	The chance of an event taking place, measured either as a proportion from 0 to 1 or as a percentage from 0 to 100% (where a value close to 0% is highly unlikely to occur by chance, whereas a value close to 100% is extremely likely to occur by chance). In statistical testing it is usually used to quantify the chance of the null hypothesis being true
Pseudoreplication	Increasing the sample size by using non-independent data points. Should be avoided where possible
Ratio data	Data on a measurement scale with an absolute zero point so that data points can be distinguished from each other in terms of both magnitude and directional differences, and ratios between points are possible
Regression analysis	Techniques for modelling the relationships between one dependent variable and one (or more) independent variable(s). Often used in predictive modelling. There are parametric and non-parametric versions
Sample	A collection of individual items (that are assumed to be random and independent) drawn from a population, measurements from which are used to estimate the true measurements (parameters) of the population
Standard deviation	Measure of the variation of normally distributed data (interval or ratio) that provides a symmetrical range around the mean within which 68.27% of data points lie (calculated as the square root of the variance)

Statistical term (alternative names)	Definition
Standard error	The standard deviation of the sampling distribution of the mean, which can be used as a measure of the reliability of the estimation of a mean, comprising a symmetrical range around the mean (± 1 standard error)
Statistic	Measurement derived from a sample (e.g. sample mean or sample standard deviation) intended to estimate the equivalent parameter of the population
Test statistic	Value calculated (or used) during a statistical test, either with a known distribution or one that approximates to a known distribution, from which the probability of the null hypothesis being true can be found
Transformation	Mathematical manipulation of all of the data points in a sample to enable them to conform to the assumptions of a test (e.g. normal distribution or a linear relationship between variables)
Treatment	The manipulation of a population in order to see whether the manipulation has an effect on a measured variable, ascertained by comparison with an unmanipulated population
Type I error	The probability of rejecting the null hypothesis when in fact it is true (i.e. producing falsely significant results)
Type II error	The probability of accepting the null hypothesis when in fact it is false (i.e. producing falsely non-significant results)
Unmatched data	Where data are independent of each other having been sampled from different individuals or groups, so that the selection of any measurement from one sample does not influence the selection of any other individual
Variable	Characteristic (comprising measurements, classifications or scores) that varies between individuals within a population
Variance	Measure of variation in a sample, standardised by the size of the sample (the square of the standard deviation)

See Everett and Skrondal (2010) for further definitions of a wide range of statistical terms.

Index

Note: Page numbers in *italics* refer to Figures; those in **bold** to Tables and Boxed text

abbreviations, use/conventions   324, 325–6, **326**
abstracts, research report   307–8
abundance measurement *see also* density estimation; population estimates
  biomass   78, 80
  littoral species scales   77–8, **78**, *79*
  microbial communities   70, **71–2**
  vegetation cover   76, **77**, 83–4
acknowledgements section, reports   308
aerial photographs   38–9, *41*
aims and objectives   7–8, 309
amphibians
  capture   187–8
    on land   189–91, *190*, *191*
    in water   188–9, *189*
  direct and indirect observation   186–7
    egg masses   187
  handling   184
  neotropical tree frogs, breeding behaviour case study   185–6
  tadpoles and juveniles   191–2
analysis of covariance (ANCOVA)   277, **278**
analysis of similarity (ANOSIM)   289–90, **290**
analysis of variance (ANOVA)   266, **267**, 345
  factors   347
  two-way   266, **268**
anemometers   44–5, *47*
ant hills (nests), distribution and sampling   27, **80**, 85
appendices, report   311, 321
aquatic multimeters   51, *52*
artificial cover traps   191–2, *192*, 198
artificial substrate samplers   131, *131*
aspect, site   55

assembly traps   163, *164*
augers, soil   48, *49*

badgers
  dung pits   221, *221*
  protection   218
  setts   67, **80**
  traps   232, *233*
Baermann funnel   129–30, *130*
Bailey's triple catch method, population estimation   254, **255**
baits
  for aquatic species   132, *132*, 174–5
  insect   **162**, 162–3, *163*, *164*
  marked dung, badgers   221, *221*
  poison   227, *228*
bar charts
  clustered and stacked types   247–8, *248*, *249*, **314**
  display of error/variation   250, **314**
bats
  bat detectors   222, *223*, **224**, 225
    triangle walk surveys   223, *223*
  capture   229, *231*
  conservation ecology, case study   230–1
  counting, based on droppings   221
  roost location and monitoring   219
beating trays   156, *156*
behavioural studies
  neotropical tree frogs, breeding behaviour case study   185–6
  problems and their avoidance   102, **102–3**
    mandrill chemical codes, case study   99–100
  recording methods   101
  study approaches   6, 98–9, 101
Belleville mosquito larvae sampler   123, *124*

belt transects   85, *85*, 189
benthic coring   128
Berger-Parker diversity index   256, **257**
bibliographies   317
Bidlingmayer sand extractors   130, *130*
bimodal distributions   246, *247*
binary data   **260**, 276, 345
   see also presence/absence data
binomial data   281, 345
binomial regression model   281, **282–4**
biological factors, environment   **37**
biological oxygen demand (BOD)   51
biotic indicators   56–7, **59–60**, *61*
birds
  capture
    handling and measurement   211, **211**
    netting techniques   *212*, 212–13
  data recording sheets   22
  direct observation
    counting parrots, case study   204–5
    distance sampling   203, 255
    hides and observation towers   201–2, *202*, *203*
    point counts   207–8
    timed species counts   203, **205**
    transect line and flush counts   208
  indirect observations
    bird song   210–11
    counting nests   210
    goose droppings survey   209, *209*
    tracks and remains   209–10
  licences   200–1, 211, **211**, 213
  ringing   213–15, *215*
  territory mapping
    breeding bird survey transects   *207*
    common birds census   206, **206**
bottle traps
  flying insects   163, *163*
  newts   189, *189*
box and whisker plots   251, *251*, **314**
Braun-Blanquet scale (vegetation cover)   76, **77**
bulb planters (soil)   48, *50*
burrows
  health and safety issues   97, 119
  monitoring techniques   109, 111, 118–19
  suction sampling   127, *128*

butterflies
  census method   112, **113**
  life cycle, case study   152–3

cage traps   229, 232, *232*, *233*
camera traps   194, *194*, 218, *218*
cannon (and rocket) nets   213, *214*
canonical correspondence analysis (CCA)   **302–3**
canonical variates analysis (CVA)   288–9, *289*
canopy cover estimates   57–8, *62*
capture-removal (population estimation)   255–6, *256*
capture techniques
  active (nets) and passive (traps)   115–16
  equipment and planning   104
  specimen preservation   104–5
censuses
  butterflies, method   112, **113**, **114**
  method limitations   6–7
central tendency   249
  comparison of samples   261–2, *262*
characterisation of sites (field work)   35, 36, **37**
chemical extraction, soil animals   137–8
chemical factors, environment   **37**
chi-square analysis   **259**
  tests for frequency association   **274**, 274–5
chi-squared distance measures   **288**
citations see references
city block (Manhattan) distance measures   **288**
clap nets   213, *214*
classification
  functional grouping   66, **66**
  species/site matrix data (cluster analysis)   287–92, 345
  systems, for vegetation communities   38, **40**
  tables, in discriminant function analysis   279, **279–80**
  taxonomic groups   62–3, **63**, 65–6
clinometers
  site slope angle   55, *57*
  tree height measurement   93, *93*
cluster analysis, hierarchical   290–1, *291*
  agglomerative methods   **292**
  dendrograms   291, *291*

clustered bar graphs   *249*, **314**
coefficient of determination   273, 281
communities
  indicator species   57
  structure, study approaches   6, 286
computer files   331
confidence intervals (limits)   250, *251*, 261, 345, 346
contents section, reports   308–9
contingency tables   259, **275**, 346
continuous data *see* interval data
correlation analysis   **30**, **260**, 269–71, 346
cover traps, artificial   191–2, *192*, 198
Craig's (Du Feu) method, population estimation   **255**
crayfish traps   132, *132*
critical values (probability)   86, 262–3, 264, **265**, 346
cross-staff, shoreline surveys   55, *57*

DAFOR abundance scale   76–7, **77**
data *see also* statistics
  independent and linked   25, 347, 348, 350
  reduction   286
  summary description   244
  types   20–2, **21**
data presentation
  figures
    display options   **314–15**
    graphs   247–8, 314, *315*, 316
    pictures, maps and diagrams   313, *313*
    kite diagrams   85, *86*
    tables   248, **249**, 311–13, **312**
    transformation and screening   244–7, *245*, *246*, *247*
data recording
  experimental or survey design   18–20
  observations and information notes   **15–16**, 36
  standardised record sheets   14–15, *22*
degrees of freedom   263, 264, 275, 346
dendrograms   291, *291*, 315
density estimation   5–6, 68, **69**, 70
  see also abundance measurement; population estimates
  census walks   112, **113**, *113*, **114**
  plotless methods   86–7, **87–8**

using quadrats and pin-frames   76, 81, 84
dependent (response) variables   272, 276–7, 279, 346
detrended correspondence analysis (DCA)   296, 298, *298*
difference, statistical testing   29, **30**
  multiple comparison tests   266, 268, **268–9**
  test statistics and interpretation   261–3, 265–6, **267–8**
discriminant function analysis (DFA)   279, **279–80**
discussion section, reports   316–17
distance measures   208, 287, **287–8**, 346
distribution pattern (spatial analysis)
  mobile species   97
  nested quadrats   81, *81*
  sampling problems   27, 87
  sampling techniques   88, **89**, 252
distributions (data) *see* frequency distribution analysis
diversity
  indices of diversity and evenness   256, **257–8**
  microorganisms, identification and counting   70, **71–2**
DNA or RNA analysis
  bar-coding   184
  microbial identification   **72**
  sample preservation   96–7
Domin scale (vegetation cover)   38, 76, **77**
drags and dredges   128–9, *129*
drift fences and nets   126–7, *190*, **190**, *191*, *198*
droppings *see* dung
dry sieving   133, *133*
Du Feu (Craig's) method, population estimation   **255**
dummy variables   276, **276**
dung
  baited, use in territory surveys   221, *221*
  diet analysis   188, 194, 217, 220
  for estimating animal density   73, 103
dynamometers (wave action)   54, *54*

eigenvalues **295–6**, 298, 299, 346
electrical extraction
 earthworms 138
 fish (electrofishing) 181
emergence traps 150, *151*, 177
environmental measurements
 biotic indicators 56–7, **59–60**, *61*
 factors affecting living organisms 36, **37**
 habitat mapping 36, 38–42, *39*, **40**
 light levels *56*, 57–8, *62*
 microclimate 42–5, *46*, *47*
 size, topography and aspect 54–5, *57*
 substrates (soil) 45–51
 water (aquatic habitats) 51–4
equipment *see also* environmental measurements
 descriptive detail, for reports 310
 development and testing 104
  barnacle larva trap, case study 105–6
 environmental multimeters 45, *47*, 51, *52*
 radio-tracking **109**
 suggested requirements, field work **13**, 14, 97, 111
ESACFORN scale *see* SACFOR abundance scales
Euclidean and squared Euclidean distance measures **287**, **288**, 289, 299
evenness indices 256, **258**
event recorders 101
experimental design 19–20, 26, 28, *28*
 Park Grass Experiment case study 74–5

field work *see also* behavioural studies; environmental measurements; sampling; surveys
 characterisation of sites 35, 36
 equipment **13**, 14, 97, **109**
 field notebooks and data recording 14–15, **15–16**
 health and safety issues 12–13, 91, 92, 186
 identification 58, 61–3, **64–5**, 65–6, 67–8
 legal aspects 11–12, 104, 107–8, 112
 research planning checklist **34**
 site selection 35–6
 working overseas 13
  insect-collecting in Costa Rica, case study 141–2

fish
 capture techniques 177, 181
  nets and traps 178–81, *179*, **179–80**
  slurp gun 174, *175*
  sport fishing 177, *178*
 direct observation 174, 176, *176*
 egg studies 176–7
 lake fish populations, case study 182–3
 larvae 181
Fisher and Ford method, population estimation 254, **255**
floatation (invertebrate extraction) 134
fogging (chemical knockdown) 156–7, *157*
forestry study methods *see* trees
frequency 347
 estimation from quadrat presence/absence 76
frequency distribution analysis **30**, 347
 association/independence testing, between distributions **274**, 274–5
 bimodal distribution *247*
 contingency tables 274, 275, **275**, 346
 skewed distributions, transformations 245
 truncation *246*
freshwater ecology
 abiotic factors, monitoring 51–4
 habitat quality, biotic indicators 56, **59–60**
 sampling stream invertebrates, case study 121–2
funnel traps 132, 163, 180–1, 190, 198

Gaussian distribution *see* normal distribution
general linear models 266, 269–73, 276–80
generalized additive models (GAM) 285, **285–6**
generalized linear models (GLM) 280–2, **282–4**
 model checking 281–2
geographical information systems (GIS) 38–9, 41, 55, 75
global positioning system (GPS) receivers *39*, 55, *58*, 73, 153
Gower's distance measure **288**

Gower's similarity coefficient 260
grabs (aquatic sampling) 128–9, *129*
gradient analyses 347
  direct 301–3
  indirect 294–301
graphs
  non-linear types *273*, 273–4
  presentation 247–8, 314, **314–15**, *315*, 316

H traps 145, *148*
habitat mapping
  landscape scale 41–2, 54–5
  Phase 1 survey 36, 38, *39*
  remote sensing methods 38–9, 41
  vegetation classification systems 38, **40**
hair traps 222, *222*, 227
half-barrel (Duffus) traps 228, *229*
harp traps 229, *231*
Harvard reference and citation
    system 317–18, **318–21**
health and safety issues (field work) 12–13,
    91, 92, 186
Hess samplers 126, *126*
Hester-Dendy multiplate samplers 131, *131*
histograms 246, **314**
humidity 42–3, *46*
hydroacoustics 182, **182**
hygrometers 43, *46*
hypotheses 347
  creation 7–8
  stating, in reports 309
  statistical testing (inferential
      statistics) 29–31, **30**, 263–5, *264*

identification
  of diet items 188, 194, 217, 220
  of ecological categories 66, **66**
  guides and keys 58, 61–2, **64–5**, 67, 114
  by indirect remains and signs 103,
      114–15, 194, 219–25
  of individuals, using markings 106–8,
      108
  of juveniles 175, 184
  specimen preservation 67–8, 104–5, 116,
      116–17
  taxonomic groupings 62–3, **63**, 65–6
inclined tray light-based separators 137, *137*

independent (predictor) variables 272,
    276–7, 347
indicator species
  data analysis methods 292–4, **293**, *293*
  monitoring habitat quality 56–7,
      59–60, *61*
IndVal (indicator value) analysis 292–4, **293**
inferential statistics *see* hypotheses, statistical
    testing
interception traps for airborne
    invertebrates 165–7, 172
interpolation **144**, 348
interval data 21, **21**, 347
introduction, reports 309
inundation and water flow 53–4
invertebrates, sampling
  airborne
    capture 158, 160–73
    census survey 112–14
    hedgerow pollinators, case study 158–9
  aquatic 121
    barnacle larva trap, case study 105–6
    stream invertebrates, case study 121–2
    substrate-living 127–32
    surface and swimming species 123–7
  associated with plants
    butterfly life cycles, case study 152–3
    capture 152, 153–7
  ground-active
    capture 139–51
    insect collection, Costa Rica case
        study 141–2
    Tarantula distribution and behaviour,
        case study 118–19
  numbers in polluted habitats 56, **59–60**
  soil-living 133
    extraction methods 133–8
  techniques for killing and
      preservation 116, **116–17**

Jaccard index of similarity **259**
Jolly–Seber method, population
    estimation 254, **255**

Kempson bowl extractors 135, *136*
keys, for choice of statistical tests
  analysis, main types **239**

keys, for choice of statistical tests (*Continued*)
  descriptive statistics 240
  hypothesis testing 241
  pattern and structure analysis 243–4
  predictive analysis 242
  terms used 238, **238–9**
kick nets and screens, aquatic sampling 124, *125*
kite diagrams 85, *86*

ladder transects 85, *85*
Latin and foreign words, writing style 328–9, **329**
Latin square sampling design 28, *28*
leaf litter mesh bags 131, *131*
legal aspects, field work 11–12, 104, 107–8, 112
lentic habitats *see* freshwater ecology
lichens
  abundance and distribution 73, 76, 77, 80
  identification 67, 84
  as sulphur dioxide level indicators 56, *61*
light meters 55, *56*
light traps (night-flying insects)
  choice factors **171**
  light types 168, *170*
  trap types 167, *167*, 168–71, *169*
  use and success factors 168
Lincoln Index (Peterson method) for population estimation 253–4
line and line intercept transects 84–5, *85*
linear regression, simple **270**, 271–3, 349
  line plot display 271, *272*, **315**
literature review
  appraised in research reports 309
  location and use of sources 8–9, 10–11
  primary and secondary literature 9
  professional ecology software 9–10
  publication bias 8
  search terms 10
littoral (shore) habitats
  environmental measurements 54, *54*, 55, *57*
  SACFOR (ESACFORN) abundance scales 77–8, *78*, **79**
  transects 84, 85
Longworth traps 225–7, *226*
lotic habitats *see* freshwater ecology

Mahalanobis generalized distance measures **288**
malaise traps 166, *167*
mammals *see also* badgers; bats
  capture 225–9, 231–3
  direct observation *217*, 217–19
    camera traps 218, *218*
    nocturnal mammals 218, 219
  indirect observation
    dung and droppings 219, 220–1, *221*
    hair sampling 221–2, *222*
    sound tracking 222–5
    tracks 219–20, *220*
  research standards and opportunities 216–17
Manhattan (city block) distance measures **288**
Manly–Parr method, population estimation 254, **255**
Mann–Whitney *U* test 266, **267**
mapping
  birds, territory mapping 206, **206**
  habitat types 36, 38–42, *39*
  landscape, habitat mosaics 41–2, 54–5
  using animal sounds 114–15, *115*
mark-release-recapture (population estimation) 106–8, 153, 252–4, **255**
mark-resight 254
marking
  assumptions, for data use 106–7
  legal and ethical issues 107, 108
  tarantula distribution and behaviour, case study 118–19
  techniques
    amphibians 192
    birds 213–15, *215*
    fish 181–2
    invertebrates 107, *108*, 117, 120
    mammals 107, 230, 233
    reptiles 199
matched (paired) data 25, 266, **267–8**, 348
maximum/minimum thermometers 42, *44*
mean 249–50
  graphical display of variation 250, *251*
median 249, 251, *251*
methodology 311
methods section, reports 310–11

metric (measurement) data 29–30, 348
  see also interval data; ratio data
microbial counting and diversity 70,
    71–2
microclimate
  effects and measurement 42, *43*, 45,
    47
  humidity 42–3, *46*
  temperature 42, *44*
  wind speed 43–5, *47*
minnow pots *179*, 181
mist nets
  bats 229
  birds *212*, 212–13
mobile organisms
  behaviour studies 98–102, **102–3**
  capture and preservation techniques 104,
    106
  direct observation 97–8, 112
  indirect evidence 103
  marking 106–8
  radio-tracking 108–10, **109**
  sampling challenges 95–7
mode 249
models *see* predictive analysis
mole traps 228, *229*
monitoring
  habitat quality 56–7, **59–60**, *61*
  scope of studies 4–6
  site characterisation 35, *36*
  traps, using transmitters 110
mosses **40**, 67, 68, 80, 85
moth traps *see* light traps
multidimensional scaling (MDS) 299–301,
    300
  stress measures **299**
multimeters, environmental
  aquatic 51, *52*
  microclimate and terrestrial
    measurements 45, *47*
multiple comparison tests **268–9**
multiple regression 276–7
multivariate analysis 348
  multivariate analysis of variance
    (MANOVA) 279
  planning and data collection 32
  simple community analyses 345

National Vegetation Classification
    (NVC) 38, **40**
nearest individual/neighbour sampling 87,
    87
Nemenyi test **267**
nested quadrats 81, *81*
nests, sampling of 27, **80**, 85, 210
nets
  airborne insects 158, 160, *160*
  aquatic invertebrates *123*, 123–7, *125*,
    126, *127*
  birds *212*, 212–13
  fish 178, *179*, **179–80**
  invertebrates associated with
    plants 154–5, *155*
nominal (categorical) data 20–1, 247,
    348
non-parametric tests 246, 266, 289, 348
nooses 197, *198*
normal distribution 30, *31*, 348
  data conformation and
    transformations 244–7, *245*
  mean and standard deviation *250*
null hypothesis 7–8, 262–3, *264*
  test statistics 263–4, 350
numbers, presentation in reports 324–5

ordinal (ranked) data 21, 29, 78, 348
  difference testing 265–6
ordination 348
  biplot display 294, **297**, *298*, 315
  comparison of data matrices/
    ordinations 301, **301**
  direct gradient analysis 301–3
    canonical correspondence analysis
      (CCA) **302–3**, *303*
  indirect gradient analysis 294–301
    correspondence analysis (CA) 296
    detrended correspondence analysis
      (DCA) 296, 298, *298*, 301
    multidimensional scaling (MDS) **299**,
      299–310, **300**
    principal components analysis
      (PCA) 294, **295–6**
    principal coordinates analysis
      (PCoA) 299
Owen traps 150, *151*

paired (matched) data   25, 266, **267–8**, 348
paired *t* test   **267**
parabolic reflectors   114–15, *115*
parameters, definition   22, **23**, 349
parametric tests   349
passive integrated transponders (PIT tags)
　invertebrates   109, 117, **119**, 120
　vertebrates   192, 199, 233
pattern analysis   286
　cluster analysis   287–92
　data collection requirements   32–3
　indicator species analysis   292–4
　ordination   294–303
Pearson's product moment correlation   269, **270**, 273
penetrometers   46, *48*
percentage (Renkonen) distance measure   **288**
percentage (Renkonen) index of similarity   **259**, *291*
permanent recording stations   73, 74–5, 94
personal communications, reports   308, **319**
Peterson (Lincoln Index) method   **253–4**
Phase 1 habitat maps   36, 38, *39*
phase-separation, invertebrate capture   134
pheromone traps   163, *164*
photography
　aerial and satellite images   38–9, 41
　camera traps   194, *194*, 218, *218*
　for canopy cover estimates   57–8, *62*
　static organism surveys   73, 76
　used in presentation   313, *313*
phutt nets   213
physical factors, environment   **37**
pie charts   247, *248*, **314**
Pielou index for evenness   256, **258**
pilot studies   3, 16, **17**, *17*, 23
pin-frames (point quadrats)   83–4, *84*
pipe traps   198, *198*
pitfall traps
　analysis of data   **144**, 144–5
　barriers, for movement analysis   145, *145*, *148*
　materials and covers   139–40, *141*, *143*
　setting   140, *142*, *143*
　use, factors for consideration   **146–7**
plankton nets   127, *127*
planning   2–3, 4, **34**

see also project design
plants *see* trees; vegetation cover
plotless sampling   86–7, **87–8**, **89**, 97
point centred quarter method (density)   87, **88**
point charts   **314**
　display of variation around mean   250, *251*
point quadrats (pin-frames)   83–4, *84*
poison bait   227, *228*
Poisson distribution   246, 281, **284**, 349
pollution
　biological indicators   56–7, **59–60**, *61*
　chemical contamination analysis   51, 53
　habitat quality monitoring   56–7, **59–60**, *61*
pond nets
　drift and plankton nets   126–7, *127*
　flat-bottomed (D-) nets   *123*, 124
　kick-sampling   124, *125*
pooters   153–4, *154*
population estimates *see also* abundance measurement
　capture-removal method   255–6, *256*
　distance sampling   203, 255
　mark-release-recapture techniques   106–8, 252–4, **255**
　　mark-resight   254
　　Peterson (Lincoln Index) method   **253–4**
　mobile species
　　direct counting   97–8
　　indirect methods   103
　from static organism counts   68, **69**
　study approaches   5–6
populations
　definition   22, **23**, 349
　fragmentation and connectivity   41
predictive analysis
　analysis of covariance (ANCOVA)   277, **278**, 279
　data collection   31–2
　discriminant function analysis (DFA)   279–80
　generalized linear model (GLM)   280–2, **282–4**
　multiple linear regression   276–7

multivariate analysis of variance
    (MANOVA)   279
  simple linear regression   **270**, 271–3, *272*
predictor (independent) variables   272,
    276–7, 347
prefixes (units)   **325**
presence/absence data
  patterns, correspondence analysis   296
  percentage frequency calculation   76, 81
  similarity measures   **259, 260**
presentation *see also* data presentation;
    writing style
  communication quality   305–6, 331
  computer files and backup   331
  equations and formulae   316
  font choice   328–9
  illustrations (tables, figures and
      graphs)   311–16
  report structures   306–7, *307*
  SI units   324, **325**
primary literature   9
principal components analysis (PCA)   294,
    295–6
  biplot production   294, **297**
principal coordinates analysis (PCoA)   299
project design (research projects)
  aims, objectives and hypotheses   7–8
  choice of topic   1–2, *3*
  experimental and survey data   18–22, *22*
  literature review   8–11
  practical aspects   11–15, **15–16**, 33–4,
      35–6
  research questions   4–7
  sampling design   22–8
  statistical analysis planning   28–33,
      235
  time-scales   16, **17**, *17*, 18
propelled nets   213, *214*
protected species   112, 143, 183, 193,
    216
  see also legal aspects, field work
  British regulations   189, 196, 200,
      218, 230
  tarantula distribution, case study   118–19
pseudoreplication   25, 349
publication bias   8
punctuation, report writing   327–8

quadrats   *75*
  biomass estimation   78, 80
  cover estimation   76–8, **77**, *78*
    subdivided quadrats   76, *77*
  density and frequency estimation   76, 81
  distribution pattern analysis   81, 88, **89**
  fixed   73
  nested   81, *81*
  placement   73, 82, *82*
  shape   82–3, *83*
  size   **80**, 80–1, *81*
  use in experimental design   73, 74–5
quotations, in reports   **319**

radio-tracking   108–10, **109**
  see also passive integrated transponders
    (PIT tags); telemetry
random numbers   26, **27**, 82, *82*
random sampling   26, *26*
  transect data limitations   86
ranging poles (slope inclination
    measurement)   55, *57*
ratio data   21, **21**, 349
Raunkiær plant life-form system   66, *66*
references   317
  citations in text   317–18, **318–19**
  reference lists   317, 318, **320**
  web material   **321**
refuges   131, *131*, 165, *165*
  see also artificial cover traps
regression *see* linear regression, simple;
    predictive analysis
relationship testing *see* hypotheses, statistical
    testing
remote sensing, imaging techniques   38–9, 41
Renkonen (percentage) distance measure   **288**
Renkonen (percentage) Index of similarity
    259, *291*
reports, research *see* presentation
reptiles
  capture techniques   195
    hand capture, risks and
        equipment   197, *198*
    traps   *198*, 198–9
  diet, case study   195–6
  direct observation   193
  indirect monitoring   194, *194*

research questions 4–7, 18–19
  choice of statistical analysis
    technique 30–1, 238, **238–44**
response (dependent) variables 272, 276–7,
    279, 346
results *see also* data presentation
  interpretation 316–17
  presentation 311–16
ringing
  bats 230
  birds 211, 213–15, *215*
rocky shore sampling *see* littoral (shore)
    habitats
rotary traps 172, *172*

SACFOR (ESACFORN) abundance
    scales 77–8, *78*, **79**
salinity 51, 53
sampling *see also* capture techniques;
    quadrats; transects
  choice of method 33–4, 70, 73
  definition of terms 22, **23**, 349
  design of sampling strategy 25–8, *26*
  fraction (sampling area) **69**
  heterogeneity and homogeneity 252, 347
  mobile organisms 95–7
  plotless methods 86–7, **87–8**, **89**
  sample size and replication 23–4,
    24, 261
  static organisms 67–8
  timing 96, 97, 101, 122
scatterplots 269, *270*, 271, **314**
scissor traps 228, *229*
search terms and key words 10
seashores *see* littoral (shore) habitats
Secchi discs 52, *53*
secondary literature 9
Shannon–Wiener diversity index 256, **257**
SI units 324, **325**
significance testing **265**
similarity measures 258, **259–60**, 261
simple matching coefficient of
    similarity **259**
Simpson's Index (diversity and
    evenness) 256, **257**, **258**
site selection 35–6
slope 55, *57*

slurp guns 174, *175*
small mammal traps 225–9, *226*, *229*
software
  digital image analysis 58, 76
  ecology, management and
    recording 10–11
  indicator species analysis 294
  population estimation, data analysis 255
  statistical analysis **236–7**
soil
  invertebrates, capture and sampling 135,
    136, *137*
    dry extraction methods 133–8
  nutrient status 48, 50
  organic content 50
  salinity 51
  sampling equipment 48, *49*, *50*
  temperature 42, *45*
  water content 46, *48*
Sørensen index of similarity (coefficient of
    community) **259**, **287**
sounds, use in surveys
  bat detectors 222–5, *223*, **224**
  bird song 210–11
  insects 114–15, *115*
spatial or temporal autocorrelation
    86, 252
Spearman's rank correlation **270**, 271
species accumulation curves 24, **24**
species names 310, 324, 328–9
species richness 5, 256, **257–8**
  see also diversity; monitoring, habitat
    quality
stacked bar graphs 248, **314**
standard deviation 250, *250*, *251*, 349
standard error 250, *251*, 261, 350
static organisms *see also* trees
  counting, for population/density
    estimates 68, **69–70**
  quadrat-based studies, uses 73–80
  scale and sampling techniques 70,
    73
  short vegetation, cover estimates
    83–4, *84*
  spatial distribution 86–9
  transect-based studies 84–6, *85*, *86*
  types and logistical problems 67–8

zoned, along environmental
    gradients   84–6, *86*
statistical analysis *see also* keys, for choice of
    statistical tests
  data collection requirements   28, 29–30,
    30
  descriptive statistics   28–9, 244, 261
  hypothesis testing   29–31, **30**, 261–5,
    264, **265**
  keys for choice of tests   238, **238–44**
  level of information in reports   310
  parametric and non-parametric   246
  pattern and structure analysis   32–3, 286
  predictive analysis   31–2, 276–86
  software   **236–7**
statistics (descriptive)
  central tendency (mean, median and
    mode)   249
  definition   22, 350
  documentation and presentation   261,
    312, **314**
  population estimates   252–6
  richness and diversity indices   256, **257–8**
  similarity coefficients   258–61
  variability measures   *250*, 250–1, *251*
sticky traps
  with chemical attractants and baits   163,
    164
  for flying insects   161, *162*
  for hair collection   221–2, *222*
  with light trapping   167
  for reptiles   199
  with sound attraction   115
stratified random sampling   26, 27–8, 68
stress measures   299, **299**
suction samplers
  aquatic burrowing invertebrates   127, *128*
  invertebrates in vegetation   148–50, *149*
  Rothamsted suction trap   160, *161*
  slurp gun   174, *175*
sulphur dioxide estimation (lichen
    zones)   56, *61*
Surber samplers   125, *126*
surveying tools   55, *57*
surveys
  design and limitations   20, 25–6
  use of photography   73

sweep nets   154–5, *155*
systematic sampling   26, 27, 112

T-square sampling method (density and
    distribution)   87, **88**, **89**
t tests   265, **267**, 280
tariff charts (timber volume)   94
telemetry *see* hydroacoustics; radio-tracking
temperature measurements   42, *44*, *45*, 52
tense, in writing   324
thermometers
  maximum/minimum   42, *44*
  soil   42, *45*
  water   52
time management   16, **17**, *17*, 18
titles of reports   307
topic selection (research project)   1–2, *3*
topography   55, *57*
tracking tunnels   194, 220, *220*
transects   84–6, *85*, *86*
  Breeding Bird Survey   *207*
  'W' shaped walk   112, *113*
transformations, data   350
  to approximate linear models   280
  skewed distributions   244–6, *245*
  truncation of percentage scale   246, *246*
transmitters, radio-tracking   108–10, **109**
trap-nests, bees and wasps   165, *165*
traps
  artificial refuge traps   131, 165, 191–2, *192*,
    198
  assembly   163, *164*
  baited   162–4, *164*, 227–8, 232
  barrier and bucket   190, *190*, *191*, 191–2
  bottle   163, *163*, 189, *189*
  cage   229, 232, *232*, *233*
  camera   194, *194*, 218, *218*
  drift fencing and funnels,
    amphibians   190, *190*, *191*
    reptiles   198
  emergence (Owen, fish egg)   150,
    151, 177
  fish funnels and pots   *179*, 180–1
  funnel   132, *132*, 190, *190*
  harp   229, *231*
  interception (window, malaise)   165–6,
    166, 167

traps (*Continued*)
  light attraction *167*, 167–71, *169*, **171**
  missing, data adjustment **144**
  pitfall 139–48
  rotary 172, *172*
  small mammals (Longworth, Trip, mole) 225–7, *226*, *229*
  sticky 125, 161, *162*
  water 172–3, *173*
trees
  age 94
  basal area 92
  branching, as ecological history marker 58
  canopy cover, estimation 57–8, *62*
  density estimation (plotless sampling) 86–7, **87–8**
  diameter at breast height (dbh) 90, 92, 93, 94
  growth and condition, case study 91–2, 94
  height 93, *93*
  timber volume 93–4
Trip-traps *226*
Tukey test **267, 268, 269**
Tullgren funnels 134–5, *135*
turbidity 52, *53*
two-way indicator species analysis (TWINSPAN) 292, *293*, 293–4
Type I and Type II errors 263, **265**, 350

Vancouver reference and citation system 318
variables **23**, 350
  see also data
  dependent (response) and independent (predictor) 272, 276–7, 346, 347
  displaying trends *271*
  dummy 276, **276**
  fixed, measured and derived 21–2, 347
  non-linear relationships *273*, 273–4
  relationship testing, statistics 263, 269–74, **270**
variance 280, 281, 350
vegetation classification systems 38, **40**
vegetation cover
  canopy cover estimates 57–8, *62*
  defined scales 76, **77**
  quantitative estimation 76, *77*
  use of pin-frames (point quadrats) 83–4, *84*

'W' shaped transect walks 112, *113*
water
  chemical analysis 51, 53
  inundation and flow 53–4, *54*
  physical factors 52, *53*
  sampling and testing 51, *52*, 53
water traps (insects) 172–3, *173*
wave action dynamometers 54, *54*
weather stations, portable 42, *43*
web (internet) references **321**
wet extraction
  aquatic invertebrates 129–30, *130*
  floatation, from soil and plants 134
whirling hygrometers 43, *46*
wind speed 43–5, *47*
window (interception) traps 165–6, *166*
Winkler samplers 135, *136*, 137
writing style *see also* presentation
  abbreviations 324, 325–6, **326**
  ambiguities 322–3
  checking and resources 321, 322–4, 329
  Latin and foreign words 328–9, **329**
  numbers and units 324–5, **325**
  punctuation 327–8
  scientific writing, good practice 321–2, 324
  tense 324
  word use **323, 329–30**